COMPUTATIONAL INTELLIGENCE IN SUSTAINABLE COMPUTING AND OPTIMIZATION

COMPUTATIONAL INTELLIGENCE IN SUSTAINABLE COMPUTING AND OPTIMIZATION

TRENDS AND APPLICATIONS

Edited by

BALAMURUGAN BALUSAMY
Shiv Nadar University, Delhi, India

VINAYAKUMAR RAVI
Prince Mohammad Bin Fahd University, Dhahran, Saudi Arabia

RAJESH KUMAR DHANARAJ
Symbiosis Institute of Computer Studies and Research (SICSR), Symbiosis International (Deemed University), Pune, Maharashtra, India

SUDHA SENTHILKUMAR
School of Computer Science and Engineering, Vellore Institute of Technology, Vellore, Tamilnadu, India

BRINDHA K.
School of Computer Science Engineering and Information Systems, Vellore Institute of Technology, Vellore, Tamilnadu, India

MK

MORGAN KAUFMANN PUBLISHERS

ELSEVIER

AN IMPRINT OF ELSEVIER

For information on all Morgan Kaufmann publications visit our website at
https://www.elsevier.com/books-and-journals

Publisher: Mara Conner
Acquisitions Editor: Chris Katsarapoulos
Editorial Project Manager: Deepak Vohra
Production Project Manager: Sharmila Kirouchenadassou
Cover Designer: Christian Bilbow

Typeset by TNQ Technologies

Contents

12. Artificial intelligence—based computational intelligence solutions for robotic automation

Dasaradharami Reddy Kandati and Anusha Sirasanambeti

13. Developing green computing awareness based on optimization techniques for environmental sustainability

A. Arivoli, B. Kalaavathi, and Chen Joy Iong-Zong

14. Bio-inspired meta-heuristic algorithm for solving engineering optimization problems based on computational intelligence

S. Mohana Saranya, S. Mohanapriya, and Dinesh Komarasamy

15. Secure sharing of health records stored in cloud using cryptographic secret sharing schemes through computational intelligence: A review

Sameera Mahammad and Usha Rani Kuruba

16. Private blockchain-based encryption framework using Computational Intelligence approach

T. Sarath, K. Brindha, Rajesh Kumar Dhanaraj, and Balamurugan Balusamy

Contributors

Sriram Anbalgan SRM TRP Engineering College, Department of Electronics and Communication Engineering, Tiruchirappalli, Tamil Nadu, India

A. Arivoli Vellore Institute of Technology, School of Computer Science and Engineering, Vellore, Tamil Nadu, India

G. Arun Sampaul Thomas Department of AI&ML, J.B. Institute of Engineering and Technology, Hyderabad, Telangana, India

Balamurugan Balusamy Shiv Ndar University, Delhi, India

Lakshmi Bheemavarapu Department of Computer Science, Sri Padmavati Mahila Visvavidyalayam (Women's University), Tirupati, Andhra Pradesh, India

K. Brindha School of Computer Science Engineering and Information Systems, Vellore Institute of Technology, Vellore, Tamil Nadu, India

Jeya A. Celin Kalasalingam Academy of Research and Education, Department of Information Technology, Krishnankoil, Tamil Nadu, India

C. Chellaswamy SRM TRP Engineering College, Department of Electronics and Communication Engineering, Tiruchirappalli, Tamil Nadu, India

Vimala Chinta Department of Computer Science, Sri Padmavati Mahila Visvavidyalayam, Tirupati, Andhra Pradesh, India

Harshita Das Modi University of Science and Technology, Lakshmangarh, Rajasthan, India

Periyamuthu Deivendran Department of IT, Velammal Institute of Technology, Chennai, Tamil Nadu, India

Rajesh Kumar Dhanaraj Symbiosis Institute of Computer Studies and Research (SICSR), Symbiosis International (Deemed University), Pune, Maharashtra, India

Ganesh Dhandapani Jain (Deemed to be) University, School of CSIT, Bengaluru, Karnataka, India

Chen Joy Iong-Zong Dayeh University, Department of Electrical Engineering, Dacun, Taiwan

B. Kalaavathi Vellore Institute of Technology, School of Computer Science and Engineering, Vellore, Tamil Nadu, India

Dasaradharami Reddy Kandati School of Computer Science Engineering & Information Systems, Vellore Institute of Technology, Vellore, Tamil Nadu, India

Dinesh Komarasamy Department of Computer Science and Engineering, Kongu Engineering College, Erode, Tamil Nadu, India

Saravanan Krishnan Department of CSE, College of Engineering Guindy, Anna University, Chennai, Tamil Nadu, India

Sampath Kumar Kuppuchamy AMET University, Department of Computer Science and Engineering, Chennai, Tamil Nadu, India

Sameera Mahammad Department of Computer Science, Sri Padmavati Mahila Visvavidyalayam (Women's University), Tirupati, Andhra Pradesh, India

Aaditri Mittal Vellore Institute of Technology, Vellore, Tamil Nadu, India

S. Mohanapriya Department of Computer Science and Engineering, Kongu Engineering College, Erode, Tamil Nadu, India

S. Mohana Saranya Department of Computer Science and Engineering, Kongu Engineering College, Erode, Tamil Nadu, India

Shaik Khaja Mohiddin Department of CSE, Koneru Lakshmaiah Education Foundation, Vaddeswaram, Andhra Pradesh, India

Vaishnavi Munusamy School of Computer Science and Engineering, Vellore Institute of Technology, Vellore, Tamil Nadu, India

S. Muthukaruppasamy Department of EEE, Velammal Institute of Technology, Chennai, Tamil Nadu, India

S. Nagaraj Alliance College of Engineering & Design Alliance University, Department of Computer Science & Engineering, Bengaluru, Karnataka, India

Sivakumar Nagarajan School of Computer Science and Engineering, Vellore Institute of Technology, Vellore, Tamil Nadu, India

Indeti Naga Padmaja Department of Computer Science, Sri Padmavati Mahila Visvavidyalayam (Women's University), Tirupati, Andhra Pradesh, India

Bhargavi Peyakunta Department of Computer Science, Sri Padmavati Mahila Visvavidyalayam, Tirupati, Andhra Pradesh, India

K. Usha Rani Department of Computer Science, Sri Padmavati Mahila Visvavidyalayam (Women's University), Tirupati, Andhra Pradesh, India

Ripal D. Ranpara Department of Computer Science, Atmiya University, Rajkot, Gujarat, India

Jayashree J. Reddy Department of Database Systems, Vellore Institute of Technology, Vellore, Tamil Nadu, India

Vijayashree Reddy Department of Database Systems, Vellore Institute of Technology, Vellore, Tamil Nadu, India

Dilipkumar S. VIT University, Department of Analytics, Vellore, Tamil Nadu, India

T. Sarath School of Computer Science Engineering and Information Systems, Vellore Institute of Technology, Vellore, Tamil Nadu, India

K. Sasikumar School of Computer Science and Engineering, Vellore Institute of Technology, Vellore, Tamil Nadu, India

Sudha Senthilkumar School of Computer Science and Engineering, Vellore Institute of Technology, Vellore, Tamil Nadu, India

Vandana Sharma CHRIST (Deemed to be University), Delhi, India

Shaik Sharmila Department of IT, VNITSW, Guntur, Andhra Pradesh, India

Jyothi Singaraju Department of Computer Science, Sri Padmavati Mahila Visvavidyalayam (Women's University), Tirupati, Andhra Pradesh, India

Anusha Sirasanambeti School of Computer Science Engineering & Information Systems, Vellore Institute of Technology, Vellore, Tamil Nadu, India

S. Srinivasan AMET University, Department of Advanced Computing Sciences, Chennai, Tamil Nadu, India

Muthuvel Subramanian AMET University, Department of Computer Science and Engineering, Chennai, Tamil Nadu, India

R. Thanuja SASTRA Deemed University, SRC Campus, Department of Computer Science and Engineering, Kumbakonam, Tamil Nadu, India

CHAPTER

1

Journey of computational intelligence in sustainable computing and optimization techniques: An introduction

Sampath Kumar Kuppuchamy[1], S. Srinivasan[2], Ganesh Dhandapani[3], S. Nagaraj[4], Jeya A. Celin[5] and Muthuvel Subramanian[1]

[1]AMET University, Department of Computer Science and Engineering, Chennai, Tamil Nadu, India; [2]AMET University, Department of Advanced Computing Sciences, Chennai, Tamil Nadu, India; [3]Jain (Deemed to be) University, School of CSIT, Bengaluru, Karnataka, India; [4]Alliance College of Engineering & Design Alliance University, Department of Computer Science & Engineering, Bengaluru, Karnataka, India; [5]Kalasalingam Academy of Research and Education, Department of Information Technology, Krishnankoil, Tamil Nadu, India

1. Introduction to computational intelligence

Computational intelligence: Computational intelligence is a branch of artificial intelligence that deals with creating algorithms and systems that can learn from data and make decisions based on what they have learned. This includes tasks such as machine learning, neural networks, and evolutionary computation. Computational intelligence is an emerging field of study that has been gaining momentum in recent years. It seeks to explore the potential for machines and computers to think, solve problems, and accomplish tasks just as humans do. This article will investigate how this technology can be used to create solutions that are more effective than traditional algorithms and methods.

Computational Intelligence in Sustainable Computing and Optimization
https://doi.org/10.1016/B978-0-443-23724-9.00001-3

1

The concept of computational intelligence builds on many existing technologies such as artificial intelligence, machine learning, data mining, and optimization algorithms. By combining these disciplines together, it enables us to develop solutions that are not only faster but also much more accurate than before. Moreover, due to its versatility, it can be applied across a wide range of industries from finance to healthcare. Computational intelligence is a field of study that seeks to understand, explain, and predict intelligent behavior. It applies the principles of computer science, mathematics, engineering, and statistics in order to create artificial systems that can solve complex problems. This discipline encompasses a wide range of methods including fuzzy logic, neural networks, learning theory, evolutionary computation, genetic algorithm, and deep learning.

Soft computing techniques such as fuzzy logic allow for the development of systems with capabilities similar to those found in biological nervous systems by utilizing heuristics instead of traditional algorithms. Artificial neural networks are modeled after the structure of human brains and used for pattern recognition or classification tasks. They have been widely used in various fields such as image processing, natural language processing, and robotics. On the other hand, evolutionary computation focuses on exploring solutions through search-based optimization algorithms over generations. Through mutation and selection operations, these techniques are able to generate new solutions from existing ones.

Computational intelligence (CI) is the theory, design, application, and development of biologically and linguistically motivated computational paradigms. Traditionally, the three main pillars of CI have been neural networks, fuzzy systems, and evolutionary computation. However, in time, many natures inspired computing paradigms have evolved. CI plays a major role in developing successful intelligent systems, including games and cognitive developmental systems. Over the last few years, there has been an explosion of research on deep learning, in particular deep convolutional neural networks (CNNs). Nowadays, deep learning has become the core method for artificial intelligence. In fact, some of the most successful AI systems are based on CI. The existence computational machine intelligence is the ability of a computer to acquire a particular process or job from any type of real-time information or results of some experimental projects.

In general, computational machine intelligence is the art of computational intelligence, which consists of integrated nature-inspired computational techniques and machine learning styles to report multifaceted real-world issues to which classical mathematical based method can be unusable for a several reasons: the procedures may be too multivariant complexity for mathematical-based computational reasoning; it includes several uncertainties while proceeding the procedure, or may be stochastic process; several real-time implemented issues cannot be converted into machine language for computers to understand and process it. Therefore, the computational machine—based intelligence gives different solutions for various issues in real world.

There are various techniques used which are closed to the live human method of reasoning, that is, it uses inaccurate and imperfect mastery mind, and it may produce control movements in a flexible mode. Computational machine-based intelligence uses an amalgamation of different complementary methods. The concept of fuzzy logic in artificial intelligence, which makes the computer to recognize natural language processing, artificial intelligence—based neural networks that allows the computer to understand real world on experimental data by biological-based evolutionary computing. Presently, the biologically inspired methods such as swarm intelligence and artificial intelligence—based robotic immune systems that can be a part of evolutionary computation, data mining, image processing, natural language processing, machine learning, and artificial intelligence that inclines to be confused with computational based machine intelligence. Even though both computational machine intelligence and artificial intelligence finds parallel goals which minor distinctions among themselves. Computational-based machine intelligence acts like robotics instead of replication of behavior human.

Computational machine intelligence is the strategy, machine learning—based application and expansion of biologically and communicative motivational measurable paradigms. In general, neural network, traditional computation, and fuzzy logic—based network are major themes of computational machine intelligence.

2. Goals of computational intelligence

Computational intelligence is a rapidly evolving field of computer science that seeks to develop algorithms and techniques to enable machines to solve complex problems. It involves the use of machine learning, neural networks, fuzzy systems, swarm intelligence, and probabilistic methods in order to extract meaningful information from data.

The main goals of computational intelligence are to create models that can accurately represent and predict real-world phenomena, as well as produce solutions which are capable of autonomously adapting their behavior according to changing environments. This requires the development of membership functions and artificial neural networks to understand data patterns and generate accurate predictions or decisions. The primary applications of computational intelligence include robotics, image processing, natural language processing, autonomous navigation, medical diagnosis, and fault detection.

In addition to that computational machine intelligence focused on nature-based stimulated calculated paradigms, environment protection-based intelligence, artificial robotic machine intelligence, social and culturally based intelligence, artificial hormone creation intelligence-based networks, and entertainment intelligence—based system and intellectual information technology—based development systems. Nowadays, more researchers focused on deep learning—based neural networks; it is also one theme of computational machine intelligence. So computational intelligence is one of major domains, which have been applied on different sector in real-time environment as shown in Fig. 1.1.

2.1 Fuzzy logic sets

Fuzzy logic is an efficient method which allows variable for processing numerous probable truth values to be administered by the same variable. Fuzzy logic mainly to solve critical issues with open standard, indefinite range of facts from heuristically based information that provide a collection of exact true inferences. The Boolean logic allows variable for processing two truth values (0/1 or yes/no) to be administered by the same variable.

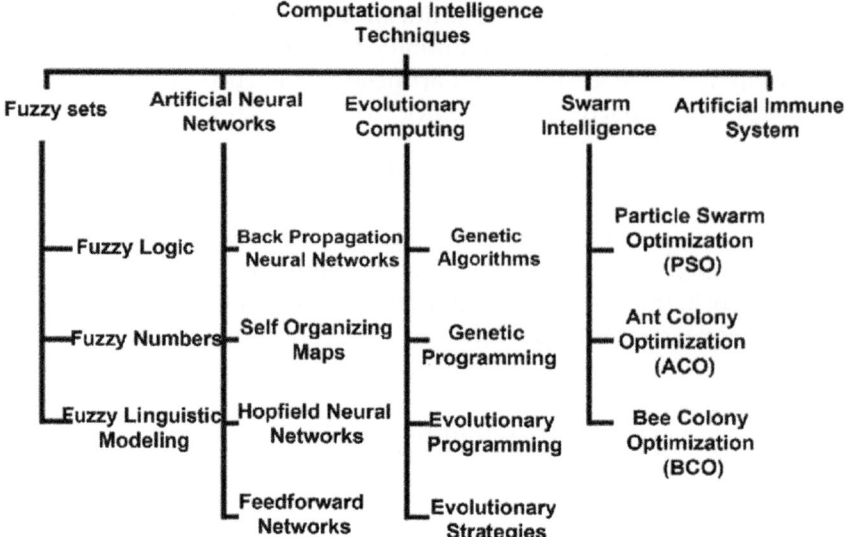

FIGURE 1.1   Computational machine intelligence techniques.

2.1.1 Evolution of fuzzy logic

In a 1965, Lotfi Zadeh projected fuzzy logic for the Journal Information and Control. He insisted the fuzzy sets and its implications in real world. Zadeh tried to bring the concept and implementation of different types of real facts used in IT data processing and originate the primary logical rubrics for these various types of data set. Zadeh elaborates that the classes of objects come across in the real world might not vaguely defined principles of membership. But still roughly defined classes appeared in human behavior—based computational machine learning, artificial intelligence implication—based pattern recognition, information communication—based technologies, and various types of abstraction.

2.1.2 Understanding the concept of fuzzy logic

Fuzzy logic—based branches enforce multivalued logic in various applications. The difference between general logic and fuzzy logic is dealing with single value or multiple values on various facts. For example, "Is this object yellow?" declaration indicates perfect single value truth that is general logic. But fuzzy logic narrates reasonable set of addresses or values with proper relations or meaning like shortest or smallest and build or construct. This type of fuzzy logic tries to analyze the real-world scenarioor problems and make decisions that depend on vague values. In some scenario, this type of constructs allows limited values based on "true" condition. This creates a prospect for algorithms to make decisions based on ranges of values as opposed to one detached value. Presently fuzzy logic has been implemented in different areas of applications like automatic control-based aerospace engineering, collaborative business understanding with decision-making, highly productive-based industrial engineering process, and efficient automotive traffic control in real-time environment using some of the basic elements of fuzzy logic as shown in fuzzy logic system represented in Fig. 1.2.

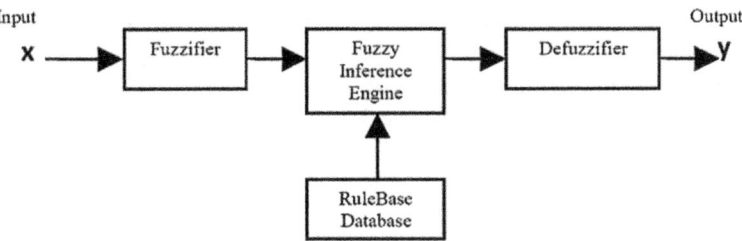

FIGURE 1.2 Some basic elements of fuzzy logic.

All components of fuzzy logic induce the concepts of artificial intelligence, machine learning, medical image processing, and so on. Some of the basic components of fuzzy logic are fuzzification, defuzzification, inference engine, and so on. We have to know the concept of the above basic mentioned components as represented below.

2.1.2.1 Fuzzification

Fuzzification is defined as the progression in which exact value which may be random input is transformed into approximate grade of relationship of fuzzy sets. Crisp sets are the group of values used in our real life. The goal of crisp set is to seek an element is either a participant of the set or not. Therefore, the crisp members are transformed into fuzzy sets. Internet-of-Things (IoT) which uses crisp set by means of crisp precise inputs are measured by different types of IoT components like sensors and which are passed into the fuzzy control system for further processing. There are several categories of fuzzifiers, which have been basic elements of fuzzification of fuzzy logic are singleton, Gaussian, and trapezoidal fuzzifiers.

The rule base in fuzzy logic is a group of rules which involve conditional statements like "If.Then" with in them. The rule base is otherwise called as fuzzy rules, which propose conditions, based on the value of condition or linguistic values, the business experts make a better decision-making in their system.

2.1.2.2 Defuzzification

Defuzzification in fuzzy login converts the conclusions of fuzzy into complete output fuzzy defuzzification values. The defuzzification elements transformed the set of values used in fuzzy through the inference engine into a crisp set. Some of specific expert system use defuzzification methods for purpose of reduce the error in their system.

2.1.2.3 Inference engine

The inference engine in fuzzy logic matches the degree of the random exact input with corresponding to each rule base. The inference engine selects the specific rule as per the input value. These specific rules are merged to form fuzzy inference control actions.

2.2 Fuzzy logic set and decision trees

Fuzzy logic assists the experts to make a better decision by decision tree. Many expert people analyze their business value in broader scale with the help of fuzzy logic rule–based inferences with analysis of decision tree through artificial intelligence—based programmed by rule-based inferences.

In general, fuzzy logic uses huge number of case studies which can be developed through decision tree analysis, which enforces fuzzy logic rules induces artificial intelligence rule-based programming. In this complex fuzzy logic set, code developers widen the logic rules used to govern the variables used in their rule-based programming.

2.2.1 Fuzzy control

It is a technique to embody human-like thinkings into a control system. It may not be designed to give accurate reasoning, but it is designed to give acceptable reasoning. It can emulate human deductive thinking, that is, the process people use to infer conclusions from what they know. Any uncertainties can be easily dealt with the help of fuzzy logic.

2.2.2 Merits of fuzzy logic system

This system can work with any type of inputs whether it is imprecise, distorted, or noisy input information. The construction of fuzzy logic systems is easy and understandable. Fuzzy logic comes with mathematical concepts of set theory and the reasoning of that is quite simple. It provides a very efficient solution to complex problems in all fields of life as it resembles human reasoning and decision-making. The algorithms can be described with little data, so little memory is required.

2.2.3 Demerits of fuzzy logic systems

Many researchers proposed different ways to solve a given problem through fuzzy logic, which leads to ambiguity. There is no systematic approach to solve a given problem through fuzzy logic. Proof of its characteristics is difficult or impossible in most cases because every time we do not get a mathematical description of our approach. As fuzzy logic works on precise as well as imprecise data, so most of the time accuracy is compromised.

2.2.4 Application of fuzzy logic control system

It is used in the aerospace field for altitude control of spacecraft and satellites. It has been used in the automotive system for speed control, traffic control. It is used for decision-making support systems and personal evaluation in the large company business. It has application in the chemical industry for controlling the pH, drying, and chemical distillation process. Fuzzy logic is used in natural language processing and various intensive applications in artificial intelligence. Fuzzy logic is extensively used in modern control systems such as expert systems. Fuzzy logic is used with neural networks as it mimics how a person would make decisions, only much faster. It is done by aggregation of data and changing it into more meaningful data by forming partial truths as Fuzzy sets represented in Fig. 1.3.

2.3 Fuzzy semantics in artificial intelligence

The semantic means interpretation of words in different meaning with respect to various context. Semantics can be used in fuzzy logic so we called as fuzzy semantics. Integration of

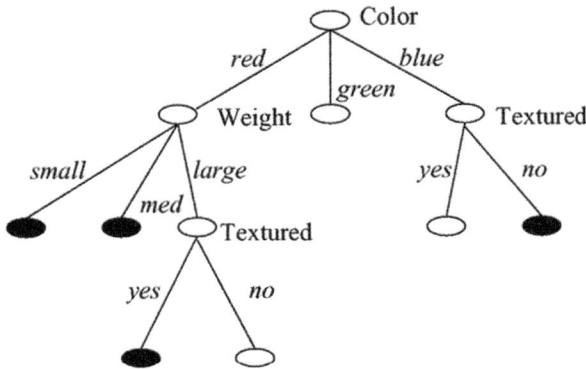

FIGURE 1.3 Example for fuzzy logic set and decision tree.

fuzzy logic and fuzzy semantics is the main element to artificial intelligence—based programming. Artificial intelligence methods and tools provide solutions with expand to across a range of subdivisions as the programming abilities from fuzzy semantics and fuzzy logic also enlarged as shown in Fig. 1.4.

Artificial Intelligence—Fuzzy Logic Systems.

IBM's Watson is artificial intelligence systems through different variations of fuzzy semantics and fuzzy logic. Fuzzy logic and fuzzy semantics are used in artificial intelligence—based machine learning systems supporting perfect output value of investment through business intelligence in financial sectors.

2.3.1 Examples of fuzzy logic

In advanced business trading models, software systems can use rule-based programming fuzzy set to analyze millions of protections in real-time environment. Fuzzy logic is often used when a business people find to create multiple effects for consideration. Business people may also have the ability to program a set of rules for passing trades. Two examples are depicted below:

Rule 1: If the moving average is low and the Relative Strength Index (RSI) is low, then sell.

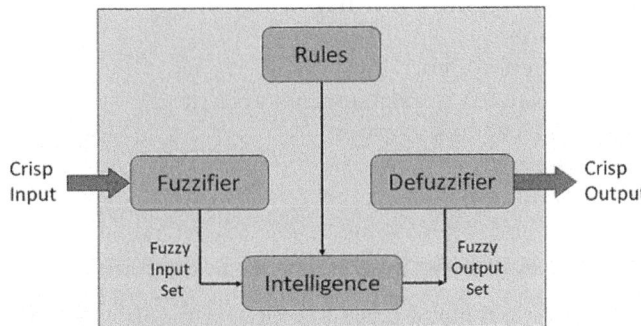

FIGURE 1.4 Fuzzy semantics in artificial intelligence.

Rule 2: If the moving average is high and the Relative Strength Index (RSI) is high, then buy.

In the above example fuzzy semantic and fuzzy logic allows a business people to program their own subjective rule inferences on low and high to reach at their private automated artificial intelligence-based trading signals.

2.4 Applications of fuzzy logic and fuzzy semantics

Computational machine—based intelligence fuzzy logic has been implemented in gear selection of automobile industry; it is based on aspects like engine load, road conditions, and so on. In home appliances also the fuzzy logic rule—based intelligence has implemented, which able to find the washing strategy and amount of power required to operate the washing machine depends on number of vessels and the range of food reside on the vessel or dishes. In small-scale commercial business like Xerox or paper copier machine shops, the automated computational machine-based intelligence is applied, and it is used to adjust drum voltage based on paper and image density and power or temperature.

The fuzzy logic concepts and expert system foundations are applied in aerospace industry, which is able to maintain altitude control for artificial intelligence—based machine learning satellites and dynamic spacecraft based on global environmental aspects. In medical diagnosis also, artificial intelligence—based expert system provide automated robotics computer-assisted diagnosing system during operation theater in many best class global hospitals in real world.

The computational intelligence—based fuzzy logic rule base and fuzzy semantics applied in the following major areas are represented below:

- Automatic control system in automobile industry.
  - Selection of automatic gear control
  - Design and implementation of four-wheeler steering control mechanism
  - Detection and protection of pollution control in automobile industry on vehicle maintenance system
  - Maintenance of electronic goods used by the customer
- Commercial high-fidelity control systems
- Xerox or paper copier automated machine control system
- Controlling of motion pictures and videos
- Audio—video with well-equipped smart television system.
- Utilization of goods in home
- Automated microwave oven system
- Automatic temperature changes in refrigerators system
- Perfect dust detection and vacuum cleaners
- Seventh sense air conditioners
- Automated dryers and heaters

2.5 Advantages of fuzzy semantic and fuzzy logic systems

- Fuzzy logic applied in mathematical concepts like set theory, reasoning, etc. is quite moderate.

- Manipulation of fuzzy logic rules is easy to applied in real-time applications.
- Fuzzy logic applied in medicine, automobile, and so on, it is able to find better a solution to complex problems and make decision-making related to the particular field.
- Fuzzy logic and fuzzy semantic are easily understandable structure.
- Fuzzy logic implemented many algorithms on distinct data, which required less memory space and produce high power processing output on different applications
- Fuzzy semantics and fuzzy logic provide perfect results with imprecise data.

2.6 Disadvantages of fuzzy semantic and fuzzy logic systems

- There is no organized method to design of fuzzy logic system.
- It will concentrate on less accuracy related problems.
- Fuzzy logic provides ambiguity with respect to provide solutions to a problem due to unorganized method.

2.7 AI in immunology—machine learning algorithms for immunological data analysis

Machine learning algorithms have revolutionized immunology by enabling the analysis of complex immunological data applied in various research areas, including the analysis of high-dimensional immunological data, prediction of immune responses, biomarker discovery, and analysis of large-scale immunological datasets. In the analysis of high-dimensional immunological data, machine learning techniques like clustering and dimensionality reduction algorithms have been used to identify distinct cell populations and characterize their phenotypes based on marker expression. This approach enhances our knowledge of immune cell subsets and their roles in diseases.

Machine learning models have been developed to predict immune responses, such as vaccine efficacy or disease progression, by training algorithms on immunological and clinical data. Support vector machines, random forests, and artificial neural networks are commonly used algorithms for building predictive models that optimize treatment plans for individual patients. Biomarker discovery has been significantly advanced by machine learning algorithms that integrate multiomics data. Feature selection and classification algorithms aid in identifying robust biomarkers for diagnostics, disease monitoring, and targeted therapy development.

Machine learning techniques, including hidden Markov models and sequence alignment algorithms, have been applied to large-scale immunological datasets, such as immune repertoire sequencing data, to uncover patterns and features related to immune responses. Noteworthy machine learning tools in immuno-genetics research include scikit-learn, an open-source Python library with a wide range of algorithms for data preprocessing and model evaluation. Deep learning techniques, such as CNNs and recurrent neural networks, extract meaningful features from immuno-genetics data. So, machine learning algorithms have revolutionized immuno-genetics research by analyzing complex data, predicting immune responses, discovering biomarkers, and uncovering the intricacies of immune repertoire sequencing. These advancements contribute to our understanding of immune system dynamics, disease mechanisms, and personalized medicine, ultimately improving diagnostics, treatments, and patient outcomes in immunology.

3. Artificial neural networks

Neural network is sub-domain of artificial intelligence. Neural network is one of the computing systems, which is constituted by biological neural networks of living animal brains. Imposing artificial intelligence on neural networks, which is a collection of interconnected nodes named as artificial neurons, which is basically a biological brain. Each nodes passing signals or values to other nodes in biological brain—based neural networks. An artificial neural network node obtains the values or input or signals and then processes it and produces output. The output of neuron is calculated through nonlinear function of the summation of its neuron input value or signals. Edges are connections, which connect nodes among neural networks. Each edge has some weight, which is adjustable as learning procedure process value may increase or decrease based on the signal strength at each connection. Each neuron has a threshold value, based on the value of threshold; the signals may cross the nodes if the cumulative signal crosses that value of threshold. In general, neurons are accumulated into layers. Each layer can do different transformations on their inputs value or signals. The input value or signal passes from the initial layer called as input layer to final layer called output layer possibly after navigating the layers many times.

3.1 Structure of artificial neural network

The structure of artificial neural network contains a huge number of artificial intelligence neurons arranged in ordered sequence of layers shown in.

The sequence of layers in artificial neural network is mentioned in below:

Artificial neural network input layer: It get inputs in various formats given by experts and AI programmer.

Artificial neural network hidden layer: This layer connects both input and output layer. It is otherwise called as the intermediate layer. It calculates hidden features and artificial intelligence patterns.

Artificial neural network output layer: The various sequence of transformation done on hidden layer which collect input from input layer and produce output values or results in output layer.

The artificial intelligence neural network collects the input and calculated the summation of the weighted value available in input layer with a bias. The calculation is in the form of a transmission or transfer function as:

$$\sum\nolimits_{i=0}^{n} wi * xi + b$$

3.1.1 Merits of artificial intelligence neural network

Similar processing ability:

Artificial intelligence neural networks have an algebraic numerical value, which performs more than one job concurrently.

Retrieval and storing information on the artificial neural network:

Artificial intelligence neural networks processing the input value and storing the output value in artificial neural network.

With having a memory dispersal:
Artificial intelligence neural network obtain output with accurate results in different formats and stored in different locations in large memory specified by the experts and programmer.

3.1.2 Demerits of artificial intelligence neural network

Anonymous behavior of artificial intelligence neural network:
It is the utmost substantial problem of this network, when this network produces a proper testing and provides solution, but it does not the processing logics, which results in diminutions trust in the neural network.

Computer hardware dependence:
Artificial intelligence neural networks require high-speed processors with similar processing power with respect to artificial neural network structure. Due to this impact, the comprehension of the devices or equipment is dependent.

3.2 Working principle of artificial intelligence neural networks

Artificial intelligence neural network is a weighted directed graph, which is connected with nodes as artificial neurons and edges as weighed values. There relation between neuron inputs and output can be represented as unidirectional edges with appropriate weights. The artificial intelligence neural network obtains the input values or signal from the various sources from external part, which may be vector as pattern or image and then inputs are represented by the notations as $x(n)$ for each n number of inputs value.

3.2.1 Types of artificial intelligence neural networks

Feedforward artificial intelligence neural network: It is basic foundational artificial intelligence network, which indicates the movement of signals passed in unidirectional neural network. The signal or data pass through input layer, hidden layer, and output layer and features of front-spread signals only.

Convolutional artificial intelligence neural network: It is little artificial intelligence network, which is similar to feedforward network, but the connections between nodes on different layer have different weights. This network proposed a complication operation on the input signals and then passes the output of each node on different layer to the next layer. This convolutional artificial intelligence neural network applied in speech recognition and image processing, which is used in computer vision.

Modular artificial intelligence neural network: This network includes a combination of many distinguish artificial neural networks. Each network in this modular artificial intelligence neural network is independently work and produces output without any interaction among them. Each of the different artificial neural networks can do various sub-job by getting inputs from different networks.

Radial basis function (RBF) artificial intelligence neural network: It is functional-based neural network, which focuses the distance between two nodes with respect to center point using two layers. The input value or signal is assigned to the *RBFs* in the hidden layer, and then the output layer calculated the output based on radial functional values.

Recurrent artificial intelligence neural network: It is otherwise called back propagation recurrent artificial intelligence neural network, which saves the output value of a layer and pass this output value or signal back to the input layer, which propose to predict the outcome of the layer.

3.2.2 *Uses of artificial intelligence neural networks*

Interactive social media: Artificial intelligence—based neural networks applied in social media. Many categories of people know the features of Facebook. By using the Facebook social media, everyone updates their profile and analyze others profile, seeking others area of interests, inviting relatives and friends.

Digital marketing and sales: Amazon, Flipkart may seek your interest on various products with assistance of history of user stored in browser. Similarly, Zomato, Swiggy, will find customer likely foods and based on customer feedback or interest they provided restaurant recommendations. Similarly in movies are also rated in this manner.

Protective healthcare: Artificial intelligence neural networks applied in medical fields like oncology to train methodologies or algorithmic technique of any fields of computer science, which recognize and seek cancerous diseases at measuring minute level at the identical precision trained by well skilled health-based physicians. Some diseases physical characteristics can be recognized in their early phases through pattern recognition like facial detection on the various patient images.

Personal digital assistants: The personal digital assistants interact with natural language processing, artificial intelligence—based neural network managing the language format, formal speech recognition, semantics, and so on.

3.2.2.1 Role of artificial intelligence neural network

Nowadays most businesses and companies make use of these technologies to solve complex problems like facial recognition, which helps the companies to have tight security.

Having facial recognition means no outsider can enter the company without the identified person. Due to their parallel architecture, these are especially suited for real-time systems, as they respond quickly.

There are several other applications that include speech-to-text transcription, data analytics, handwriting recognition, weather prediction, etc. The most fascinating feature of neural networks is the possibility of developing "conscious" networks in the future.

These networks have the potential to analyze the raw data and reveal new insights for which they might not even be trained. It can also learn and improve over time based on the user's behavior.

For example, consider a neural network that automatically suggests music by analyzing your music taste. Let us assume that the model was trained to play Rock and Metal genres songs. However, if you frequently listen to Jazz, the neural network can automatically learn and start suggesting types of songs that you usually like to listen to.

Neural networks can always prove helpful in the banking and business sector to identify frauds. Big start-ups like Uber and Swiggy use artificial neural networks to identify frauds and prevent losses.

3.3 Advantages and disadvantages of artificial neural network

The advantages are listed below

(1) A neural network can perform tasks that a linear program cannot.
(2) When an element of the neural network fails, its parallel nature can continue without any problem.

(3) A neural network learns, and reprogramming is not necessary.

(4) It can be implemented in any application.

(5) It can be performed without any problem.

The disadvantages are described below

(1) The neural network needs training to operate.

(2) The architecture of a neural network is different from the architecture of microprocessors. Therefore, emulation is necessary.

(3) Requires high processing time for large neural networks.

3.4 Neural network architecture types

The neural network architecture types are as follows:

3.4.1 Perceptron model

Neural network is having two input units and one output unit with no hidden layers. These are also known as "single-layer perceptrons."

Types of perceptron models.

Based on the layers, perceptron models are divided into two types. These are as follows: Single-layer perceptron model multilayer perceptron model.

3.4.1.1 Single layer perceptron model

This is one of the easiest artificial neural network (ANN) types. A single-layered perceptron model consists of feed-forward network and also includes a threshold transfer function inside the model. The main objective of the single-layer perceptron model is to analyze the linearly separable objects with binary outcomes. In a single-layer perceptron model, its algorithms do not contain recorded data, so it begins with inconstantly allocated input for weight parameters. Further, it sums up all inputs (weight). After adding all inputs, if the total sum of all inputs is more than a predetermined value, the model gets activated and shows the output value as +1. If the outcome is same as predetermined or threshold value, then the performance of this model is stated as satisfied, and weight demand does not change. However, this model consists of a few discrepancies triggered when multiple weight inputs values are fed into the model. Hence, to find desired output and minimize errors, some changes should be necessary for the weights input. "Single-layer perceptron can learn only linearly separable patterns."

3.4.1.2 Multilayered perceptron model

Like a single-layer perceptron model, a multilayer perceptron model also has the same model structure but has a greater number of hidden layers. The multilayer perceptron model is also known as the backpropagation algorithm, which executes in two stages as follows:

Forward stage: Activation functions start from the input layer in the forward stage and terminate on the output layer. Backward stage: In the backward stage, weight and bias values are modified as per the model's requirement. In this stage, the error between actual output and demanded originated backward on the output layer and ended on the input layer. Hence, a multilayered perceptron model has considered as multiple artificial neural networks having various layers in which activation function does not remain linear, similar to a single layer perceptron model. Instead of linear, activation function can be executed as sigmoid, TanH, ReLU, etc., for deployment.

A multilayer perceptron model has greater processing power and can process linear and nonlinear patterns. Further, it can also implement logic gates such as AND, OR, XOR, NAND, NOT, XNOR, and NOR.

3.4.1.3 Advantages of multilayer perceptron

A multilayered perceptron model can be used to solve complex nonlinear problems. It works well with both small and large input data. It helps us obtain quick predictions after the training. It helps to obtain the same accuracy ratio with large as well as small data.

3.4.1.4 Disadvantages of multilayer perceptron

In multilayer perceptron, computations are difficult and time-consuming. In multilayer perceptron, it is difficult to predict how much the dependent variable affects each independent variable. The model functioning depends on the quality of the training.

3.4.2 *Radial basis function*

These networks are similar to the feed-forward neural network, except the *RBF* is used as these neurons' activation function.

The popular type of feed-forward network is the *RBF* network. It has two layers, not counting the input layer, and contrasts from a multilayer perceptron in the method that the hidden units implement computations. Each hidden unit significantly defines a specific point in input space, and its output, or activation, for a given instance based on the distance between its point and the instance, which is only a different point. The closer these two points, the better the activation. This is implemented by utilizing a nonlinear transformation function to modify the distance into a similarity measure. A bell-shaped Gaussian activation service of which the width can be different for each hidden unit is generally used for this objective. The hidden units are known as RBFs because the points in the instance area for which a given hidden unit makes a similar activation form a hypersphere or hyperellipsoid. The output layer of an RBF structure is similar to that of a multilayer perceptron. It takes a linear set of the outputs of the hidden units and in classification issues passage it through the sigmoid function.

The parameters that such a network understands are the centers and widths of the RBFs and the weights used to design the linear set of the outputs acquired from the hidden layer. An essential benefit over multilayer perceptrons is that the first group of parameters can be decided independently of the second group and make accurate classifiers. One method to decide the first group of parameters is to use clustering. The simple k-means clustering algorithm can be applied, clustering each class independently to obtain k-basis functions for each class. The second group of parameters is understood by keeping the first parameters constant. This includes learning a simple linear classifier using one of the approaches such as linear or logistic regression. If there are long fewer hidden units than training instances, this can be done fastly. The limitation of RBF networks is that they provide each attribute with a similar weight because all are considered equally in the distance computation unless attribute weight parameters are contained in the complete optimization process. Therefore, they cannot deal efficiently with inappropriate attributes, against multilayer perceptrons. Support vector machines share similar issues. Support vector machines with Gaussian kernels (i.e., "RBF kernels") are a definite method of RBF network, in which one function is centered on each training instance, all basis functions have a similar width, and the outputs are merged linearly by calculating the maximum-margin hyperplane. This has the result that some of the RBFs have a nonzero weight the ones that define the support vectors.

3.4.3 *Multilayer perceptron*

A multilayer perceptron is a type of feedforward neural network consisting of fully connected neurons with a nonlinear kind of activation function. It is widely used to distinguish data that are not linearly separable. MLPs have been widely used in various fields, including image recognition, natural language processing, and speech recognition, among others. Their flexibility in architecture and ability to approximate any function under certain conditions make them a fundamental building block in deep learning and neural network research. Let's take a deeper dive into some of its key concepts.

3.4.3.1 Input layer

The input layer consists of nodes or neurons that receive the initial input data. Each neuron represents a feature or dimension of the input data. The number of neurons in the input layer is determined by the dimensionality of the input data.

3.4.3.2 Hidden layer

Between the input and output layers, there can be one or more layers of neurons. Each neuron in a hidden layer receives inputs from all neurons in the previous layer (either the input layer or another hidden layer) and produces an output that is passed to the next layer. The number of hidden layers and the number of neurons in each hidden layer are hyperparameters that need to be determined during the model design phase.

3.4.3.3 Output layer

This layer consists of neurons that produce the final output of the network. The number of neurons in the output layer depends on the nature of the task. In binary classification, there may be either one or two neurons depending on the activation function and representing the probability of belonging to one class, while in multiclass classification tasks, there can be multiple neurons in the output layer.

3.4.3.4 Weights

Neurons in adjacent layers are fully connected to each other. Each connection has an associated weight, which determines the strength of the connection. These weights are learned during the training process.

3.4.4 *Recurrent*

A recurrent neural network (RNN) is a deep learning model that is trained to process and convert a sequential data input into a specific sequential data output. Sequential data are data—such as words, sentences, or time-series data—where sequential components interrelate based on complex semantics and syntax rules. An RNN is a software system that consists of many interconnected components mimicking how humans perform sequential data conversions, such as translating text from one language to another. RNNs are largely being replaced by transformer-based artificial intelligence (AI) and large language models (LLM), which are much more efficient in sequential data processing.

RNNs are often characterized by one-to-one architecture: one input sequence is associated with one output. However, you can flexibly adjust them into various configurations for specific purposes. The following are several common RNN types.

3.4.4.1 One-to-many

This RNN type channels one input to several outputs. It enables linguistic applications like image captioning by generating a sentence from a single keyword.

3.4.4.2 Many-to-many

The model uses multiple inputs to predict multiple outputs. For example, you can create a language translator with an RNN, which analyzes a sentence and correctly structures the words in a different language.

3.4.4.3 Many-to-one

Several inputs are mapped to an output. This is helpful in applications like sentiment analysis, where the model predicts customers' sentiments like positive, negative, and neutral from input testimonials.

3.4.5 Long short-term memory neural network

The type of neural network in which memory cell is incorporated into hidden layer neurons is called an LSTM network.

Long short-term memory is an improved version of recurrent neural network designed by Hochreiter & Schmidhuber. LSTM is well-suited for sequence prediction tasks and excels in capturing long-term dependencies. Its applications extend to tasks involving time series and sequences. LSTM's strength lies in its ability to grasp the order dependence crucial for solving intricate problems, such as machine translation and speech recognition. The article provides an in-depth introduction to LSTM, covering the LSTM model, architecture, working principles, and the critical role they play in various applications.

3.4.5.1 Working principle of LSTM

A traditional RNN has a single hidden state that is passed through time, which can make it difficult for the network to learn long-term dependencies. LSTMs address this problem by introducing a memory cell, which is a container that can hold information for an extended period. LSTM networks are capable of learning long-term dependencies in sequential data, which make them well-suited for tasks such as language translation, speech recognition, and time series forecasting. LSTMs can also be used in combination with other neural network architectures, such as CNNs for image and video analysis.

The memory cell is controlled by three gates: the input gate, the forget gate, and the output gate. These gates decide what information to add to, remove from, and output from the memory cell. The input gate controls what information is added to the memory cell. The forget gate controls what information is removed from the memory cell. And the output gate controls what information is output from the memory cell. This allows LSTM networks to selectively retain or discard information as it flows through the network, which allows them to learn long-term dependencies.

3.4.5.2 Bidirectional LSTM

Bidirectional LSTM (Bi LSTM/BLSTM) is recurrent neural network (RNN) that is able to process sequential data in both forward and backward directions. This allows Bi LSTM to learn longer-range dependencies in sequential data than traditional LSTMs, which can only process sequential data in one direction. Bi LSTMs are made up of two LSTM networks, one that processes the input sequence in the forward direction and one that processes the input sequence in the backward direction. The outputs of the two LSTM networks are then combined to produce the final output. Bi LSTM have been shown to achieve state-of-the-art results on a wide variety of tasks, including machine translation, speech recognition, and text summarization. LSTMs can be stacked to create deep LSTM networks, which can learn even more complex patterns in sequential data. Each LSTM layer captures different levels of abstraction and temporal dependencies in the input data.

3.4.5.3 Advantages and disadvantages of LSTM

3.4.5.3.1 The advantages of LSTM are as follows   Long-term dependencies can be captured by LSTM networks. They have a memory cell that is capable of long-term information storage. In traditional RNNs, there is a problem of vanishing and exploding gradients when models are trained over long sequences. By using a gating mechanism that selectively recalls or forgets information, LSTM networks deal with this problem. LSTM enables the model to capture and remember the important context, even when there is a significant time gap between relevant events in the sequence. So where understanding context is important, LSTMS are used, for example, machine translation.

3.4.5.3.2 The disadvantages of LSTM are as follows   Compared to simpler architectures like feed-forward neural networks, LSTM networks are computationally more expensive. This can limit their scalability for large-scale datasets or constrained environments. Training LSTM networks can be more time-consuming compared to simpler models due to their computational complexity. So training LSTMs often require more data and longer training times to achieve high performance. Since it is processed word by word in a sequential manner, it is hard to parallelize the work of processing the sentences.

3.4.5.4 Applications of LSTM

Some of the famous applications of LSTM include:

Language modeling: LSTMs have been used for natural language processing tasks such as language modeling, machine translation, and text summarization. They can be trained to generate coherent and grammatically correct sentences by learning the dependencies between words in a sentence.

Speech recognition: LSTMs have been used for speech recognition tasks such as transcribing speech to text and recognizing spoken commands. They can be trained to recognize patterns in speech and match them to the corresponding text.

Time series forecasting: LSTMs have been used for time series forecasting tasks such as predicting stock prices, weather, and energy consumption. They can learn patterns in time series data and use them to make predictions about future events.

Anomaly detection: LSTMs have been used for anomaly detection tasks such as detecting fraud and network intrusion. They can be trained to identify patterns in data that deviate from the norm and flag them as potential anomalies.

Recommender systems: LSTMs have been used for recommendation tasks such as recommending movies, music, and books. They can learn patterns in user behavior and use them to make personalized recommendations.

Video analysis: LSTMs have been used for video analysis tasks such as object detection, activity recognition, and action classification. They can be used in combination with other neural network architectures, such as CNNs, to analyze video data and extract useful information.

3.4.6 Hopfield network

A fully interconnected network of neurons is in which each neuron is connected to every other neuron. The network is trained with input patterns by setting a value of neurons to the desired pattern. Then its weights are computed. The weights are not changed. Once trained for one or more patterns, the network will converge to the learned patterns. It is different from other neural networks.

Hopfield network is a special kind of neural network whose response is different from other neural networks. It is calculated by converging iterative process. It has just one layer of neurons relating to the size of the input and output, which must be the same. When such a network recognizes, for example, digits, we present a list of correctly rendered digits to the network. Subsequently, the network can transform a noise input to the relating perfect output.

A Hopfield network which operates in a discrete line fashion or in other words, it can be said the input and output patterns are discrete vector, which can be either binary 0, 1 or bipolar $+1$, -1 in nature. The network has symmetrical weights with no self-connections, that is, $wij = wji$ and $wii = 0$.

3.4.7 Boltzmann machine neural network

These networks are similar to the Hopfield network, except some neurons are input, while others are hidden in nature. The weights are initialized randomly and learned through the backpropagation algorithm. Boltzmann machines is an unsupervised DL model in which every node is connected to every other node. That is, unlike the ANNs, CNNs, RNNs, and SOMs, the Boltzmann machines are undirected (or the connections are bidirectional). Boltzmann machine is not a deterministic DL model but a stochastic or generative DL model. It is rather a representation of a certain system. There are two types of nodes in the Boltzmann machine—visible nodes—those nodes which we can and do measure, and the Hidden nodes—those nodes which we cannot or do not measure. Although the node types are different, the Boltzmann machine considers them as the same and everything works as one single system. The training data is fed into the Boltzmann machine and the weights of the system are adjusted accordingly. Boltzmann machines help us understand abnormalities by learning about the working of the system in normal conditions.

3.4.8 Convolutional neural network

Get a complete overview of it through our blog Log Analytics with Machine Learning and Deep Learning. A CNN is a type of deep learning neural network architecture commonly used in computer vision. Computer vision is a field of artificial intelligence that enables a computer to understand and interpret the image or visual data. When it comes to machine learning, artificial neural networks perform really well. Neural networks are used in various datasets like images, audio, and text. Different types of neural networks are used for different purposes, for example, for predicting the sequence of words we use recurrent neural networks more precisely an LSTM, similarly for image classification we use convolution neural networks. In this blog, we are going to build a basic building block for CNN.

In a regular neural network, there are three types of layers:

Input layers: It is the layer in which we give input to our model. The number of neurons in this layer is equal to the total number of features in our data (number of pixels in the case of an image). Hidden layer: The input from the input layer is then fed into the hidden layer. There can be many hidden layers depending on our model and data size. Each hidden layer can have different numbers of neurons which are generally greater than the number of features. The output from each layer is computed by matrix multiplication of the output of the previous layer with learnable weights of that layer and then by the addition of learnable biases followed by activation function which makes the network nonlinear. Output layer: The output from the hidden layer is then fed into a logistic function like sigmoid or softmax, which converts the output of each class into the probability score of each class. The data are fed into the model and output from each layer is obtained from the above step is called feedforward, we then calculate the error using an error function, some common error functions are cross-entropy, square loss error, etc. The error function measures how well the network is performing. After that, we backpropagate into the model by calculating the derivatives. This step is called backpropagation which basically is used to minimize the loss.

3.4.9 Modular neural network

It is the combined structure of different types of it like multilayer perceptron, Hopfield networks, recurrent neural networks, etc., which are incorporated as a single module into the network to perform independent subtasks of whole complete. A modular neural network is an artificial neural network characterized by a series of independent neural networks moderated by some intermediary. Each independent neural network serves as a module and operates on separate inputs to accomplish some subtask of the task the network hopes to perform. The intermediary takes the outputs of each module and processes them to produce the output of the network as a whole. The intermediary only accepts the modules' outputs it does not respond to, nor otherwise signal, the modules. As well, the modules do not interact with each other.

3.4.10 Physical neural network

In this type of artificial neural network, electrically adjustable resistance material is used to emulate synapses instead of software simulations performed in the neural network. Physical neural network is a type of artificial neural network in which an electrically adjustable material is used to emulate the function of a neural synapse or a higher-order (dendritic) neuron model. "Physical" neural network is used to emphasize the reliance on physical hardware used to emulate neurons as opposed to software-based approaches. More generally, the term is applicable to other artificial neural networks in which a memristor or other electrically adjustable resistance material is used to emulate a neural synapse.

3.5 Applications of neural networks

Neural networks are regulating some key sectors including finance, healthcare, and automotive. These artificial neurons function in a way similar to the human brain. They can be used for image recognition, character recognition, and stock market predictions. Let us understand the diverse applications of neural networks.

3.5.1 *Facial recognition*

Facial recognition systems are serving as robust systems of surveillance. Recognition systems match the human face and compare it with the digital images. They are used in offices for selective entries. The systems thus authenticate a human face and match it up with the list of IDs that are present in its database. CNNs are used for facial recognition and image processing. Large number of pictures are fed into the database for training a neural network. The collected images are further processed for training. Sampling layers in CNN are used for proper evaluations. Models are optimized for accurate recognition results.

3.5.2 *Stock market prediction*

Investments are subject to market risks. It is nearly impossible to predict the upcoming changes in the highly volatile stock market. The forever changing bullish and bearish phases were unpredictable before the advent of neural networks. But well what changed it all? Neural networks of course. To make a successful stock prediction in real time a multilayer perceptron (MLP, class of feedforward artificial intelligence algorithm) is employed. MLP comprises multiple layers of nodes, each of these layers is fully connected to the succeeding nodes. Stock's past performances, annual returns, and nonprofit ratios are considered for building the MLP model.

3.5.3 *Social media*

Artificial neural networks are used to study the behaviors of social media users. Data shared everyday via virtual conversations are tacked up and analyzed for competitive analysis. Neural networks duplicate the behaviors of social media users. Postanalysis of individuals' behaviors via social media networks the data can be linked to people's spending habits. Multilayer perceptron ANN is used to mine data from social media applications. MLP forecasts social media trends; it uses different training methods like mean absolute error (MAE), root mean squared error (RMSE), and mean squared error (MSE). MLP takes into consideration several factors like user's favorite Instagram pages, bookmarked choices, etc. These factors are considered as inputs for training the MLP model. In the ever-changing dynamics of social media applications, artificial neural networks can definitely work as the best fit model for user data analysis.

3.5.4 Aerospace

Aerospace engineering is an expansive term that covers developments in spacecraft and aircraft. Fault diagnosis, high-performance auto piloting, securing the aircraft control systems, and modeling key dynamic simulations are some of the key areas that neural networks have taken over. Time delay neural networks can be employed for modeling nonlinear time dynamic systems. Time delay neural networks are used for position-independent feature recognition. The algorithm thus built based on time delay neural networks can recognize patterns. (Recognizing patterns are automatically built by neural networks by copying the original data from feature units.) Other than this TNN are also used to provide stronger dynamics to the NN models. As passenger safety is of utmost importance inside an aircraft, algorithms built using the neural network systems ensure the accuracy in the autopilot system. As most of the autopilot functions are automated, it is important to ensure a way that maximizes the security.

3.5.5 Defense

Defense is the backbone of every country. Every country's state in the international domain is assessed by its military operations. Neural networks also shape the defense operations of technologically advanced countries. The United States of America, Britain, and Japan are some countries that use artificial neural networks for developing an active defense strategy. Neural networks are used in logistics, armed attack analysis, and for object location. They are also used in air patrols, maritime patrol, and for controlling automated drones. The defense sector is getting the much needed kick of artificial intelligence to scale up its technologies. CNNs are employed for determining the presence of underwater mines. Underwater mines are the underpass that serves as an illegal commute route between two countries. Unmanned airborne vehicle (UAV) and unmanned undersea vehicle (UUV), these autonomous sea vehicles use CNNs for the image processing. Convolutional layers form the basis of CNNs. These layers use different filters for differentiating between images. Layers also have bigger filters that filter channels for image extraction.

3.5.6 Healthcare

CNNs are actively employed in the healthcare industry for X-ray detection, CT Scan and ultrasound. As CNN is used in image processing, the medical imaging data retrieved from aforementioned tests is analyzed and assessed based on neural network models. Recurrent neural network (RNN) is also being employed for the development of voice recognition systems. Voice recognition systems are used these days to keep track of the patient's data. Researchers are also employing generative neural networks for drug discovery. Matching different categories of drugs is a hefty task, but generative neural networks have broken down the hefty task of drug discovery. They can be used for combining different elements which forms the basis of drug discovery.

3.5.7 Signature verification and handwriting analysis

Signature verification, as the self-explanatory term goes, is used for verifying an individual's signature. Banks and other financial institutions use signature verification to cross check the identity of an individual. Usually, a signature verification software is used to examine the signatures. As cases of forgery are pretty common in financial institutions, signature verification is an important factor that seeks to closely examine the authenticity of signed documents.

Artificial neural networks are used for verifying the signatures. ANNs are trained to recognize the difference between real and forged signatures. ANNs can be used for the verification of both offline and online signatures. For training an ANN model, varied datasets are fed in the database. The data thus fed help the ANN model to differentiate. ANN model employs image processing for extraction of features. Handwriting analysis plays an integral role in forensics. The analysis is further used to evaluate the variations in two handwritten documents. The process of spilling words on a blank sheet is also used for behavioral analysis. CNNs are used for handwriting analysis and handwriting verification.

3.5.8 Weather forecasting

The forecasts done by the meteorological department were never accurate before artificial intelligence came into force. Weather forecasting is primarily undertaken to anticipate the upcoming weather conditions beforehand. In the modern era, weather forecasts are even used to predict the possibilities of natural disasters. MLP, CNN, and RNNs are used for weather forecasting. Traditional ANN multilayer models can also be used to predict climatic conditions 15 days in advance. A combination of different types of neural network architecture can be used to predict air temperatures. Various inputs like air temperature, relative humidity, wind speed, and solar radiations were considered for training neural network based models. Combination models (MLP + CNN), (CNN + RNN) usually work better in the case of weather forecasting.

4. Evolutionary intelligence computing

Evolutionary artificial computational intelligence is a group of methodologies and algorithms for computerized local and global mathematical-based optimization stimulated through biological computing evolutional models. These methodologies and algorithm enforce a group of families of population-based experimental and error issues with solvers using a metaheuristic algorithm. Evolutionary computing enforces strategies, programming with respect to evolutionary intelligence computing and genetic concepts are represented in Fig. 1.5.

Group of participants solutions are created and iteratively recorded in evolutionary intelligence computing method. In biological-based genetic algorithm terms, a group of population measurement solutions is exposed to genetic network operations such as selection, mutation, and so on. Fitness function in genetic algorithm measured the generation of populations with increase of growth in population with respect to selected evaluation of fitness function of the genetic algorithm.

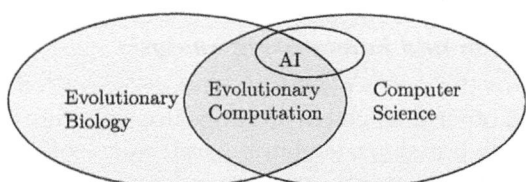

FIGURE 1.5   Evolutionary artificial intelligence computing.

Evolutionary computational intelligence methods can generate extremely mathematical-based optimal solutions in a widespread variety of different problem in information technology-based computer science concepts like data structures and so on.

Evolutionary computational intelligence techniques mostly involve in the following fields.

- Agent-based computational model
- Artificial intelligence biological human immune systems
- Coevolutionary intelligence algorithms and methods
- Estimation of probability distribution problems and algorithms
- Evolutionary artificial intelligence programming, algorithms, and strategies.
- Genetic programming, algorithms, and strategies.
- Data mining–based machine learning classification methods
- Unit swarm mathematical optimization and swarm intelligence

4.1 Evolutionary computational intelligence algorithms

Evolutionary algorithm is a subsection of evolutionary computational intelligence network, which provides methods and procedures stimulated by biological neural network operations like computational reproduction, automated mutation, neural network–based natural selection, and evaluation of fitness in real entities. The evolutionary algorithms played role in the area of genetic algorithm with programming and evolutionary strategies and programming as shown in Fig. 1.6.

4.1.1 Evolutionary computational intelligence algorithms and biological-based neural network

To predict any states of the any biological model through artificial intelligence computational genetic algorithms with the theory of dynamical systems. This way of predicting the states of any biological neural network highly-controlled, well-ordered, and perfect structured development in biological system of any neural network with assistance of dynamical systems. The biological systems are acts like artificial intelligence–based evolutionary computational machines accepts input value pass to next node in neural network

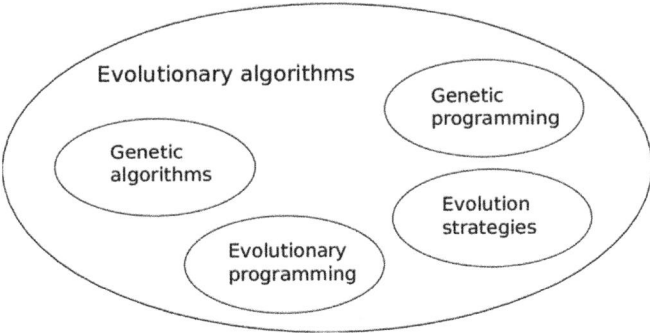

FIGURE 1.6 Involvement of evolutionary algorithms.

and then compute next states through dynamically of any biological model of any system as shown in Fig. 1.7.

Artificial intelligence evolutionary computational system involves finite deterministic automata, a simplification of computerized-Turing machines, which help to examine the characteristics of biological-based artificial intelligence neural network which obtain perfect output or results on expressiveness of artificial intelligence evolutionary computation in any type of neural network.

4.2 Working principle of evolutionary computation

An initial batch of possible solutions is created with the start of an evolutionary computation. The solutions once tried are refined as weaker solutions are stochastically removed, and small random changes are introduced to successive generations. As the generations pass the solutions become increasingly refined. In the end, the solutions produced by evolutionary computation can be tightly optimized, even though in the beginning, the approach is not understood.

4.2.1 Evolutionary algorithms function

Evolutionary algorithms are typically used to provide good approximate solutions to problems that cannot be solved easily using other techniques. Many optimization problems fall into this category. It may be too computationally intensive to find an exact solution, but sometimes a near-optimal solution is sufficient. In these situations evolutionary techniques can be effective. Due to their random nature, evolutionary algorithms are never guaranteed

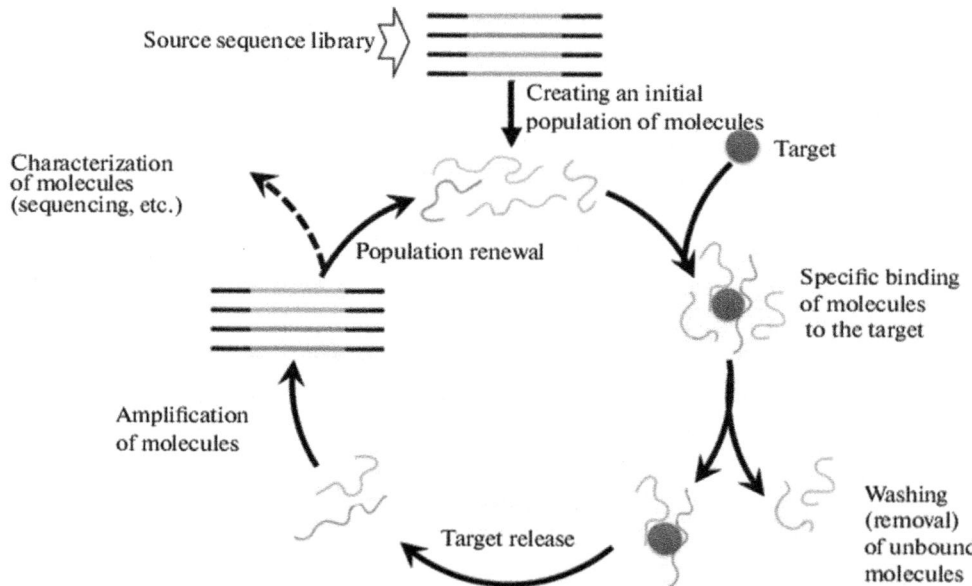

FIGURE 1.7   Artificial intelligence-based evolutionary computational biological system.

to find an optimal solution for any problem, but they will often find a good solution if one exists.

One example of this kind of optimization problem is the challenge of timetabling. Schools and universities must arrange room and staff allocations to suit the needs of their curriculum. There are several constraints that must be satisfied. A member of staff can only be in one place at a time, they can only teach classes that are in their area of expertise, rooms cannot host lessons if they are already occupied, and classes must not clash with other classes taken by the same students. This is a combinatorial problem and known to be NP-Hard. It is not feasible to exhaustively search for the optimal timetable due to the huge amount of computation involved. Instead, heuristics must be used. Genetic algorithms have proven to be a successful way of generating satisfactory solutions to many scheduling problems.

Evolutionary algorithms can also be used to tackle problems that humans do not really know how to solve. An EA, free of any human preconceptions or biases, can generate surprising solutions that are comparable to, or better than, the best human-generated efforts. It is merely necessary that we can recognize a good solution if it were presented to us, even if we do not know how to create a good solution. In other words, we need to be able to formulate an effective fitness function.

4.3 Types of evolutionary algorithms

There are various types of evolutionary algorithms. Here are the most significant ones:

Genetic algorithms (GA), genetic programming (GP), evolutionary programming (EP), evolutionary strategies (ES), and many more evolutionary algorithms also exist. These include gene expression programming, differential evolution, learning classifier systems, and neuroevolution.

4.3.1 *Genetic algorithms*

Genetic algorithms are the most popular type of evolutionary algorithms. They find solutions to problems as strings of numbers. Most of these strings are binary, but the most effective ones tend to show something about the problem in question. These algorithms make use of operators like mutation and recombination. Sometimes, they use both operators together. One of the uses of genetic algorithms is selecting the right combination of variables to build a predictive model. Selecting the right subset of variables is essentially a combinatory and optimization problem. The advantage of genetic algorithms is that it makes it possible for the best solution to emerge from the best of prior solutions. It improves the selection over time. The whole idea behind genetic algorithms is to combine the different solutions generation after generation so that it can extract the best genes or variables from each solution. It helps creating better fitted individuals. Genetic algorithms are also used for hyper-tuning parameters, finding the maximum or minimum of a function, or the search for a correct neural network architecture (Neuroevolution). It is also used in feature selection. The idea of GA is to generate a few random possible solutions that represent different variables and then combine the best possible solutions in an iterative process. The basic genetic algorithm operations are selection (picking the most fitted solutions in a generation), cross-over (creating two new individuals, based on the genes of solutions), and mutation (changing a gene randomly in an individual).

4.3.2 Genetic programming

Here, the solutions to problems are computer programs. The ability of these computer programs to solve computational problems is what determines their fitness. GP is essentially an automatic programming technique, which favors the evolution of computer programs that solve or at least approximately solve problems. It involves essentially "breeding" programs by continuously improving an initially random set of programs. Improvements are made by stochastic variation of programs and selection in line with some predefined criteria for judging the quality of a solution. Programs of genetic programming systems essentially evolve to solve predescribed automatic programming and machine learning problems. In its essence, genetic programming is a heuristic search technique that is commonly called "hill climbing." It involves searching for an optimal or at least a suitable program among the space of all programs.

4.3.3 Evolutionary programming

This is not too different from genetic programming. However, in evolutionary programming, the programs that need to be optimized have a fixed structure, while the numeric parameters can evolve. This evolutionary algorithm paradigm was first used by Lawrence J. Fogel in 1960 in an attempt to use simulated evolution as a learning process seeking to create artificial intelligence. He used finite-state machines as predictors and evolved them. Right now, evolutionary programming is a wide evolutionary computing dialect that has no fixed structure or representation. It is becoming increasingly difficult to differentiate evolutionary programming from evolutionary strategies. The main operator of evolutionary programming is mutation. In evolutionary programming, members of the population are seen as part of a specific species rather than members of the same species. Every parent generates an offspring by using a $(\mu + \mu)$ survivor selection.

4.3.4 Evolutionary strategies

Evolutionary strategies usually work by making use of self-adaptive mutation rates. They work with vectors of real numbers as representations of solutions. Evolutionary strategies are optimization techniques that are based on the ideas of evolution. They use natural problem-dependent representations and mainly make use of mutation and selection as search operators. The operators are applied in a loop, an iteration of which is known as a generation. The sequence of generations continues till a termination criterion is met. Most evolutionary algorithms work on a genotype level, but evolutionary strategies work on a behavioral level. Since the physical expression is coded directly, an individual's genes are not mapped to its physical expression. This approach is followed to give rise to a strong causality so that a small change in the coding gives rise to a small change in the individual and a large change in the coding causes a large change in the individual.

4.4 Real-time applications of evolutionary algorithms

Evolutionary algorithms have a wide range of real-world applications across various domains. While they may not be directly used for predicting future sales of products, they can be utilized in certain aspects of sales forecasting or related tasks. Here are some examples of real-world applications of evolutionary algorithms:

Optimization problems: Evolutionary algorithms excel in solving optimization problems where finding the best solution within a large search space is crucial. They can be used for optimizing resource allocation, production scheduling, logistics planning, portfolio management, and other similar tasks.

Engineering and design: Evolutionary algorithms are commonly employed in engineering and design processes. They can optimize parameters for complex systems, such as aircraft design, circuit design, antenna configuration, and architectural layouts. By iteratively evolving and improving designs, evolutionary algorithms can find optimal or near-optimal solutions.

Data mining and feature selection: Evolutionary algorithms can assist in data mining tasks by selecting relevant features and reducing dimensionality. They can be used to identify important variables, optimize feature subsets, and improve the efficiency and accuracy of machine learning models.

Machine learning and neural network optimization: Evolutionary algorithms can optimize the parameters and structure of machine learning models, including artificial neural networks. They can help find optimal weights, architecture, activation functions, and other hyperparameters, improving the performance of predictive models.

Game theory and strategy optimization: Evolutionary algorithms are used in game theory to optimize strategies for competitive scenarios. They can evolve strategies for games such as chess, poker, and strategic board games, finding optimal or near-optimal approaches.

Resource allocation and scheduling: Evolutionary algorithms can optimize the allocation of resources and scheduling of tasks in various fields. They can be applied to workforce management, project scheduling, transportation routing, and supply chain optimization, among others.

5. Swarm intelligence and unit swarm mathematical optimization

Automated swarm intelligence is the united behavior of ordered, well-organized systems, artificial intelligence-based expert system with respect to the context of bio-cellular computerized robotic systems. This system includes population of boids interacting with one another locally with in a stipulated controlled environment based biological neural network. The boids or agents follow rules and regulations governed by swarm intelligence system. According to rules, the boids should behave locally with respect to a certain degree random value, interactions between agents which lead to the development of intelligent agent, which shows moderated behavior that may acted globally in this swarm intelligence system.

Swarm robotics applies swarm principles and rules to automated robots, which have some set artificial intelligence-based algorithms which focus weather forecasting issues. Swarm robotics are worked together with genetically revised organisms in synthetic cooperative artificial intelligence. Swarm intelligence implemented in natural systems such as hawks hunting, bacterial growth, and so on.

5.1 Principles of swarm intelligence

Consciousness: Every participant must be aware of their environment and aptitudes.

Self-rule: Each participant member must be as autonomous in nature.

Extensibility: The swarm intelligence must permit dynamic enlargement where participant is effortlessly gathered.

Resiliency: When participant is evacuated the swarm system must be self-healing.

5.1.1 Applications of swarm intelligence

The swarm intelligence focused two major areas are:

Element swarm optimization: It is most known intelligence optimization method, which focused the common behavior of animals and insects. In this situation, each participant is called an element. Collaboration and learning culture is required for swarm intelligence which allows the cooperative intelligence of these disseminated elements.

Ant cluster optimization: The ant cluster optimization approach is a crucial constituent of swarm intelligence which is based on natures of actual ants. This approach forces their ant community to.

5.1.2 Particle or element swarm optimization

Particle or element swarm optimization is the one of artificial intelligence method which seeks estimated solutions to complex issues or problems as shown in Fig. 1.8. Element swarm mainly focuses population-based algorithm with the assistance of genetic algorithm. Set of individuals called elements traverse throughout entire area. At every traversal movement, this element swarm method assesses the objective function at every element in specified area. After the assessment, this method decides on the new speed of every element. Whenever the element moves, this method performs re-evaluation on each element in the specified area. The encouragement for the element swarm is gathers of birds or insects crowded. Every element is attracted to level of value, which may be called as degree to the nearest

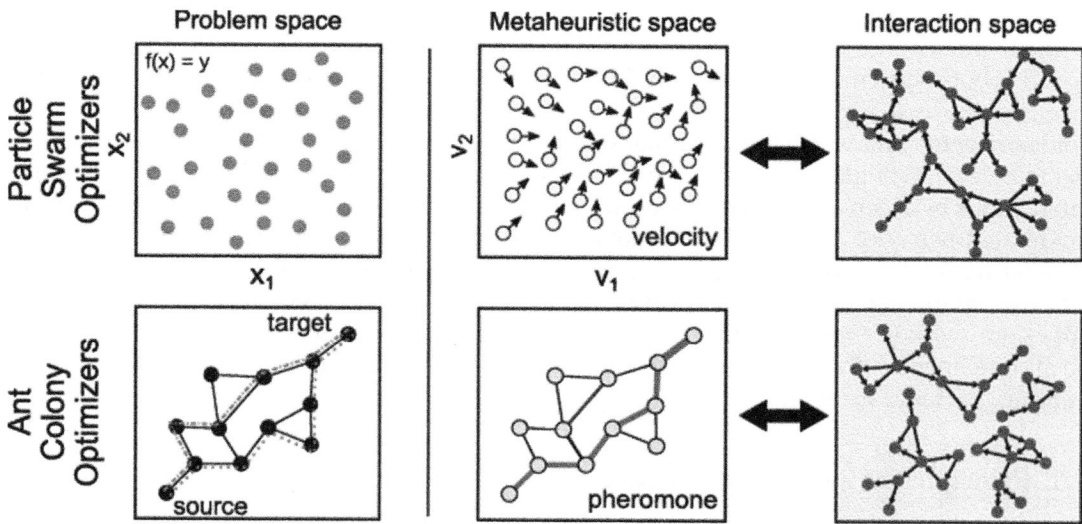

FIGURE 1.8   Element swarm optimizers versus antiassociation optimizers.

comfortable place it has found so far, and also to the good place any participants of the swarm has found. After few traversals, the population can join around single place or can merge around a rare place, or can continue to traversal.

Element swarm applied in various applications under different domain such as power control system, fuzzy logic, weather forecasting, bio-technology and bio-medicine, image preprocessing, audio and signal processing, data science and information analysis, etc.

5.2 Uses of swarm intelligence

Used in military services, NASA is generating the idea to use swarm intelligence for planetary mapping. Used in data mining, used in business to reach better financial decisions, etc.

5.3 Clustering behavior of ants

Ants build cemeteries by collecting dead bodies into a single place in the nest. They also organize the spatial disposition of larvae into clusters with the younger, smaller larvae in the cluster center and the older ones at its periphery. This clustering behavior has motivated a number of scientific studies. Scientists have built simple probabilistic models of these behaviors and have tested them in simulation. The basic models state that an unloaded ant has a probability to pick up a corpse or a larva that is inversely proportional to their locally perceived density, while the probability that a loaded ant has to drop the carried item is proportional to the local density of similar items. This model has been validated against experimental data obtained with real ants. In the taxonomy, this is an example of natural/scientific swarm intelligence system.

5.4 Nest building behavior of wasps and termites

Wasps build nests with a highly complex internal structure that is well beyond the cognitive capabilities of a single wasp. Termites build nests whose dimensions (they can reach many meters of diameter and height) are enormous when compared to a single individual, which can measure as little as a few millimeters. Scientists have been studying the coordination mechanisms that allow the construction of these structures and have proposed probabilistic models exploiting stigmergic communication to explain the insects' behavior. Some of these models have been implemented in computer programs and used to produce simulated structures that recall the morphology of the real nests. In the taxonomy, this is an example of natural/scientific swarm intelligence system.

5.5 Flocking and schooling in birds and fish

Flocking and schooling are examples of highly coordinated group behaviors exhibited by large groups of birds and fish. Scientists have shown that these elegant swarm-level behaviors can be understood as the result of a self-organized process where no leader is in charge, and each individual bases its movement decisions solely on locally available information: the distance, perceived speed, and direction of movement of neighbors. These studies have inspired a number of computer simulations that are now used in the computer graphics industry for the realistic reproduction of flocking in movies and computer games. In the taxonomy, these are examples respectively of natural/scientific and artificial/engineering swarm intelligence systems.

5.6 Ant colony optimization

Ant colony optimization is a population-based metaheuristic that can be used to find approximate solutions to difficult optimization problems. It is inspired by the above-described foraging behavior of ant colonies. In ant colony optimization (ACO), a set of software agents called "artificial ants" search for good solutions to a given optimization problem transformed into the problem of finding the minimum cost path on a weighted graph. The artificial ants incrementally build solutions by moving on the graph. The solution construction process is stochastic and is biased by a pheromone model, that is, a set of parameters associated with graph components (either nodes or edges) the values of which are modified at runtime by the ants. ACO has been applied successfully to many classical combinatorial optimization problems, as well as to discrete optimization problems that have stochastic and/or dynamic components. Examples are the application to routing in communication networks (see also the Swarm-based network management section below) and to stochastic version of well-known combinatorial optimization problem, such as the probabilistic traveling salesman problem. Moreover, ACO has been extended so that it can be used to solve continuous and mixed-variable optimization problems. Ant colony optimization is probably the most successful example of artificial/engineering swarm intelligence system with numerous applications to real-world problems.

5.7 Particle swarm optimization

Particle swarm optimization (PSO) is a population based stochastic optimization technique for the solution of continuous optimization problems. It is inspired by social behaviors in flocks of birds and schools of fish. In PSO, a set of software agents called particles search for good solutions to a given continuous optimization problem. Each particle is a solution of the considered problem and uses its own experience and the experience of neighbor particles to choose how to move in the search space. In practice, in the initialization, phase each particle is given a random initial position and an initial velocity. The position of the particle represents a solution of the problem and has therefore a value, given by the objective function. While moving in the search space, particles memorize the position of the best solution they found. At each iteration of the algorithm, each particle moves with a velocity that is a weighted sum of three components: the old velocity, a velocity component that drives the particle toward the location in the search space where it previously found the best solution so far, and a velocity component that drives the particle toward the location in the search space where the neighbor particles found the best solution so far. PSO has been applied to many different problems and is another example of successful artificial/engineering swarm intelligence system.

5.8 Swarm-based network management

The first swarm-based approaches to network management were proposed in 1996 by Schoonderwoerd et al., and in 1998 by Di Caro and Dorigo. Schoonderwoerd et al. proposed Ant-based control (ABC), an algorithm for routing and load balancing in circuit-switched networks; Di Caro and Dorigo proposed AntNet, an algorithm for routing in packet-switched networks. While ABC was a proof-of-concept, AntNet, which is an ACO algorithm, was compared to many state-of-the-art algorithms and its performance was found to be competitive especially in situation of highly dynamic and stochastic data traffic as can be observed in Internet-like networks. An extension of AntNet has been successfully applied to ad-hoc networks. These algorithms are another example of successful artificial/engineering swarm intelligence system.

5.9 Cooperative behavior in swarms of robots

There are a number of swarm behaviors observed in natural systems that have inspired innovative ways of solving problems by using swarms of robots. This is what is called swarm robotics. In other words, swarm robotics is the application of swarm intelligence principles to the control of swarms of robots. As with swarm intelligence systems in general, swarm robotics systems can have either a scientific or an engineering flavor. Clustering in a swarm of robots was mentioned above as an example of artificial/scientific system. An example of artificial/engineering swarm intelligence system is the collective transport of an item too heavy for a single robot, a behavior also often observed in ant colonies.

6. Artificial intelligence immune system

Artificial intelligence immune systems is sub-part of artificial intelligence, which perform the role of group of computational intelligent, artificial intelligence rule-based learning systems encouraged through the ideologies and procedures of the craniate intelligence immune system. This system is typically demonstrated after the artificial intelligence immune system's features of machine learning, rule-based system, and in-memory for use in creation and solving problems in artificial intelligence and robotics.

The artificial intelligence immune systems are concerned with conceptualizing the architecture and purpose of the immune system to artificial intelligence computational models and examining the different categories of application of artificial intelligence immune systems move toward mathematical-based computational problems, engineering design, artificial intelligence—based biological inspired computational methods and machine learning algorithms, and computer network as shown in Fig. 1.9.

Artificial intelligence immune systems are adaptive intelligence systems encouraged by theoretical and practical-based immunological and biology concepts and its functions, ideologies, and methods, which enforce problem solving in artificial intelligence. Artificial intelligence immune systems are implemented in a fertile ground for encouragement, DNA computing, etc.

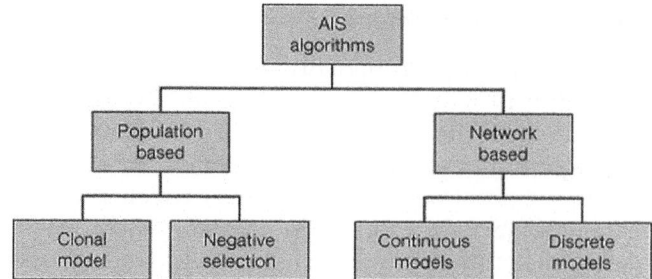

FIGURE 1.9   Categories of artificial intelligence immune systems.

6.1 History of artificial intelligence immune systems

The origins of AIS may be traced back to the 1974 work of Jerne. His work demonstrates the philosophical component of the working of the immune network, which posits that immune system cells and molecules may recognize foreign substances, respond to foreign chemicals, and regulate one another. His hypothesis is known as immunological network theory. The researchers compared immune networks to brain networks. Ishida's paper describes the first attempt to employ the immunological network in problem-solving. It centered on creating distributed diagnosis systems based on immunological network interactions.

Inspired by how the immune system distinguishes between self (normal) and nonself (abnormal), an NSA was proposed. The work announced the path from immunology to computing. More AIS algorithms began to emerge due to the considerable advantages of immune-inspired algorithms while handling various issues in many application domains. The algorithms associated with the proposition are the clonal selection algorithm (CLONAG), artificial immune network algorithm (AINE), danger theory-inspired algorithms, and dendritic cell algorithm (DCA).

6.1.1 Applications of artificial intelligence immune system

Duplicate or copy selection algorithm: A group of algorithms or methods encouraged by the clonal selection method of attained artificial intelligence immunity system that describes the way of B and T lymphocytes enhance their answers to antigens over a period of time called affinity maturation focused by Darwinian attributes of the theory through the affinity of antigen—antibody connections, replica is stimulated by cell partition, and difference is encouraged by somatic hypermutation. The duplicate or copy selection method or algorithms are implemented in optimized pattern recognition domains, hill climbing techniques, and genetic operations in genetic algorithm.

Nonpositive selection process method: Nonpositive selection indicates identify the T cells, which may select for and attack self-tissues be deleted. This nonpositive algorithms perform classification and pattern recognition with knowledge representation.

Intelligence-based immune neural network: This type of methods or algorithms encouraged by the idiotypic neural network evolutionary computational theory by Niels Kaj Jerne enforced the regulation of the intelligence-based immune system through anti-idiotypic antibodies. This method focuses on the neural network graph structures, which take antibodies cell as nodes and the artificial intelligence-based training algorithm involves pruning edges between the nodes in intelligence-based immune neural network.

Bioinspired dendritic cell algorithms: The bioinspired intelligence dendritic technique a multi-layer approach based on an intellectual perfect of dendritic cells. The dendritic cell algorithm inspecting and designing various aspects of cell function from cellular based neural networks with respect to cell behavior showed by a people of cells.

6.2 Significance of artificial immune systems

The significance of artificial immune systems in the domain of AI is underscored by their capacity to instill adaptive and self-regulating attributes within AI systems. AIS empowers AI solutions to exhibit a dynamic response to novel stimuli, fostering an intrinsic resilience that mirrors the biological immune system's ability to combat diverse challenges. This paradigm shift in AI design fosters the creation of robust and context-aware systems, revolutionizing the capabilities of AI technology.

6.3 Advantages and disadvantages of the artificial immune system

The application of artificial immune systems in AI landscape presents a spectrum of benefits and limitations, underscoring its multifaceted implications.

Advantanges:

Adaptive resilience: AIS fosters the creation of adaptive and resilient AI systems, capable of dynamically responding to evolving dynamics. Anomaly detection: AIS excels in anomaly detection, augmenting the capacity of AI systems to proactively identify and address irregularities. Robust learning: The autonomous learning mechanisms inherent in AIS engender robust AI models, enabling continuous adaptation and refinement.

Disadvantages:

Complexity: The integration of AIS introduces complexity into AI systems, necessitating a nuanced understanding of immunological and computational frameworks. Resource Intensiveness: The implementation of AIS may demand substantial computational resources, potentially posing challenges in resource-constrained environments. Ethical Considerations: AIS raises ethical considerations pertaining to autonomous decision-making and adaptive responses within AI systems, necessitating robust ethical frameworks.

6.4 Survey on artificial intelligence immune system

Farmer (1980) initiates merging immune system with the existing artificial intelligence article. Packard and Perelson (1986) and Bersini and Varela (1990) focus on immune networks issues and implications. Forrest et al. enforces the genetic algorithm operations such as selection, mutation etc. in artificial intelligence immune system. Dasgupta conducted experimental extensive revisions on selection operation under genetic algorithms in immune system. Hunt and Cooke (1990) depicts some improvements on existing immune neural network. De Castro et al. (2002) illustrates clonal selection in foundations of immune system.

Further researchers enforced danger theory and methods encouraged by the essential artificial immune system. Some developments happened in exploration of degeneracy in artificial intelligence immune system, which is motivated by its theorized part in open-source learning methods.

6.5 Techniques involved in artificial immune system

Specific immunological hypotheses that describe the function and behavior of the mammalian adaptive immune system inspired the prevalent approaches.

6.5.1 Clonal selection algorithm

Affinity maturation is a set of algorithms influenced by the clonal selection hypothesis of acquired immunity that explains how B and T cells enhance their antigen response over time. These algorithms concentrate on the Darwinian aspects of the theory, such as selection being influenced by antigen—antibody affinity, reproduction by cell division, and variation through somatic hyper mutation. Clonal selection methods, some of which resemble parallel hill climbing and the genetic algorithm without the recombination operator are most widely used in the optimization and pattern recognition areas.

6.5.2 Negative selection algorithm

T cell tolerance is inspired by the positive and negative selection processes that occur during the development of T cells in the thymus. Negative selection is the process of identifying and eliminating self-reacting cells or T cells that may select or attack own tissues. This class of algorithms is commonly utilized in problem domains like as classification and pattern recognition.

6.5.3 Immune network algorithms

Algorithms based on Niels Kaj Jerne's idiotypic network theory, which describes how anti-idiotypic antibodies regulate the immune system. This class of algorithms focuses on network graph architectures involving antibodies as nodes, with the training method creating between nodes based on affinity. Immune network techniques are similar to artificial neural networks and have been utilized in clustering, data visualization, control, and optimization.

6.5.4 Dendritic cell algorithms

The dendritic cell algorithm (DCA) is an example of a multiscale immune-inspired algorithm. This algorithm is based on a dendritic cell model that is abstract. The DCA is abstracted and implemented through a process of investigating and modeling numerous elements of DC activity, ranging from the molecular networks found within a single cell to the behavior of a population of cells as a whole. Information is granulated at multiple tiers inside the DCA, which is accomplished using multiscale processing.

7. Foundations of sustainable computing

Sustainable computing is the design, development, use, and disposal of computing systems in an environmentally sustainable manner. It involves the use of energy-efficient hardware and software, the reduction of electronic waste, and the implementation of sustainable practices in data center operations. Sustainable computing is essential to reducing the negative impact on the environment caused by the rapid growth of technology and increasing demand for computing power. By adopting sustainable computing practices, data centers can lower their carbon footprint and energy consumption while improving their bottom line.

Sustainable artificial intelligence computing represents consumption of artificial intelligence—based computing services and resources with respect to any type of artificial intelligence—based energy consumption, intelligence ecosystems, and automated pollution control system. The sustainable artificial intelligence computing compute total cost of any ownership in different business, the major environmental factors or influences and the benefits of artificial intelligence techniques.

Computational sustainable artificial intelligence is a developing arena in computing technologies which applied in social, environmental, and technological resources for the upcoming well-being of various civilization through different from computer science, mathematics and various fields shown in Fig. 1.10.

Different innovative applications by computational sustainable artificial intelligence are intelligence—based biodiversity management and species safety system, renewable resources, and online-offline storage technologies, which control and reduce the expenditure with respect to the system.

Examples of sustainable artificial intelligence computing

- Automated wind energy.
- High power saver solar energy.
- Efficient water flows and fixtures in agriculture
- Green energy

7.1 Green computing

Green computing, or sustainable computing, is the practice of maximizing energy efficiency and minimizing environmental impact in the ways computer chips, systems, and

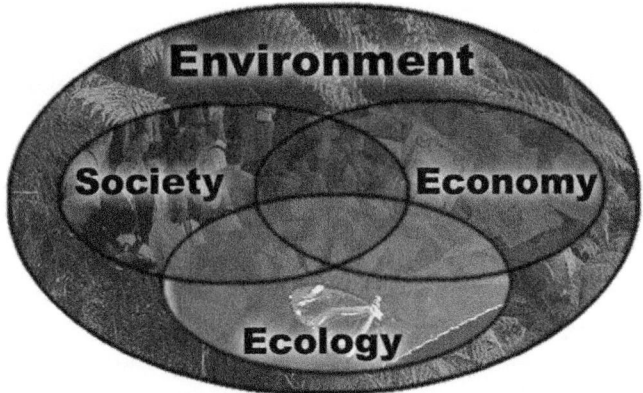

FIGURE 1.10 Implementation areas of computational sustainable artificial intelligence.

software are designed and used. Also called green information technology, green IT or sustainable IT, green computing spans concerns across the supply chain, from the raw materials used to make computers to how systems get recycled. In their working lives, green computers must deliver the most work for the least energy, typically measured by performance per watt.

7.1.1 Important of green computing

Green computing is a significant tool to combat climate change, the existential threat of our time. Global temperatures have risen about 1.2°C over the last century. As a result, ice caps are melting, causing sea levels to rise about 20 cm and increasing the number and severity of extreme weather events. The rising use of electricity is one of the causes of global warming represented in Fig. 1.11.

Computers have become an inseparable part of the modern generation. It saves our time and effort and makes lives easier. But it also consumes significant energy, generates heat, and has several other negative impacts on the environment.

- Due to high energy consumption, computers and other electronic devices impact our environment and depletes water, air, and land. Consequently, it brings climate change, air toxicity, acid rain, etc.
- The increasing emissions of harmful greenhouse gases such as carbon dioxide result in pollution and deteriorate the environment. It increases global warming and climate change.
- Computers and their peripherals, data centers, networking devices, etc., produce carbon dioxide in abundance. In addition, some parts of a computer are nonbiodegradable or recyclable.
- Manufacturing computer products involves toxic chemicals for fire protection, electrical insulation, and soldering. Exposure to these chemical fumes over a prolonged period can cause miscarriages, cancer, etc.
- Improper disposal of these devices can be harmful to the environment as they have certain hazardous chemicals and materials like lead, cadmium, mercury, etc., in them.

This is a cry for help. And green computing is a way to reduce these harmful impacts and encourage the practice of environmentally responsible computing.

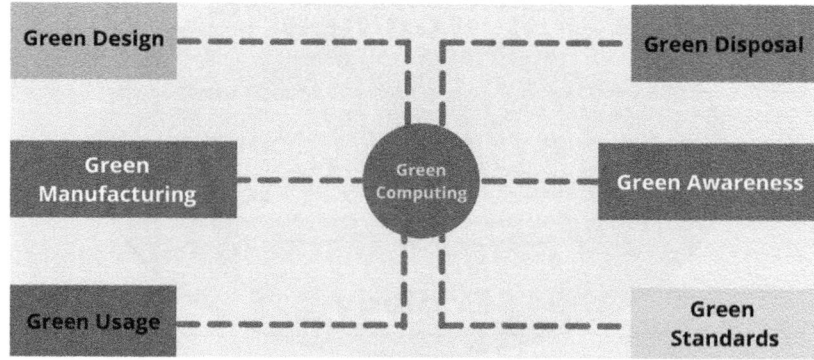

FIGURE 1.11   Green computing importance and design.

7.2 History of green computing

Green computing hit the public spotlight in 1992, when the U.S. Environmental Protection Agency launched Energy Star, a program for identifying consumer electronics that met standards in energy efficiency. A logo for energy efficient systems Fig. 1.12. The Energy Star logo is now used across more than three dozen product groups.

A 2017 report found nearly 100 government and industry programs across 22 countries promoting what it called green ICTs, sustainable information and communication technologies. One such organization, the Green Electronics Council, provides the Electronic Product Environmental Assessment Tool, a registry of systems and their energy-efficiency levels. The council claims it has saved nearly 400 million megawatt-hours of electricity through use of 1.5 billion green products it's recommended to date. Work on green computing continues across the industry at every level.

7.3 Applications of green computing

- Connected heating and lighting: The usage of connected devices for heating and lighting (through IoT), such as smart thermostats and LED technologies, is on the rise. This helps reduce energy consumption as you can control them easily with an application.
- Electric vehicles: Instead of depleting fossil fuels like petroleum, vehicles can now use electricity for running cars. And this application of green computing is in high demand these days, which explains why lots of options are available from different manufacturers such as Toyota, Tesla, etc.
- Data centers: Data centers consume high power, and maintaining them is an overhead. World's top tech giants like Google, Amazon, and Apple strive to build energy-efficient data centers with minimal environmental impact and cost.
- Alternative energy: Big tech firms are now implanting alternative energy sources like wind energy, solar power, etc., to fuel their data centers.
- Recycling devices: Computers, mobile phones, and other electronic devices contain harmful chemicals, rare metals, and whatnot. Manufacturing them on a large scale is hazardous, which is why companies have started recycling these wastes.

FIGURE 1.12   Logo for energy efficient systems.

7.4 Advantages and challenges of green computing

As with other things, green computing also has advantages and disadvantages. Let us see its advantages first.

7.4.1 Advantages
- As green computing emphasizes low energy consumption, it helps reduce fossil fuel utilization and greenhouse gas emissions.
- It aims to lower heat generation from computers and electronic devices to protect the environment.
- It promotes the effective usage of natural resources and their preservation.
- This approach emphasizes the usage of nontoxic substances reduces health hazards.
- Green computing encourages recycling and reusability of materials to reduced electronic waste, hence, lower land pollution.
- Implementing green computing makes us responsible for the environment and toward leading a sustainable future.

7.5 Challenges

- Every concept comes with inherent challenges that people need to overcome. Though green computing seems like a breakthrough, there are problems that companies face while implementing it. Some of the challenges of green computing are:
- Due to a lack of concern and huge market competition, companies resist this change. Consequently, everyone has to suffer the impact. Thought leadership and educating people about the dangers can increase awareness and adapt to eco-friendly ways.
- Frequent technology changes can confuse them to decide on how to proceed.
- The initial cost for implementing green computing is high.
- Companies find it difficult to make decisions due to fragmented data. They must put more effort into collecting and analyzing data and conclude that the method they use is environment-friendly yet profitable for their business.

7.6 Origins and motivations of computational sustainable

The field of computational sustainability has been motivated by Our Common Future, a 1987 report from the World Commission on Environment and Development about the future of humanity. Researchers in computational sustainability have primarily focused on addressing problems in areas related to the environment (e.g., biodiversity conservation), sustainable energy infrastructure and natural resources, and societal aspects (e.g., global hunger crises). The computational aspects of computational sustainability leverage techniques from mathematics and computer science, in the areas of artificial intelligence, machine learning, algorithms, game theory, mechanism design, information science, optimization (including combinatorial optimization), dynamical systems, and multiagent systems (Fig. 1.13).

FIGURE 1.13   Goals of sustainable development.

7.7 Biodiversity and conservation of computational sustainable

Computational sustainability researchers have advanced techniques to combat the biodiversity loss facing the world during the current sixth extinction. Researchers have created computational methods for geospatially mapping the distribution, migration patterns, and wildlife corridors of species, which enable scientists to quantify conservation efforts and recommend effective policies. In addition to scientific research contributions, the computational sustainability community has also contributed technologies that support citizen science conservation initiatives. An example is the creation of eBird, which enables citizens to share sightings of birds and crowd-source the creation of a global bird distribution database for researchers. An example of successful application of eBird database is the Nature Conservancy's Bird Returns program (2013), where farmers are compensated for their effort of maintaining suitable habitats for birds during the migratory periods. Another example is iNaturalist, which enables citizens to crowd-source the creation of databases about animals, plants, and other organisms, to support scientific research.

7.8 Sustainable artificial intelligence

Sustainable artificial intelligence is the use of AI systems that operate in ways contingent with sustainable business practices. Many current enterprise AI systems have a reputation for being detrimental to the environment, but doable practices can mitigate this. Those with firsthand knowledge of developing and implementing AI models understand that the arduous process of training AI models requires enormous amounts of energy, leading to unsustainable emissions and air pollution. Certain studies over the years have shown that the process of training just one machine learning (ML) model can match the carbon emissions of multiple cars combined.

Traditionally, ML algorithms attempt to parse through every insight they gather from huge data sets. However, sparse models can operate on small amounts of data with a narrower focus and are easier to train due to the need for less data. When optimizing the hardware components used for developing and training AI models, enterprises should note that developers and researchers have worked on emerging devices that deliver faster computing with lower energy consumption rates.

7.9 Current and future use cases of sustainable AI

In addition to sustainable operations, the power of AI systems can also be harnessed within many sustainability initiatives relating to pollution, natural disasters, and climate change. These applications include the following:

- Precision agriculture driven by AI.
- Weather predictions.
- Natural disaster response.
- Preparedness and strengthening infrastructure, such as power suppliers.

AI analytics can be used along with drones and other similar technologies to monitor and reduce the effects of natural disasters such as floods. Similarly, climate and risk assessments can be done with these AI tools, which work together with internet of things devices to protect crops as well as other critical assets and infrastructure. Also, predictive AI is crucial in forecasting future emissions and air pollution based on existing data and trends. In addition, many of these AI-based technologies are used in reforestation efforts in the Amazon. AI systems can analyze data from sensors, drones, and satellite imagery to monitor forests and how they evolve or devolve over time. With their predictive capabilities, ML algorithms can examine past data to determine likely future deforestation scenarios represented in Fig. 1.14.

When implemented and executed correctly, sustainable AI will mitigate the negative effects that traditional AI and ML model training and optimization practices have on the environment without sacrificing the quality of its predictions and outputs. In addition, it is applied to various use cases to help people protect their communities and natural resources. Sustainable AI is considered synonymous with the term green AI, which also describes efforts to improve these negative effects and ensure the future of AI brings with it more benefits than drawbacks.

Using computational and AI innovation to achieve the sustainable development goals

- The rapid improvements in digital technology nowadays are truly astonishing.
- Humans will be able to use massive amounts of data to harness AI innovation and achieve ground-breaking improvements in sectors like healthcare, agriculture, education, and transportation.
- Additionally, we are witnessing how AI-enhanced computing enables doctors to reduce medical errors, boost grower yields, customize student training, and identify researchers.

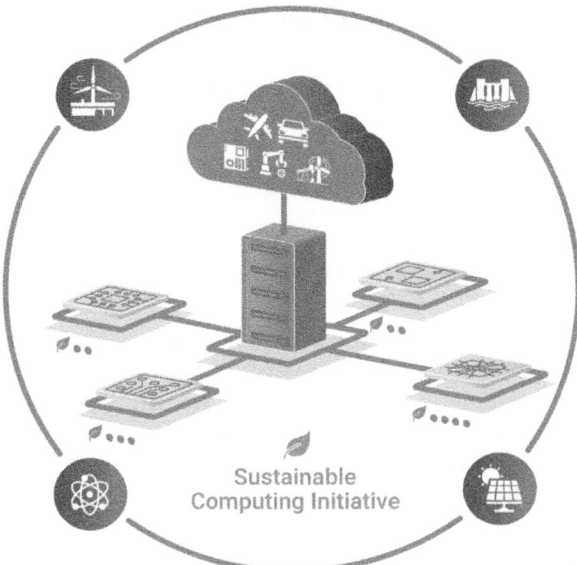

FIGURE 1.14 Sustainable computing initiative.

- Recently, there has been a lot of interest in the relationship between artificial intelligence and climate change.
- AI can be used, for instance, to manage environmental consequences and changes in a variety of economic contexts and settings.
- These are just a few examples of potential uses, which also include safer supply chains, environmental control and regulation, and weather forecasting. Renewable distributed electricity networks coupled with AI are another.

7.10 Optimization techniques

Optimization techniques are a powerful set of tools for solving complex real-world problems. The application domain has grown manifold in recent years, covering almost all engineering and science disciplines. On the technological front, many new algorithms are designed by taking inspirations from different natural phenomena, resulting in some of the popular algorithms such as GA based on Darwin's principle of survival of the fittest, ACO based on the foraging behavior of ants, PSO based on the behavior of birds flocking in swarms and many more. On the application front, irrespective of the source of inspirations, nature-inspired algorithms are suitable for most of the challenging applications dealing with current complexities of life in this 21st century. Fields of interest are as follows:

- Linear Optimization
- Meta-heuristics
- Nature-inspired Optimization
- Applications

7.10.1 Optimization techniques application

7.10.1.1 Linear programming
- Interior point methods for semi-infinite programming calculations
- Matrix shadow costs for multilinear programs
- A linear programming approach to on-line constrained optimal terrain tracking systems
- Recursive interior—point linear programming algorithm based on Lie-Brockett flows

7.10.1.2 Nonlinear programming
- Numerical experience with limited-memory quasi-Newton and truncated Newton methods
- Deriving quasi-Newton updates via the ABS approach
- Symmetric rank-one update and quasi-Newton methods
- A gradient projection variable metric method for solving the nonlinear programming
- Formulae-variable for conjugate gradient strategies
- A dual-active-set algorithm for positive semi-definite quadratic programming
- A smoothing technique for nonsmooth optimization problems
- On a class of complex nonlinear programming problems
- On the numerical solution of generalized polynomial equations using an optimization-based algorithm
- Curved-line search using ordinary differential equations
- The discrete recursive quadratic programming with artificial intelligence strategy
- Graphical optimization using HGRAM graphical format

7.10.1.3 Combinatorial optimization
- Recent developments in integer programming
- Location-search-based heuristic methods for set covering problems
- A new approach to game scheduling with preassignments
- A hybrid approach to the multiple-choice knapsack problem
- Original methods of optimization on combinatorial sets
- New progress on optimal partitioning problems
- A weighted graph partitioning problem with side constraints
- A decomposition process for the recognition of series-parallel subgraphs
- Toward nonalgorithmic optimization: a practical and general purpose method of formulating and solving dynamic programming problems

7.10.1.4 Network optimization
- Implementing optimization methods in network crashing: an illustration
- An algorithm for solving quadratic cost network flow optimization problems
- An algorithm of the shortest path for the network with negative arcs
- Algorithms for the minimum weight spanning tree with bounded diameter problem

7.10.1.5 Multi-criteria optimization
- An interactive method for multicriteria decision making
- A comparative evaluation of multiple objective linear programming methods
- An interactive approach to multiple objective transportation problems

- Major optimality and major efficiency in multicriteria optimization
- Project crashing with earliness and tardiness penalties: a goal programming model
- The multiobjective programming model of appropriate development goal of eighth 5-year plan ("8.5") for Chengdu

7.10.1.6 Optimal control
- A computational approach to an optimal control problem with a cost on changing control
- Algorithms for unconstrained optimization: a new approach using control theory
- A sufficient condition for near optimal stochastic controls and its applications to an HMMS model under uncertainty
- Multi-mode controllers for linear discrete-time systems with general state and control constraints

7.10.1.7 Large-scale optimization
- Testing the reliability and robustness of optimization codes for large-scale optimization problems
- Multi-step, multi-directional parallel algorithms for unconstrained optimization
- Modeling and analysis of large systems: some approaches
- An optimization procedure for decomposing changes in industrial energy consumption
- Parallel design optimization of large scale structural systems subjected to multiple loading and multiple constraints

7.10.1.8 Scheduling/sequencing/transportation
- Toward reducing computational demands in calculating lower bounds in permutation flow shop scheduling
- Complexity of the flow shop with multiple processors scheduling problem, and some dominance conditions
- Preference-based scheduling via constraint satisfaction
- A parametric model for staff scheduling problems with flexible demand constraints
- New heuristic algorithms for precedence constrained scheduling
- The design and development of a microcomputer-based vehicle routing and scheduling package

7.11 Optimizing software

The best ways to optimize software and deploy it are.

- *Virtualization:* It refers to the process of abstracting or dividing computer resources such as processors, storage, memory, etc., into virtual computers or virtual machines (VMs). Here, two or more virtual instances run on a single, robust physical system. It helps conserve system resources by eliminating the requirement for original hardware while reducing cooling and power consumption. It also reduces the need to manufacture more hardware. Virtualization helps distribute work effectively so that servers do not consume energy when not in use. They are either running or in sleep mode. As this technology is energy efficient, many service providers are offering software packages to enable virtual computing.

- *Creating efficient algorithms:* Efficient algorithms can be one of the factors influencing the number of resources needed for a computing function. Algorithm changes, for instance, making a search algorithm faster, can help reduce resource utilization substantially. Hence, IT companies must make sure the programmers write better, efficient code.
- *Allocating resources strategically:* IT teams can use algorithms to route data to a data center with less expensive electricity available. It saves the cost. Similarly, they can also route traffic away from a data center that experiences warmer weather. This approach allows them to shut down systems and avoid air conditioning while reducing energy usage.
- *Using terminal servers:* Terminal servers are servers or hardware devices providing terminals like PCs, smartphones, tablets, printers, etc., that share a common connection to the terminal server. Using terminal servers eliminates the need for terminals to have a network interface, modem, or card of their own. If you implement terminal servers, you can save energy consumption as well as cost.

7.12 Intelligent transportation systems

Intelligent transportation systems (ITS) seek to improve safety and travel times while minimizing greenhouse gas emissions for all travelers, though focusing mainly on drivers. ITS has two systems: one for data collection/relaying, and another for data processing. Data collection can be achieved with video cameras over busy areas, sensors that detect various pieces from location of certain vehicles to infrastructure that is breaking down, and even drivers who notice an accident and use a mobile app, like Waze, to report its whereabouts.

Advanced public transportation systems (APTS) aim to make public transportation more efficient and convenient for its riders. Electronic payment methods allow users to add money to their smart cards at stations and online. APTS relay information to transit facilities about current vehicle locations to give riders expected wait times on screens at stations and directly to customers' smart phones. Ramp meters regulate the number of cars entering highways to limit backups. Traffic signals use algorithms to optimize travel times depending on the number of cars on the road. Electronic highway signs relay information regarding travel times, detours, and accidents that may affect drivers' ability to reach their destination. With the rise of consumer connectivity, less infrastructure is needed for these ITS to make informed decisions.

Google Maps uses smartphone crowdsourcing to get information about real-time traffic conditions allowing motorists to make decisions based on toll roads, travel times, and overall distance traveled. Cars communicate with their manufacturers to remotely install software updates when new features are added or bugs are being patched. Tesla Motors even uses these updates to increase their cars efficiency and performance. These connections give ITS a means to accurately collect information and even relay that information to drivers with no other infrastructure needed. Future ITS systems will aid in car communication with not just the infrastructure, but with other cars as well.

7.13 Electrical grid

The electrical grid was designed to send consumers electricity from electricity generators for a monthly fee based on usage. Homeowners are installing solar panels and large batteries to store the power created by these panels. A smart grid is being created to accommodate the new energy sources. Rather than just electricity being sent to a household to be consumed by the various appliances in the home, electricity can flow in either direction. Additional sensors along the grid will improve information collection and decreased downtime during power outages. These sensors can also relay information directly to consumers about how much energy they are using and what the costs will be.

7.14 Power management

Effective power management is a big step toward green computing. And each company, no matter how big or small, can implement this and protect the environment. That said, there is an open industry standard called the Advanced Configuration & Power Interface (ACPI), which allows systems to turn off their components like hard drives and monitors automatically after certain periods of being inactive. You can also hibernate the systems when most of their components like RAM and CPU turn off. In addition, some programs enable you to adjust CPU voltages manually, reducing electricity consumption and heat generation. Solid-state drives (SSDs) store data in DRAM or flash memory. As there are no moving parts in them, they consume even less power. In addition, IT companies must manage power in GPUs, one of the biggest power consumers in computers. Use energy-efficient ways like using no graphics cards instead of a shared terminal, desktop sharing client, etc. You can also utilize motherboard video results or choose a GPU that consumes less power when idle.

7.15 Material recycling

Another excellent way to embrace green computing is to recycle materials in computing devices. It will prevent harmful substances like mercury, lead, cadmium, etc., from reaching landfills. You can also replace some equipment, instead of manufacturing them all over again, reducing emissions and saving energy. Additionally, IT companies can repurpose or donate computers that they no longer use to nonprofits and charities. Also, parts of outdated systems and supplies like paper, batteries, and printer cartridges are recycled. However, care must be taken while recycling old computers due to privacy issues as they can contain some data of the previous users that needs destruction before recycling.

7.16 Cloud, edge, and parallel computing

Cloud computing helps IT companies adopt green computing by addressing issues like resource consumption and energy usage. The approaches like virtualization, energy-efficient data centers, multitenancy, etc., enable cloud computing to reduce energy usage and carbon emissions. Using cloud computing eliminates the need to have energy-consuming data centers. Parallel computing, similarly, helps reduce energy consumption as multiple small-size computations or functions can run simultaneously on multiple processors that interact through shared memory, instead of running on separate hardware consuming more power.

7.17 Telecommuting

Remote work is rising high, especially after the COVID-19 breakout. Implementing teleconferencing and working from anywhere come with many benefits like reduced emissions from travel, the convenience of workers, and higher profit margins due to low office maintenance.

7.18 Real-life applications of optimization

- Traveling salesman problem (TSP): This is one of the most common combinatorial optimization problems in real life that can be solved using genetic optimization. The main motive of this problem is to find an optimal way to be covered by the salesman, in a given map with the routes and distance between two points. If genetic algorithms are used in finding the best route structure, we don't get the solution only once. After each iteration, we can generate offspring solutions that can inherit the qualities of parent solutions. TSP has a variety of applications like planning, logistics, and manufacturing (Fig. 1.15).
- Vehicle routing problem (VRP): The basic VR can be considered as a generalization of the TSP problem which is also a combinatorial optimization problem. In this problem we find an optimal weight of goods to be delivered or find an optimal set of delivery routes when other things like distance, weights, depot points are constrained or have any kind of restrictions. Genetic approaches are competitive with tabu search and simulated annealing algorithms in terms of solution time and quality.
- Financial markets: In the financial market, using genetic optimization, we can solve a variety of issues because genetic optimization helps in finding an optimal set or combination of parameters that can affect the market rules and trades. For example, in the stock market, any rule is a popular tool for analysis, research, and deciding to buy or sell shares. In this example, the success of trading depends on the selection of optimal values for all parameters and combinations of parameters. Genetic algorithms

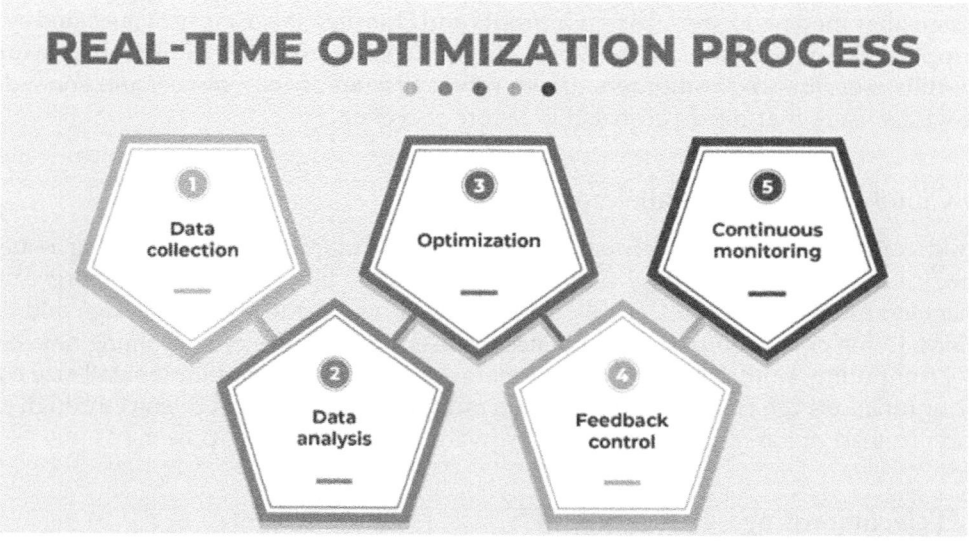

FIGURE 1.15   Real-time optimization process.

can help in finding the optimal and sub-optimal combinations of parameters. Also by genetic optimization, we can find out the near-optimal value from the set of combinations.

- Manufacturing system: One of the major applications of genetic optimization is to minimize a cost function using the optimized set of parameters. In manufacturing we can see various examples of cost function and finding an optimal set of parameters for this function can be performed by following the genetic optimization. In many cases, we can find the application of genetic optimization in product manufacturing (variation of production parameters or comparison of equipment layout). The main motive behind applying genetic optimization is to achieve an optimum production plan by taking into consideration dynamic conditions like inventories, capacity, or material quality.

- Mechanical engineering design: In many designing procedures of mechanical components, we can also find the application of genetic optimization. We can take aircraft wing design as an example where we are required to improve the ratio of lift to drag for a complex wing. This kind of designing problem can be considered as a multidisciplinary problem, the fitness function in genetic optimization can be altered by considering some specific requirement of the design.

- Data clustering and mining: Data clustering can be considered an unsupervised learning process where we try to segment data based on the characteristic of data points. One of the major parts of the procedure is to find out the center point of the clusters and we know that genetic algorithms have great capability of searching for an optimal value. In data clustering and mining we can use genetic algorithms to find a data center with an optimal error rate.

- Image processing: There are various works and researches which show the use cases of genetic optimization in various image processing tasks. One of the major tasks related to genetic approach in image processing is image segmentation. Although these genetic optimizations can be utilized in various areas of image analysis to solve complex optimization problems. Using genetic optimization in an integrated manner with image segmentation techniques can make the whole procedure an optimization problem.

- Neural networks: Neural networks in machine learning are one of the biggest areas where genetic algorithms have been used for optimization. One of the simplest examples of use cases of genetic optimization in neural networks is finding the best fit set of parameters for a neural network. Instead of these, we can find the use of genetic algorithms in neural network pipeline optimization, inheriting qualities of neurons, etc.

- Wireless sensor networks: The wireless sensor network is a network that includes spatially dispersed and dedicated centers to maintain the records about the physical conditions of the environment and pass the record to a central storage system. Some notable parameters are the lifetime of the network and energy consumption for routing which plays key roles in every application. Using the genetic algorithms in WSN we can simulate the sensors and also a fitness function from GA can be used to optimize, and customize all the operational stages of WSNs.

- Medical science: In medical science, we can find many examples of use cases of genetic optimization. The generation of a drug to diagnose any disease in the body can have the application of genetic algorithms. In various examples, we find the use of genetic optimization in predictive analysis like RNA structure prediction, operon prediction, and protein prediction, etc. also there are some use cases of genetic optimization in process alignment such as bioinformatics multiple sequence alignment, gene expression profiling analysis, protein folding, etc.

- Data collection: The first step is to collect data from sensors or other sources. This data can include variables, such as temperature, pressure, flow rate, or other variables, depending on the system or process being optimized.
- Data analysis: The next step is to analyze the data using advanced algorithms and models. That may involve using ML, AI, or other techniques to identify patterns or trends in the data.
- Optimization: Based on the analysis, the system can make real-time decisions or recommendations to optimize performance. That may involve adjusting parameters like temperature, pressure, or flow rate to achieve optimal performance.
- Feedback control: Optimization typically uses feedback control systems that continuously make adjustments based on the data. That ensures the system is constantly optimizing performance based on the latest information.
- Continuous monitoring: Finally, optimization involves continuously monitoring the system to ensure it performs optimally. That may involve setting up alarms or other notifications to alert operators if the system deviates from its optimal performance.
- Chemical manufacturing: Optimization can be used to optimize the production of chemicals by adjusting parameters, such as temperature, pressure, and flow rate, to achieve the optimal yield, reduce waste, and improve efficiency.
- Power generation: Optimization can be used to optimize the operation of power plants by adjusting parameters, such as fuel consumption, boiler temperature, and turbine speed, to maximize efficiency and minimize emissions.
- Transportation: Optimization can be used to optimize traffic flow by adjusting traffic signals in real-time to reduce congestion and improve safety.
- Agriculture: Optimization can be used to optimize crop production by adjusting parameters, such as irrigation, fertilization, and pest control, to maximize yield and minimize waste.
- Healthcare: Optimization can be used to optimize patient care by analyzing real-time data, such as vital signs, medication administration, and medical records, to make informed decisions and improve outcomes.

7.18.1 Benefits of optimization

Optimization can provide a range of benefits in different industries and applications. Here are some of the main benefits of optimization.

- Improved efficiency: Optimization can help optimize the use of resources, regardless of type—energy, raw materials, or labor—leading to improved efficiency in the production process.
- Increased productivity: Optimization can increase productivity by minimizing downtime, reducing errors, and streamlining processes.
- Cost savings: Optimization can help reduce costs by optimizing the use of resources and minimizing waste, leading to lower production costs.

- Improved quality: Optimization can help improve the quality of products or services by ensuring that processes are operating at their optimal levels.
- Enhanced safety: Optimization can help enhance safety in the workplace by monitoring processes in real-time and detecting any issues or anomalies that could pose risks.
- Greater flexibility: Optimization can provide greater flexibility by adapting processes quickly to changing conditions or requirements.
- Improved sustainability: Optimization can help improve sustainability by optimizing the use of resources and minimizing waste, leading to reduced environmental impact.

8. Computational learning theory

Computational learning theory (CoLT) is a branch of AI concerned with using mathematical methods or the design applied to computer learning programs. It involves using mathematical frameworks for the purpose of quantifying learning tasks and algorithms. It seeks to use the tools of theoretical computer science to quantify learning problems. This includes characterizing the difficulty of learning specific tasks. Computational learning theory can be considered to be an extension of statistical learning theory or SLT for short that makes use of formal methods to quantify learning algorithms.

Computational learning theory (CoLT): Formal study of learning tasks. Statistical learning theory (SLT): Formal study of learning algorithms.

8.1 Importance of computational learning theory

Computational learning theory provides a formal framework in which it is possible to precisely formulate and address questions regarding the performance of different learning algorithms. Thus, careful comparisons of both the predictive power and the computational efficiency of competing learning algorithms can be made. Three key aspects that must be formalized are:

The way in which the learner interacts with its environment, the definition of success in completing the learning task, and a formal definition of efficiency of both data usage (sample complexity) and processing time (time complexity). It is important to remember that the theoretical learning models are abstractions of real-life problems. Close connections with experimentalists are useful to help validate or modify these abstractions so that the theoretical results reflect empirical performance. The computational learning theory research has therefore close connections to machine learning research. Besides the model's predictive capability, the computational learning theory also addresses other important features such as simplicity, robustness to variations in the learning scenario, and an ability to create insights to empirically observed phenomena.

8.2 Computational learning theory in machine learning

These are sub-fields of machine learning that a machine learning practitioner does not need to know in great depth in order to achieve good results on a wide range of problems. Nevertheless, it is a sub-field where having a high-level understanding of some of the more prominent methods may provide insight into the broader task of learning from data. Theoretical results of computational learning in machine learning mainly deal with a type of inductive learning called supervised learning. In supervised learning, an algorithm is given samples that are labeled in some useful way. For example, the samples might be descriptions of mushrooms, and the labels could be whether or not the mushrooms are edible. The algorithm takes these previously labeled samples and uses them to induce a classifier. This classifier is a function that assigns labels to samples, including samples that have not been seen previously by the algorithm. The goal of the supervised learning algorithm is to optimize some measure of performance such as minimizing the number of mistakes made on new samples. In addition to performance bounds, computational learning theory studies the time complexity and feasibility of learning. In computational learning theory, a computation is considered feasible if it can be done in polynomial time.

There are two kinds of time complexity results:

Positive results—Showing that a certain class of functions is learnable in polynomial time. Negative results—Showing that certain classes cannot be learned in polynomial time. Negative results often rely on commonly believed, but yet unproven assumptions, such as:

Computational complexity $-$ P $\neq$ NP (the P vs. NP problem); Cryptographic $-$ One-way functions exist.

9. Probabilistic models

Machine learning algorithms today rely heavily on probabilistic models, which take into consideration the uncertainty inherent in real-world data. These models make predictions based on probability distributions, rather than absolute values, allowing for a more nuanced and accurate understanding of complex systems. One common approach is Bayesian inference, where prior knowledge is combined with observed data to make predictions. Another approach is maximum likelihood estimation, which seeks to find the model that best fits observational data.

Probabilistic models are an essential component of machine learning, which aims to learn patterns from data and make predictions on new, unseen data. They are statistical models that capture the inherent uncertainty in data and incorporate it into their predictions. Probabilistic models are used in various applications such as image and speech recognition, natural language processing, and recommendation systems. In recent years, significant progress has been made in developing probabilistic models that can handle large datasets efficiently.

9.1 Categories of probabilistic models

These models can be classified into the following categories:
Generative models discriminative models. Graphical models.

9.1.1 *Generative models*

Generative models aim to model the joint distribution of the input and output variables. These models generate new data based on the probability distribution of the original dataset.

Generative models are powerful because they can generate new data that resembles the training data. They can be used for tasks such as image and speech synthesis, language translation, and text generation.

9.1.2 Discriminative models

The discriminative model aims to model the conditional distribution of the output variable given the input variable. They learn a decision boundary that separates the different classes of the output variable. Discriminative models are useful when the focus is on making accurate predictions rather than generating new data. They can be used for tasks such as image recognition, speech recognition, and sentiment analysis.

9.1.3 Graphical models

These models use graphical representations to show the conditional dependence between variables. They are commonly used for tasks such as image recognition, natural language processing, and causal inference.

9.2 Importance of probabilistic models

Probabilistic models play a crucial role in the field of machine learning, providing a framework for understanding the underlying patterns and complexities in massive datasets. Probabilistic models provide a natural way to reason about the likelihood of different outcomes and can help us understand the underlying structure of the data. Probabilistic models help enable researchers and practitioners to make informed decisions when faced with uncertainty. Probabilistic models allow us to perform Bayesian inference, which is a powerful method for updating our beliefs about a hypothesis based on new data. This can be particularly useful in situations where we need to make decisions under uncertainty.

9.3 Advantages of probabilistic models

Probabilistic models are an increasingly popular method in many fields, including artificial intelligence, finance, and healthcare. The main advantage of these models is their ability to take into account uncertainty and variability in data. This allows for more accurate predictions and decision-making, particularly in complex and unpredictable situations. Probabilistic models can also provide insights into how different factors influence outcomes and can help identify patterns and relationships within data.

9.4 Disadvantages of probabilistic models

There are also some disadvantages to using probabilistic models. One of the disadvantages is the potential for overfitting, where the model is too specific to the training data and doesn't perform well on new data. Not all data fits well into a probabilistic framework, which can limit the usefulness of these models in certain applications. Another challenge is that probabilistic models can be computationally intensive and require significant resources to develop and implement.

Designing computational intelligence techniques based smart framework for sustainable computing

G. Arun Sampaul Thomas[1], S. Muthukaruppasamy[2], Periyamuthu Deivendran[3] and Saravanan Krishnan[4]

[1]Department of AI&ML, J.B. Institute of Engineering and Technology, Hyderabad, Telangana, India; [2]Department of EEE, Velammal Institute of Technology, Chennai, Tamil Nadu, India; [3]Department of IT, Velammal Institute of Technology, Chennai, Tamil Nadu, India; [4]Department of CSE, College of Engineering Guindy, Anna University, Chennai, Tamil Nadu, India

1. Introduction

Cities throughout the globe are rapidly becoming hubs of human activity and creativity, and this trend is accelerating. There is a growing need to address issues of sustainable development, effective resource management, and enhanced quality of life in our urbanized world. As a solution, the idea of "smart cities" has evolved, which centers on the utilization of IoT (Internet of Things), AI (Artificial Intelligence), and big data to propel data-driven urban growth. This chapter digs into the revolutionary potential of smart cities, emphasizing their role in influencing the future of cities and bettering the lives of their inhabitants (no date a) (Zhou, 2022). ICT is used to sense, integrate, store, process, analyze, predict, and respond to key information in all aspects of urban operation to provide intelligent responses and aid decision-making for urban livelihoods, environmental protection, public safety, urban services, industrial and commercial activities, and urban residents' quality of life. Hardware, software, administration, computing, data analysis, and other services are used in "Smart Cities" in metropolitan areas. AI, big data, cloud computing, IoT, mobile internet, and intelligent sensing terminals power smart cities' urban operating systems (no date b) (Rabe et al., 2021).

China's smart city development started slowly but is currently accelerating. Almost every province, municipality, and autonomous territory in China has a smart city pilot plan, and that number will expand. Guangdong, Shandong, and Shanghai are China's smart city innovation centers, with over 600 smart city pilots established since 2012. China's urban development has shifted toward human-centered, high-quality structures. A human-centered, high-tech urban construction ecosystem is emerging as the government, developers, integrators, service operators, and third-party service agencies collaborate to build a more efficient, complex, and intelligent governance model. Smart transport, security, and communities are becoming the main application scenarios because of legislative support, mature technology, and infrastructure (no date c) (Zhou, 2022).

2. The prospects of high-tech smart cities

Projections show that by 2050, the vast majority of the world's population will live in urban settings, giving the notion of smart cities great relevance. To put it simply, a smart city is one in which state-of-the-art technology is used to improve transportation, energy efficiency, waste management, and public safety. AI, IoT, and Big data Analytics with block chain are the catalysts for this shift, and they completely alter the way people live in cities (Zhou, 2022; Rabe et al., 2021).

3. IoT and information amassment

The Internet of Things (IoT), a network of linked devices outfitted with sensors and software, is fundamental to smart city architecture. These gadgets gather and transmit information from a wide range of sources, providing instantaneous insights on the city's operations. Imagine a metropolis where sensors are installed to track and record data on things like traffic, air quality, energy use, and more. These statistics provide a holistic picture of city operations, allowing officials to make educated choices and respond rapidly to problems (no date a) (Arun Sampaul Thomas et al., 2023).

4. AI and data analysis

When combined, IoT and AI can unleash the full potential of smart cities. Artificial intelligence processes the enormous amounts of data that IoT devices produce, and it then extracts insightful patterns. Predicting traffic congestion, optimizing energy distribution, and spotting abnormalities that demand rapid attention are all within the capabilities of AI systems. The end result is a city that can quickly adjust to new situations, improving efficiency and cutting down on waste.

5. Blockchain solutions

Traditional smart city data silos are disrupted by blockchain. Blockchain facilitates data interchange across organizations and systems without needing system modifications by circulating each system's original data or data fingerprints. Using carrier aggregation (CA) technology, the anonymous blockchain will be trusted, with rights and duties assigned to systems and persons through identity authentication and consensus authority settings. Data processing leaves fingerprints on the blockchain, which can be monitored due to its immutability, transparency, and strict time stamp mechanism. Big data analysis and AI algorithms have also enabled more precise and personalized service delivery and management (no date c). Urban planning is changing with the promise of "smart cities" leveraging technology to increase liveability, productivity, and sustainability. Data-driven solutions and digital technologies enhance infrastructure, services, and resource management in these cities. This chapter explores the smart city movement's core principles, the efforts made in this area, and the barriers that have stopped the smart city model from reaching its full potential (no date d).

6. Concrete applications of smart cities

6.1 Challenges of establishing smart cities

During smart city building, many IoT smart terminal devices are exposed in public, increasing the cyberattack risk. Data and systems are hacked at any moment, causing huge expenses and losses for important firms. This security hazard hinders smart city development (Tamilmani et al., 2023; Qose and Fregan, 2022). Information silos are another issue since they prevent natural collaboration between different initiatives. There are several issues with communication that lead to higher costs of operation and maintenance, including inconsistencies in data structures and circulation interfaces, a lack of standardization, and a significant lack of connectivity. High-quality analysis-based prediction are now beyond the reach of one-dimensional data analysis. Data silos are a growing problem that threatens to undermine the early warning and preventive capabilities of many different systems by preventing them from communicating with one another in times of crisis.

Globally, smart cities are already altering the cityscape. Some concrete ways in which the AI, IoT with Blockchain, and data-driven urban planning are improving city dwellers' lives are listed below.

Smart transport: metropolitan areas suffer from traffic congestion often. To alleviate traffic congestion and shorten travel times, smart cities use Internet of Things (IoT) sensors and Artificial Intelligence (AI) analysis to streamline traffic flow. Thanks to the Internet of Things, traffic in Barcelona is now 20% less congested than before.

Smart energy management is a solution to the inefficiency and pollution of many conventional energy use practices. Sensors connected to the Internet of Things track energy use in smart cities, allowing for more efficient distribution and less waste. Thanks in large part to London's energy monitoring measures, the city has successfully reduced its energy use by 10%. Overflowing landfills and sloppy trash collection are problems that affect many urban areas. By enabling authorities to track garbage volumes and optimize collection routes using the Internet of Things, we can reduce both costs and environmental impacts.

Disposal expenses in San Francisco have gone down by 20% because of IoT-based trash management. The use of artificial intelligence (AI) in public safety initiatives enhances police work and emergency response times. Internet of Things sensors able to detect gunshots and other emergencies, allowing for quicker and more accurate responses. The installation of a gunshot detection system in Chicago has led to a 20% drop in gun deaths. The Internet of Things and artificial intelligence are changing the face of healthcare.

The use of Internet of Things (IoT) devices for remote patient monitoring enables preventive treatments, decreasing the need for hospitalization. Hospitalizations in Singapore have fallen by 10% thanks to healthcare efforts fueled by Internet of Things sensors. Every smart terminal device will own a unique set of public and private keys, which can be verified by the blockchain, ensuring the security of the overall smart city system's development. By keeping track of terminal IDs via smart contracts, the blockchain system can prevent rogue terminals from accessing the network and prevent data pollution. Consequently, the blockchain has created a reliable data-sharing infrastructure. Furthermore, local governments install blockchain nodes locally, validate data localization, identify the data source and validity, and guarantee the data's integrity from its origin to the user's terminal (Shen and Pena-Mora, 2018; Athira Jayavarma et al., 2022).

6.2 The potential of blockchain-enabled smart cities

Constructing a novel kind of "smart city" is a methodical endeavor that calls for the simultaneous use of several different technologies while making full use of their mutual benefits. Therefore, municipal planners can't just use blockchain. To support the intelligent growth of city administration and services, they should aggressively integrate new technologies like big data, cloud computing, and artificial intelligence. This would speed up the exploration of urban application scenarios under technology integration and improve resident outcomes. Blockchain technology improves central systems in new smart cities, intermediate large data resource centers, and sophisticated intelligent applications. Logistics for temperature-sensitive goods, such as pharmaceuticals, perishable foods, and beverages, is a prime example of a use case for blockchain technology. The supply chain begins with the suppliers and continues via logistics providers, customers, and end users. Data are not shared publicly or easily in the conventional transportation process, and only logistics providers have access to all transportation data. Logistical firms collect one-of-a-kind data because process errors cannot be precisely described.

Blockchain and the internet of things can verify all perishable food and other items in the cold chain, from the warehouse to consumers. IoT acquisition chips can collect data at every point and moment, and after the authorized signature, the item's temperature, humidity, geographical location, lighting, and other parameters can be reported and uploaded to the blockchain.

Blockchain's potential applications are currently being investigated. There are more possibilities today since rules from 1024 nations provide guidance and the public is increasingly gaining a more thorough grasp of blockchain technology. CyberVein has contributed to the food industry's supply chain by working with seafood suppliers (Qose and Fregan, 2022).

Large, long-distance fishing vessels sometimes spend anywhere from a month to 6 months at sea to catch their maximum quota of fish before returning to port. The only way for the folks on shore to get in touch with the sailors is via the captain's communication equipment, which is highly costly to use and has no reception out at sea. They would have to bring all their catch back to the harbors, tabulate it, and then store it until they could sell it. Because of this, the seafood exchange platform can now collect data on the catch directly from the vessels that caught the fish. This has reduced the time and money spent by both parties waiting for the items to be delivered after landing at the port. Image recognition software instantly add images of recently caught fish to the blockchain. This ensures the data on seafood is accurate, which is important since some fishermen try to deceive customers by giving them false information using the more traditional method. As the location of the fish's native habitat will affect its price, the GPS system also records the area in which the seafood is captured. Since the cost of some seasonal seafood items varies greatly throughout the year, the exact time of catch is also recorded. Companies and eateries in the food service business that are part of the exchange network use the blockchain data to determine the fair market value of the seafood they are interested in purchasing. The fisherman do all of this while out on the water or on their way home. The blockchain has made it easier for the fishermen and the purchasers to establish confidence, which has led to more equitable pricing and a more streamlined transaction procedure. If a fisherman's catch isn't sold right away, the platform can connect them with third-party service providers who can store it until it's sold, as well as logistics providers who can transport it to the buyers; information about the process of storage and delivery is recorded and made transparent via the blockchain (no date c).

7. The road ahead for smart cities

Through the IoT, towns all around the globe upgraded to "smart cities," ushering in a whole new age of urbanization. Future city inhabitants will reap the advantages of higher security, less traffic, decreased pollution, more efficient energy usage, and an improved quality of life.

The notion of "smart cities" has been around for around 10 years, but it has evolved significantly since its inception and is now set to drastically transform city life due to the advent of key enablers like the Internet of Things. There's little doubt that adding sensors to common gadgets that gather and send data is a useful tool for both business and personal life. This new trend shows how powerful IT systems propel smart city development to heights that its creators couldn't have dreamed of a decade ago.

With sensors installed in almost every vehicle, gadget, or piece of equipment used in a city daily, IoT technologies will enable future smart cities to completely transform the way we live and do business. The Internet of Things generates vital data for a wide variety of business intelligence systems, including those used for crime prevention, emergency services, parking management, and many more. In addition, Smart America estimates that local governments will spend more than $41 trillion on infrastructure improvements over the next 2 decades so that they can take advantage of the IoT. The world's brightest cities are currently implementing the following program.

- Intelligent roadways, which monitor traffic conditions and adjust accordingly (Sharma et al., 2020)
- Energy, lighting, and resource optimization in smart buildings (Arun Sampaul Thomas et al., 2023; Muthukaruppasamy et al., 2018)

- Adaptive streetlights and other forms of smart lighting
- High-tech garbage monitoring and collection systems
- Smart grids can monitor and allocate energy use based on the current situation.

Since traffic management and parking 1 day be computer-controlled, these are only two examples of the numerous metropolitan infrastructures that will need updating as IoT technology progresses. Many futurists, for instance, envision a day when self-driving cars would drop off city dwellers at their destination, then drive themselves to the next parking spot and send a notification to the owner. While this is an intriguing possibility, it will need substantial infrastructure upgrades, such as new highways and improved communication networks, to become a reality (Sharma et al., 2020).

This is only one way in which the Internet of Things (IoT) promises to improve people's day-to-day lives in the cities of the future (no date b). As the world rushes toward a future where smart cities rule, several themes are shaping their future: (no date a) The uplift of Internet of Things (IoT) devices will lead to the development of a massive network of linked systems, which in turn will improve the efficiency with which data can be collected and analyzed. The growth of AI will enable smart cities to enhance their prediction abilities, automate routine jobs, and make better decisions. The use of big data will help cities better understand their residents' habits, requirements, and preferences, paving the way for more strategic urban planning and design.

Internet users greatly benefit from blockchain technology since it can be used to fortify digital rights, verify digital access, and secure digital assets. Still, widespread adoption is a long way off, and the future of blockchain is clouded by a brew of discouraging economic reports, inane commercial marketing, and a nebulous libertarian ideology that struggles to win converts while promising "magic money." To the extent that the blockchain adoption as an urban developer and scientist, I do so with the goal of creating a system that makes utilities more widely available. Blockchain makes me think of a city where people join in droves if they were motivated to do so, since cities have always been about pooling resources. Blockchain technology is used to make a city's shared resources more accessible to the public (no date e) (Shen and Pena-Mora, 2018).

The public spaces and private homes that make up a city are inseparable. They can be found in every major city. The well-off rely on the city's shared services, while the poor carry their belongings on their backs. There are endless artifacts in the urban infrastructure, which is an overlay of technological and social networks interlaced with an accumulation of physical substance structured for human comfort. The urban infrastructure seen as a system of transient buildings that can easily adjust to new circumstances. The blockchain strikes me as an innovative piece of ubiquitous sociotechnical urban infrastructure that used to improve current urban environments. With this clarification, blockchain technology used not only as a digital replacement for traditional banking but also as a decentralized system of storage lockers for tangible objects.

7.1 Free and available urban technology

While industrialization has been the driving force behind societal changes in the last two centuries, cities have long served as political and social hubs, amassing wealth in the form of objects, treasures, virtual capitals, knowledge, techniques, cultural wonders, works of art, and monuments ever since they were first built. However, because of industrialization, capital was redirected from real estate to other, more productive investments. At the same time as industrialization allowed for the nearly limitless spread of exchange values, banking made wealth movable and built exchange networks that allowed the movement of money. The contemporary era has shown the many ways in which consumer goods function as a logic, a language, and a universe unto themselves.

Since the active people fled the city for the industrial agglomeration around or apart from it, it is assumed that industry was the primary mobile for urbanization. At the same time as many historic downtown areas degenerated or erupted, ancient agricultural systems collapsed, forcing many uprooted peasants to relocate to cities in search of jobs and a means of survival. Offices have supplanted homes in urban cores as wealth is redistributed to more remote residential or productive peripheries. These hubs deteriorate into "poor" ghettos when they are left to the "poor." Urban centers have supported an intensive, deteriorated "style of life" throughout history, and this has never changed. Urban centers adapt to survive in the face of erosion and incorporation by urbanization processes. They can stay in business because of their dual function as a consumer destination and a destination for consumers. Central cities are always in the process of becoming something new, and they serve as hubs for social interaction, value exchange, information dissemination, organizational capacity, and institutional decision-making. Facts, representations, and pictures from the past are used in the theoretical construction of the city, but this construction is always evolving and refining itself.

7.2 Blockchain technology to facilitate ongoing evolution

The term "open-source urbanism" was used to describe the decade-long fusion of urban theory with hacker culture and the sharing economy. The IT sector and the do-it-yourself (DIY) movement both contributed to a proliferation of prototypes for urban innovation, many of which sought to influence established social institutions. This goal is embedded in the concept of the internet itself, which undermines the traditional systems of social control. Open-source initiatives rely on a global community to share knowledge and resources. As a result, the stakes and frameworks for property and government are shifting. Instead of seeing open-source urban technology as an expansion or complement to existing social skills, we should view them as manifestations of this cyborg "beta" urbanization. In truth, infrastructure is not an extra but rather a fundamental human right to choose and reorder one's own social existence. Most prototypes in today's age of open technologies will never be finalized, yet they continually branch and make possible innovative extensions of themselves. Instead of striving for perfection, prototypes are more interested in spreading throughout existing infrastructure by building, breaking down, and rebuilding. One way to see the prototype is as a form of "infrastructural being," a shifting collection of people and objects whose holding processes are "in suspension" of private effects and public sway. An open ledger that records possessions and associated digital identities is well suited to sustain a society of this size given the fluid nature of short-term property and individual well-being.

7.3 Blockchain to organize urban chaos

There are essentially three eras in the development of contemporary urban areas: First, urban reality as it was before industrialization is deconstructed and its practices and philosophy transformed into trade value. The second is the worldwide development of cities and the rise of urban society. There is an economic and social acceptance of a new urban awareness. Thirdly, urban cores suffer from their own destruction in both theory and practice when they strive to restore centrality. So, urban ideas are either conceived or reconceived. Separated from urban reality, the city appears to have a certain degree of social reality, including direct personal and interpersonal ties. A clear understanding of the global dynamics (economic, social, political, and cultural) that have molded urban space is essential for urban analysts. Everyday life's shifts alter cityscapes without residents' active participation. Establishing or reestablishing coherence amid chaotic chaos is an essential part of restoring centrality and redesigning urban centers. In order to properly analyze urban phenomena, one must go from the broadest possible understanding to an intimate familiarity with each unique urban condition. It necessitates the fusion of breaks in the continuity of urban formations, social relations, and relations between individuals and groups that have been tracked over time and space, always translated on the ground and etched into the urban landscape by forces both past and future.

In his 1968 book The Right to the City, Henri Lefebvre claimed that the city was "a projection of society on the ground, not only on the actual site but at a specific level, perceived and conceived by thought, which determines the city and the urban." Without privileged places and times of meeting unencumbered by the market, apart from the means of exchange value and the ties that determine profits, the multiplication and complexity of exchanges are impossible. Blockchain developers seemed unable to provide remedies to the urban issue and to act otherwise than through technological performances that simply protected the status quo, because of their focus on planning industries and organizing businesses. Marketers' misunderstanding of the socialization of technology' has slowed widespread adoption. The anarchist rhetoric and the pirate hacker character have frightened off the nestlings and squelched their zeal. Urbanization is synonymous with socialization, which refers to widespread acceptance by society. Blockchain, which provides a standard for the digital recording and transfer of everything, seems to be the most helpful tool for bringing order to the chaos of cities and so increasing their overall worth.

7.4 Blockchain's potential to improve urban life

The logic of money and trade value is inherent in the retail world and transcends all boundaries. The logic of goods and services is not the same as that of urban society, which is a collection of temporal acts privileging a space and being favored by it. The value of usage underpins urban planning. A decade of widespread internet use and the rise of social media have shifted the emphasis of human interactions onto digital platforms. Urban societies' reality nevertheless fights and protects the use value of a city, which has been subjected for millennia to trading value. Residents' initiatives to preserve city centers in the face of certain tensions—between use value and exchange value; between the mobilization of private wealth and unproductive public spending; by regulating the distribution of activities and time allocation toward streets and neighborhoods—have helped spread the concept of bike-sharing, car-sharing, house-sharing, coworking, and community-operated spaces (Gopal et al., 2021).

Recent decades have seen an increase in the degree to which cities' infrastructure and residents are technologically advanced. In metropolitan regions, information technologies are often accessible to everyone who wants to use them, but they need complicated infrastructures of hardware, software, and resources to function. Rules for access, extraction, and exclusion must be put into open technologies to maintain the long-term viability of the shared resource. In their economic models, digital ownership—that is, property over one's own data, one's user profile, and one's digital collection represents values that used to motivate users to adopt the most responsible behavior toward a shared resource. Any city greatly benefits from blockchain technology if it were used to tokenize real-world assets and incorporate them into a token economy in which crypto users are given preferred privileges or benefits for adopting the best behavior for society at scale (Thomas and Robinson, 2020; Thomas et al., 2020).

7.5 Personal belongings in the information age

In the previous 10 years, we saw a dramatic transition from analogue to digital products. Digital content has supplied not just media needs such as books, music, and movies but also software-enabled products such as phones, vehicles, and medical equipment. There is mounting evidence in the judicial system, the marketplace, and our personal lives that the ability to legally own, manage, repair, and utilize the objects we purchase is significantly impacted by whether those products are analogue or digital. Analogue and digital personal possessions have vastly distinct legal regimes. Having been applied for hundreds of years to all forms of portable property, the rules of analogous ownership seem self-evident. In the case of digital products, the rules are different: you do not "own" the e-books you pay for, but rather a "license" to use them.

The issue of digital ownership extends well beyond the realm of books. Most of us purchase items and utilize services with the expectation that we have certain rights regarding their usage, but these rights are really established by a nonnegotiable agreement. Manufacturers' and merchants' weak interpretation of consumers' rights amounts to a claim to ownership. The most obvious consequence of not owning anything is the loss of many important rights. Selling something you don't own is illegal. It's not something you can loan out or gift. Unauthorized media players and readers are prohibited. The technology you rely on is not repairable or open source. There are compelling arguments in favor of relinquishing such rights, such as the availability of free, unrestricted material in any location. However, buyers aren't fully aware of the differences between ownership and licensing, and they also forego reliability and longevity without even realizing it. Blockchains provide a permanent, distributed ledger of digital assets with rules for ownership that can't be changed. It assigns unique identifiers to digital goods, making them just as readily movable, tradable, and inheritable as tangible possessions.

After the worldwide economic crisis of 2008, the concept of the blockchain was proposed as a means of providing an alternative to the centralized banking system. To encourage positive user behavior, the blockchain first focused on cryptocurrency before branching out into other industries. Ethereum's blockchain has been a major driver of the rise in popularity of smart contracts since sometime in 2016, since it offers a transparent and traceable platform on which parties engage in trust less transactions with one another without the need for middlemen. Nonfungible tokens allow crypto tokens to represent anything other than fiat money, such as admission to a movie, loyalty points, ownership in a firm, or a license to use a piece of software. Many people still think that our present understanding of blockchain is inadequate and that there is a lack of information on the areas in which blockchain technology deliver the social benefits, despite the efforts of blockchain developers. One of the most unrealized benefits of blockchain technology is NFTs, which were overlooked since its digital ownership paradigm was seen as a fleeting internet fad.

In 2021, when the NFT market was experiencing record sales, public interest in them skyrocketed. To have a more accurate statistical grasp of the NFT market, academic research analyzed data from June 23, 2017, to April 27, 2021 (Hemenway Falk, Tsoukalas and Zhang, 2022; Nadini et al., 2021), covering 6.1 million transactions of 4.7 million NFTs on the Ethereum and WAX blockchains. While art NFT transactions account for just 10% of worldwide exchanges, their historical dominance of the NFT market volume makes them a useful vehicle to hold value for certain niche markets. Like works of art themselves, art NFTs are purchased by admirers or investors who want to hold on to them for the long haul rather than quickly flip them. Since the digital abundance of NFTs in games, for example, led to a large reduction in value, the most traded NFTs belong to games and collections for which NFTs were found to be bad speculative assets. NFTs cost an average of $15, with the costliest one's going for an average of $1500. The average selling price for the top 1% is above $12,000 for utility NFTs. Another realization is that NFTs have their own speculative bubble, which is difficult to anticipate and hence heavily reliant on hype cycles, even if their prices are driven by bitcoin values. As a result, the reliance on hype has generated concerns about the rise of wash trading, in which the same firm buys and sells the same financial instrument to create a false impression of market activity.

The NFT market is very concentrated, as shown by a network analysis of trading in NFTs. The top 10% of traders account for 85% of all transactions and trade 97% of all assets in the crypto market, which specializes in NFTs. Many dealers focus on only a few collections of aesthetically similar items. At least 73% of their business is conducted in their primary collection, and 82% in their primary and secondary collections combined. Traders often focus on one set of assets and transact with other traders who deal in the same set of NFTs when buying and selling. The most active 10% of purchasers and vendors account for 90% of all business. The promise of NFTs to decentralize wealth, expertise, and power has not been realized, at the very least (Tamilmani et al., 2023).

The long-term viability of the embedded creator royalty concept is frequently questioned since prudent investors typically remove the royalty from their speculation before making a purchase. In the minds of prospective buyers and sellers, the resale royalty rate is a tax (a transaction cost) on future revenues from the secondary sale of the NFT, which slow down resales. In practice, owners of NFTs migrate their NFTs from the architecture of the original minting platform to an off-platform wallet and then sell their NFTs on a marketplace platform that does not employ the architecture and smart contract protocols of the original minting platform. Investors shouldn't dismiss the embedded creator royalties as a simple tax on speculative income because of the several reasonable benefits they provide. For the sake of the entire NFT economy, it is important to reinforce embedded royalties at the smart-contract level for at least three reasons: creator royalties allow risk sharing of future price volatility; creator royalties redistribute future revenues from better-informed speculators or market influencers; and creator royalties allow a distributed source of funding (Athira Jayavarma et al., 2022).

While NFTs (nonfungible tokens) used in conjunction with blockchain technology to safely hold an unforgeable asset on a global ledger, they are still vulnerable to plagiarism. To safeguard the rights of NFT producers and NFT owners, researchers have experimented with an approximate pattern-matching technique. Data can be protected against being copied immediately using hash algorithms; however, this security does not extend to data that has been even slightly modified. This indicates that NFT technology does not provide a fool proof solution to the issue of uniqueness. For instance, if you change only one pixel in the picture's transparent layer, you'll end up with two distinct hashes for the identical image. For an NFT to be "minted," according to one proposed NDFA-based method, the digital data linked with it must first pass a plagiarism check. It is recorded as "minted" in the ledger after verification is complete. This would cause a delay in the coining process and needs to be fixed (A New High-Performance Approach to Approximate Pattern-Matching for Plagiarism Detection in Blockchain-Based Non-Fungible Tokens, 2022).

An innovative semifungible token, halfway between the fungible and nonfungible forms, has been suggested and tested. In contrast to FTs and NFTs, which require the deployment of a new smart contract for each kind of token, SFTs provide a more scalable method for token manufacturing thanks to their single smart contracts' support for an infinite number of tokens. Fees can't be avoided with SFTs since they need a smart contract that locks in the underlying NFT asset. Also, the smart contract's purchasing functionality incorporates explicit coding for the distribution of payments to numerous recipients. There is no built-in way to purchase without paying the charge, and there is no way to trade the SFT outside of the initial smart contract. If a creator or investor wishes to swap an NFT's income model for anything else, they simply redeem and externalize the NFT. With SFTs, the NFT holder relax this restriction as soon as it releases the entire quantity to the market, secure in the knowledge that it will save money on future transactions. Fractional selling is immune to revenue stream remodeling and fee avoidance strategies if the NFT holder retains at least one share in the fractionalized NFT. While traditional NFTs are unable to deviate from their established protocol, SFTs endlessly reassembled and fractionalized to accommodate changing market conditions (Zanella et al., 1991).

8. Smart city peer-to-peer system

Businesses all throughout the globe, from Airbnb to Uber, have developed systems to put idle property to good use. It has been suggested that homeowners of automobiles, apartments, and even Wi-Fi hotspots rent them out to others in exchange for a fee. To capitalize on their monopoly position in a particular exchange market, the giants of the sharing economy have evolved into matchmaking platforms that aggregate all the available offers for shared goods on a given market, charge fees to make connections between buyers and sellers, and collect data on both parties. These p2p enterprises become more valuable as their networks expand, leading to a winner-take-all scenario in which a small number of companies end up controlling the market and drawing in most customers. If a peer-to-peer platform is trying to gain a critical mass of users, it offers discounts or reduced commissions in the beginning, only to turn around and increase prices and commissions once it has established itself as the market leader.

To create a universal sharing network with the potential to become a potent source for adding use value to cities, we must link all types of physical assets to the blockchain, including cars, houses, and more, and use the network infrastructure to rent, sell, share, or swap them in a peer-to-peer fashion. The web3 version of the sharing economy will rely less on the intermediary platform to carry out p2p transactions and more on a shared, transparent backbone, in contrast to the web2 giants. A smart contract will make costs more transparent and predictable, while an incentive scheme designed to encourage users and service providers to operate in the ecosystem's best interests will clarify their responsibilities and ensure that data privacy is protected.

8.1 Blockchain city planning: A tactic for the future

The city is seen as a functional social level by urban planners because it provides a localized but scalable diversity of human groupings with the opportunity for feedback control through social engineering. Rapid urbanization throughout the globe is a result of the population boom, which has led to several problems such as increased traffic, pollution, the depletion of nonrenewable resources, and widening socioeconomic disparities. As part of their larger attempts to create a better urban future, several cities throughout the globe are initiating blockchain projects. Blockchain cities have been proposed by researchers as the next great urban innovation that will help alleviate the problems caused by rapid urbanization.

The fundamental information and communication technologies that support cities are essentially similar to blockchain technology. So, blockchain cities readily connect to what has been researched for 3 decades under the general title Smart City, and blockchain research can benefit from the backdrop of 3 decades of sustainable and smart-city research frameworks. Smart cities are evaluated in part by how well they have performed in the past when it comes to sustainability and quality of life, with the inclusion of several cutting-edge technology components. When it comes to automating a city, the blockchain as well be the ideal infrastructure, as it offers a transparent and verifiable platform on which parties can conduct trust-free transactions through smart contracts and eliminate the need for middlemen. Crypto token models on permission less blockchains provide the way for dispersed and trust less people to better align economically, share interests, and coordinate efforts. Inevitably, token holders on the blockchain care deeply about the development of the blockchain technology that underpins their token and the utility or security value it represents.

The blockchain's potential applications in cities divided in half:

- Integrating blockchain technology into already existing municipal procedures to increase confidence, visibility, and command.
- Trying out different models of ownership and administration with the use of blockchain technology.

8.2 Blockchain, IoT, and AI together to make cities smarter and greener

Blockchain cities can apply the usual smart city categories to classify their early use cases into existing urban innovation categories: governance, economy, transportation, energy, construction, and sustainability.

Using the typical smart city categories, such as governance, economics, transportation, energy, construction, and sustainability, blockchain cities categorize their early use cases into pre-existing urban innovation categories.

Governance: A method of recording governmental decisions in a public ledger that cannot be altered and can be checked at any time. It is a mechanism through which companies provide sensitive information to the government while protecting themselves from legal repercussions. It provides secure mechanism for direct document exchange between government entities electronic voting that protects voters' privacy while still being open and honest. It deals with a taxation system in which the tax authorities can track VAT invoices and maintain a permanent record of taxable dealings. Governance based on consensual procedure for prioritizing urban requirements submitted by residents, used by authorities to formulate policy, and voted on by citizens as a means of resolving policy conflicts.

Economy: It is a network of manufacturers and consumers where the latter post requests for services and the former provide them. It is based on a method of tracking product ownership that uses the last link in the supply chain—the buyer—to identify the origin of fake items. automated vending machine blockchain records, providing consumers with instant access to machine status updates. It is an automated power payment system that requires no human intervention and consists of a smart wire and a smart socket. a cryptographically secure and blockchain-recorded smart lock that checks the legitimacy of an entered code. It is defined as a protocol for the anonymous rental of commonplace physical objects between strangers, in which no personal information is exchanged between the renter and the tenant. It deals with a decentralized marketplace for exchanging surplus foreign cash that help recycle old bills.

9. AI, IOT, blockchain-based cities uses cases

Since 2011, I've been actively involved in several urban innovation initiatives, and I've seen some concepts that are like experiments I've done in the past. Before providing you with my proposal of a strategic strategy for blockchain cities, I will analyze the ones in which I have invested the most and share my personal experience with you. Starting with a neighborhood forum called "Immobilizer Participative" in 2011, which spawned the creation of a FabLab in 2013 dedicated to DIY urban innovation, which was followed by the organization of a monthly open innovation meetup in my hometown in 2015 to showcase the solutions made up by local inhabitants to improve our city, and finally by academic research between 2018 and 2021 for my Ph.D. thesis about "Citizen Engagement in Urban Innovation," I have invested a great deal of time and energy into various experiences of In my experience, political considerations permeate every aspect of citizen and stakeholder participation in collaborative urban government. In 2016, I developed a system for citizens to submit their urban needs to local authorities using a 3D map of my hometown as a digital interface for citizen participation in decision-making of local urban developments, for which I was awarded a Youth Trophy in the economic category. By that time, I'd garnered positive attention from the regional media as well as backing from academic institutions and the government. As an advocate for the widespread use of technology in government, I've been asked to speak at several community events, but my efforts to get funding for this have so far been unsuccessful. I was unable to get funding from either the corporate sector (which saw me as a political innovator) or the governmental sector (which saw me as a potential money-maker). My whole business strategy hinged on getting everyone in a city to agree to implement urban innovation, and that was contingent on getting regular people to utilize my digital tools via a combination of widespread education and financial incentives. Similar propositions of the blockchain cities include a system to record and manage product ownership, leveraging the postsupply chain consumer, and a system for sharing any kind of everyday tangible object, which reminds me a lot of my years-long promotion of participatory housing models and local barter networks, which I evolved through various research and initiatives, culminating in the creation of a nonprofit association to synchronize inhabitants in their collaborative efforts. The fact that my study culminated in a nonprofit model rather than a for-profit one speaks volumes about how unsuccessful I was at turning the lessons I learned from my time in Germany and Denmark—two countries where the interest in organizing social life through the management of common housing is more developed than in the south of Europe—into a lucrative business. One of my former business associates, Habab, set itself up as a cooperative urban developer, but, as far as I am aware, its members have never been paid; rather, they are a loose confederation of architects who try out communal living in their spare time. When applied to housing and urban infrastructure, the open-source idea requires so much time and energy from its participants that it falls short of providing the standard of reliability, comfort, and safety that most people look for when investing a sizable sum of money in a home. People I encountered in the DIY sharing economy were a mix of utopian politicians and people living on low incomes whose ultimate objective wasn't to become self-sustaining or lucrative but to get access to a public subsidy (Bhavana et al., 2022; Jiang et al., 2022).

In a social relations-based production model, individuals work together on a similar goal, provide feedback on one another's work, and come to a decentralized agreement on the worth of the final output. In fact, I've used a methodology fairly similar to this one for all of my start-up efforts, and in 2013 and 2014, I even sought to create the open-source program Better Means for the French-speaking market. To achieve this goal, I've been able to organize a low-key consensus in which team members join me on projects, contribute their talents and time to advance a budding start-up in exchange for credits, and then depart for greener pastures in a conventional firm or their own personal endeavors whenever they like. If the project has access to liquid capital, investors exchange their shares for that, or they can retain their money in the company and wait for a better selling opportunity or continue as long-term shareholders. I admire the creative approach to organization, and I think that using blockchain for accountability would make it the most safe and scalable solution, all while accurately recording everyone's efforts. However, there are already several software firms offering project management systems, such as Trello, Asana, Hubspot, and others; I have no interest in getting into this business.

This infographic shows how artificial intelligence, blockchain, and the Internet of Things intersect to provide new opportunities for checks and balances in a future where computing power is being rapidly dispersed. Blockchain can improve the transparency of AI and IoT. IoT applications, blockchain, and process automation all benefit from software and hardware intelligence. Fig. 2.1 demonstrates how the significance of responsibility, security, data access, and trust has increased in tandem with the prevalence of connectivity and algorithms across industries and sectors.

The figure used to provide insight of AI, IoT and blockchain convergence.

Considering, on the one hand, my more than a decade of Fig. 2.1 perience in urban innovation projects, I have decided to pursue the most basic, bounded, and sellable product that can be both beneficial and lucrative from the smallest possible prototype up to the largest possible scale. In contrast, I will begin developing a blockchain-controlled smart lock after realizing that "a platform where owners can lock their utility NFTs in a vault and release a series of tokens representing shares of them" is essentially the concept of the Partage app I developed this year. To me, this is the finest plan for utility NFTs to add use value to any city's urban infrastructure and the best way to bring the digital world closer to the real world.

9.1 AI, IoT with blockchain-controlled smart lock

The use of digital locks to safeguard property has risen in popularity in recent years. The usage of smartphone as a key to unlock shared utilities like an Airbnb flat, hotel room, vehicle, or scooter, or private assets like a Tesla or BMW. You can get a wide variety of them on the market, each with its own style and set of functions, for about $300. These locks did show several flaws, however, that addressed using blockchain technology. Locking and unlocking smart locks managed by the blockchain is possible depending on the value of a contract variable. The smart contract, which will also give hardware access to the Partage Protocol's programmability and robust safety features, will handle the operation of the lock. To safely allow many people to use the same set of keys to enter a shared space, vehicle, boat, or piece of sporting equipment. Incorporating a smart contract as the underlying layer to determine lock and unlock behavior increases the flexibility and security of the conventional smart lock. Many possibilities become available thanks to the almost boundless programmability and ease with which any payment method will be linked. A high security multisign lock, for instance, would need approval from a certain number of addresses before a state change is allowed. In a nutshell, blockchain-controlled locks are significantly safer than conventional smart devices while still providing all the conveniences associated with digital technology (Thomas and Robinson, 2020; Thomas et al., 2020).

There are now three methods available for controlling the smart lock from afar: The lock's state stored in the distributed ledger. The smart lock's status update and operation will be restricted to the account used to launch the contract. For the lock to be unlocked, a legitimate transaction must be performed, at which point the smart contract will provide a token to a local server, which will then execute an unlocking script. This is the next best thing to existing smart door solutions, but it led to new problems, such as increased gas costs and a longer wait time to open the door. In addition, the user is not guaranteed an unlimited number of opens and shuts within a certain time frame if the door's opening is contingent on payment.

For the purpose I have in mind, I don't believe this is the best choice. It's also possible for a smart contract to provide a random 4-digit code that is used to unlock a keypad lock for an extended period. Choose a time slot on the calendar (anything from an hour to many days),

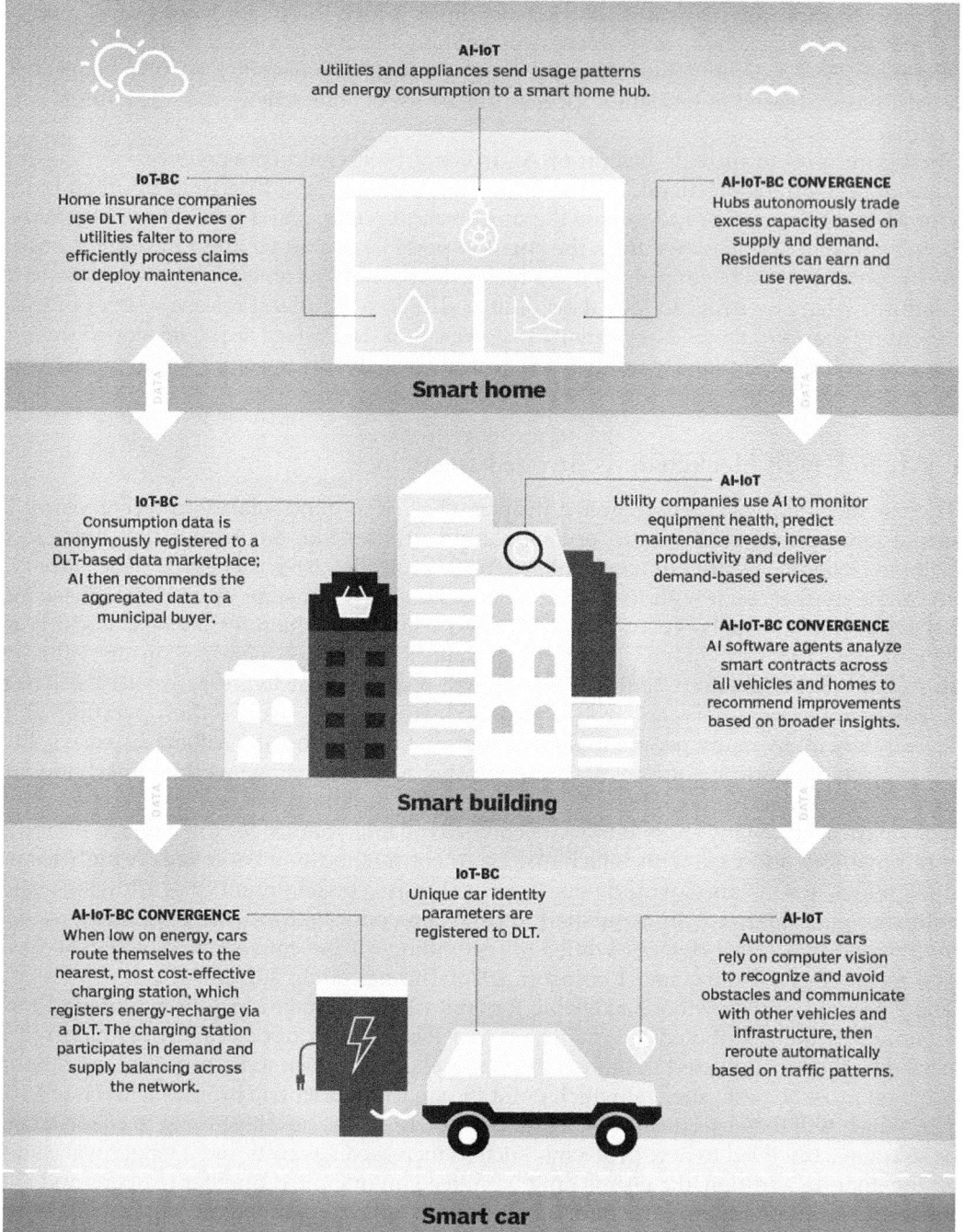

FIGURE 2.1   AI, IOT, and blockchain convergence toward sustainable smart city development.

submit cash to the owner's address, and you'll get a 4-digit code good for that period. The code will be invalid when your credit period ends, and a new code will be created at random for the next purchaser. If the buyer doesn't show up, the owner isn't inconvenienced because the door will remain locked, and no one will have access to the temporary code. On the other hand, if the property is damaged, the owner will know exactly who it did because only that one person had access to the door. Even though no one should know where the property is situated or when the code will be valid, this is not the best plan of action.

10. Conclusion

A new age of data-driven city planning has arrived with the confluence of AI, IoT, and Big Data Analytics is ushering in a new era of data-driven urban planning. The possibilities for better productivity, longevity, and general well-being are enormous. Smart cities have already shown their potential to improve urban life via programmes like smart mobility, energy management, trash reduction, and enhanced public safety and healthcare. Possible solutions include having the smart contract either transmit a token (such as an NFT) or produce a QR code (which can be read at the door's entry) to activate the lock. Since I don't believe that such door scanners are industry standards, it will be more costly and difficult to develop the initial prototypes and later to clone that solution. However, it is undoubtedly the most blockchain-centric engineering solution and likely the most marketable solution. Because of the blockchain's authentication features, buyers and fishermen feel more confident in the platform's reliability. The platform's ability to monitor the various operations within this supply chain at a lower cost has allowed it to attract new customers and contribute to the company's expansion. Innovative and potentially game-changing applications of these technologies hold great promise for the future of our cities and the development of a more live able, interconnected metropolitan area. The ultimate success of the Internet of Things and artificial intelligence will be determined by data security, not the capabilities of the underlying technology. This is nothing more than a scientific fair where the capabilities of AI and IoT devices are shown without any consideration for security. Therefore, security is the primary criterion that must be met if these technologies are to achieve widespread adoption. There is potential for these technologies to complement one another, leading to a future in which devices trade with one another using blockchain while data are analyzed using AI and machine learning.

References

A New High-Performance Approach to Approximate Pattern-Matching for Plagiarism Detection in Blockchain-Based Non-fungible Tokens. Ciprian Pungilă, Darius Galis, Viorel Negru, 2022.

Arun Sampaul Thomas, G., Muthukaruppasamy, S., Sathish Kumar, S., Saravanan, K., 2023. Green IoT use case approaches for blockchain technology taking industry 5.0 to the next level. Green Blockchain Technology for Sustainable Smart Cities 119—141. https://doi.org/10.1016/B978-0-323-95407-5.00012-8.

Athira Jayavarma, S., Preetha, P.K., Nair, M.G., 2022. Smart contract based energy trading-an overview. In: INDICON 2022—2022 IEEE 19th India Council International Conference. Institute of Electrical and Electronics Engineers Inc., India https://doi.org/10.1109/INDICON56171.2022.10039697. http://ieeexplore.ieee.org/xpl/mostRecentIssue.jsp?punumber=10038981.

Bhavana, G.B., Anand, R., Ramprabhakar, J., 2022. Designing microgrid energy markets using smart information system and blockchain technology. In: International Interdisciplinary Humanitarian Conference for Sustainability,

IIHC 2022—Proceedings. Institute of Electrical and Electronics Engineers Inc., India, pp. 12—15. https://doi.org/10.1109/IIHC55949.2022.10060599.

Gopal, Muthuselvan, Muthukaruppasamy, 2021. Model Predictive Controller—Based Quadratic Boost Converter for WECS Applications, vol 31. Wiley.

Hemenway Falk, B., Tsoukalas, G., Zhang, N., 2022. Economics of NFTs: the value of creator royalties. SSRN Electronic Journal. https://doi.org/10.2139/ssrn.4284776. Elsevier BV.

Jiang, S., Jakobsen, K., Bueie, J., Li, J., Haro, P.H., 2022. A tertiary review on blockchain and sustainability with focus on sustainable development goals. IEEE Access 10, 114975—115006. https://doi.org/10.1109/access.2022.3217683.

Muthukaruppasamy, S., Abudhahir, A., Gnana Saravanan, A., Gnanavadivel, J., Duraipandy, P., 2018. Design and implementation of PIC/FLC plus SMC for positive output elementary super lift Luo converter working in discontinuous conduction mode. Journal of Electrical Engineering and Technology 13 (5), 1886—1900. https://doi.org/10.5370/JEET.2018.13.5.1886.

Nadini, M., Alessandretti, L., Di Giacinto, F., Martino, M., Aiello, L.M., Baronchelli, A., 2021. Mapping the NFT revolution: market trends, trade networks, and visual features. Scientific Reports 11 (1). https://doi.org/10.1038/s41598-021-00053-8.

Qose, S., Fregan, B., 2022. A review of cyber security and blockchain. In: SACI 2022—IEEE 16th International Symposium on Applied Computational Intelligence and Informatics, Proceedings. Institute of Electrical and Electronics Engineers Inc., Hungary, pp. 87—92. https://doi.org/10.1109/SACI55618.2022.9919583.

Rabe, J., Ietto, B., Muth, R., Eisenhut, K., Pascucci, F., 2021. Citizens' engagement in urban development through blockchain: a human-centered design approach. In: 2021 IEEE International Conference on Technology Management, Operations and Decisions, ICTMOD 2021. Institute of Electrical and Electronics Engineers Inc., Germany https://doi.org/10.1109/ICTMOD52902.2021.9739434.

Sharma, A., Awasthi, Y., Kumar, S., 2020. The role of blockchain, AI and IoT for smart road traffic management system. In: Proceedings—2020 IEEE India Council International Subsections Conference, INDISCON 2020. Institute of Electrical and Electronics Engineers Inc., Iraq, pp. 289—296. https://doi.org/10.1109/INDISCON50162.2020.00065.

Shen, C., Pena-Mora, F., 2018. Blockchain for cities—a systematic literature review. IEEE Access 6, 76787—76819. https://doi.org/10.1109/ACCESS.2018.2880744.

Tamilmani, S., Mohan, T., Jeyalakshmi, S., Shukla, G.P., Gehlot, A., Shukla, S.K., 2023. Blockchain integrated with industrial IOT towards industry 4.0. In: 2023 International Conference on Artificial Intelligence and Smart Communication, AISC 2023. Institute of Electrical and Electronics Engineers Inc., India, pp. 575—581. https://doi.org/10.1109/AISC56616.2023.10085226.

Thomas, A.S., Robinson, H., 2020. Real-time health system (RTHS) centered internet of things (IoT) in healthcare industry: benefits, use cases and advancements in 2020. Springer's Multimedia Technologies on the Internet of Things Environment (Scopus Indexed) 978—981.

Thomas, A.S., Robinson, H., IoT, 2020. Internet of Things and Big Data Applications—Part of the Intelligent Systems Reference Library Book Series. Springer Book Chapter 180, pp. 978—983.

Zanella, G., Liu, C.Z., Choo, K.-K.R., 1991. Understanding the trends in blockchain domain through an unsupervised systematic patent analysis. IEEE Transactions on Engineering Management 70.

Zhou, J., 2022. Big data classification model and algorithm based on blockchain. In: 2022 IEEE 2nd International Conference on Electronic Technology, Communication and Information, ICETCI 2022. Institute of Electrical and Electronics Engineers Inc., China, pp. 1096—1101. https://doi.org/10.1109/ICETCI55101.2022.9832399.

Further reading

https://medium.com/@betaarrays/smart-cities-of-tomorrow-iot-ai-and-data-driven-urban-development-c933045 2d475.

https://medium.com/@bramahendramahendra1/smart-cities-concepts-implementation-and-challenges-ec0a4cf 9f694.

https://medium.com/hackernoon/smart-cities-of-the-future-powered-by-iot-59808de5c8ae#:~:text=Smart% 20cities%20of%20the%20future%20will%20allow%20IoT%20systems%20to,uses%20on%20a%20daily%20basis.

https://juliencarbonnell.medium.com/democratizing-access-to-utilities-blockchain-for-smart-cities-25eefb0348e7.

https://medium.com/cybervein/blockchain-and-smart-cities-660d20b77fa.

Multiple parameter optimization methods based on computational intelligence techniques in context of sustainable computing

Indeti Naga Padmaja, Jyothi Singaraju and K. Usha Rani

Department of Computer Science, Sri Padmavati Mahila Visvavidyalayam (Women's University), Tirupati, Andhra Pradesh, India

1. Introduction

Recently, there has been a notable surge in the field of machine learning, drawing in a large number of scholars and practitioners. It is currently one of the most well-liked research avenues and is important for a variety of applications, including recommendation systems, image recognition, speech recognition, machine translation, and more. One of the fundamental elements of machine learning is optimization. Building an optimization model and using the provided data to determine the parameters in the objective function is the fundamental process of most machine learning techniques.

The efficiency and efficacy of numerical optimization algorithms play a major role in the deployment and popularization of machine learning models in the age of massive data. To encourage the advancement of machine learning, several efficient optimization techniques.

Modern sustainability difficulties must consider coupled systems, lengthy time horizons, diverse goals, and profound uncertainty (Liu et al., 2013, 2016; Hull et al., 2015) Environmental and human systems must be taken into account in the context of sustainable ecosystem management, water planning, and climate change adaptation and mitigation, for example. These domains, or systems, are inextricably linked because modifications to the composition and behavior of the natural world frequently result in modifications to human institutions and incentives. On the other hand, the ways in which human preferences, technology, and institutions evolve greatly influence the directions in which natural resource systems grow. One or both systems frequently cease to operate sustainably if these interactions are not tracked and controlled (Hull et al., 2015; Ostrom, 2009, 2011) As an illustration, in the case of accelerated global climate change, for instance, increasing anthropogenic emissions will lead to an increase in greenhouse gas concentrations in the atmosphere, which will cause climate imbalances (e.g., altered precipitation patterns, elevated temperatures) and potentially irreversible changes in natural ecosystems (e.g., biodiversity loss) and the economy (e.g., increased inequality) (Molina-Perez et al., 2020).

It is difficult to analyze policies in the context of sustainability. First, the environmental and human domains are complex systems: vast timescales must be taken into account due to path dependencies in both domains, and their nonlinear interactions result in dynamic behavior that is challenging to predict and describe. Second, there is substantial uncertainty that impacts both domains since experts and stakeholders frequently disagree on how complex systems should be causally represented, the importance of particular analytical parameters, and the applicability of certain measures to characterize sustainability (Lempert, 2003; Marchau et al., 2019).

Opportunities are presented by contemporary computational intelligence techniques including agent-based modeling, machine learning, optimization, and data visualization (Lempert et al., 2006; Groves and Lempert, 2007; Bryant and Lempert, 2010; Kasprzyk et al., 2013; Isley et al., 2015; Kwakkel, 2017). However, when these are applied in an integrated manner, their analytical potential for sustainability sciences can be most effectively utilized.

Many areas of machine learning are significantly impacted by optimization techniques. For machine translation problems, for instance (Pajarinen et al., 2019), introduced the transformer network employing Adam optimization (Kingma and Ba, 2015; Ledig et al., 2017) suggested a generative adversarial network for superresolution images, which Adam also optimized (Wu et al., 2017). suggested the use of trust region optimization in the actor-critic model to address the issue of deep reinforcement learning in MuJoCo settings and Atari games.

Three characteristics can be distinguished between common optimization techniques when seen through the lens of gradient information: first-order optimization techniques, shown by the popular stochastic gradient techniques; Heuristic derivative-free optimization techniques, of which the coordinate descent method is a representative, and high-order optimization techniques, of which Newton's method is a typical example.

As the representative of first-order optimization methods, the stochastic gradient descent method (Robbins and Monro, 1951; Jain et al., 2018), as well as its variants, has been widely used in recent years and is evolving at a high speed. However, many users pay little attention to the characteristics or application scope of these methods. They often adopt them as black box optimizers, which may limit the functionality of the optimization methods. In this chapter, we comprehensively introduce the fundamental optimization methods. Particularly, we systematically explain their advantages and disadvantages, their application scope, and the characteristics of their parameters. We hope that the targeted introduction will help users to choose the first-order optimization methods more conveniently and make parameter adjustment more reasonable in the learning process.

High-order optimization techniques (Shanno, 1970; Hu et al., 2019) converge more quickly than first-order techniques and improve search direction due to the curvature information. While they receive a lot of attention, high-order optimizations present significant difficulties.

The operation and storage of the Hessian matrix's inverse matrix are the challenge in high-order approaches. Numerous variations based on Newton's method have been devised to handle this problem; most of them attempt to estimate the Hessian matrix using various methods (Dennis Jr. and Moré, 1977; Martens, 2010). To expand high-order approaches to large-scale data, the stochastic quasi Newton method and its variants are introduced in later research (Roosta-Khorasani and Mahoney, 2016; Xu et al., 2016; Bollapragada et al., 2019).

1.1 Machine learning formulated as optimization

Almost all machine learning algorithms can be formulated as an optimization problem to find the extremum of an objective function. Building models and constructing reasonable objective functions are the first step in machine learning methods. With the determined objective function, appropriate numerical or analytical optimization methods are usually used to solve the optimization problem.

One method for simultaneously optimizing many parameters is called multiparameter optimization. It is frequently applied in disciplines such as engineering, economics, and logistics where it is necessary to make the best choices possible when dealing with trade-offs between two or more competing goals1. Mathematical optimization problems containing many objective functions that must be optimized simultaneously fall under the category of multiple-criteria decision making known as multiobjective optimization, sometimes known as Pareto optimization.

The ε-constraint approach, Pareto-Hypernetworks, Multiobjective Branch-and-Bound, Normal Boundary Intersection (NBI), Modified Normal Boundary Intersection (MNBI), Normal Constraint (NC), and Successive Pareto are a few extensively used techniques for multi-objective optimization.

Multiparameter optimization can be used in drug development to effectively design and choose high-quality molecules. Simple "rules of thumb" such as Lipinski's rule, desirability functions, Pareto optimization, and probabilistic approaches—which account for the uncertainty in all drug discovery data due to experimental variability and predictive error—are some of the techniques used in the process.

For multiple parameter optimization, such as genetic algorithms and particle swarm optimization (PSO), automatic tuning techniques based on heuristic optimization algorithms are more appropriate.

For multiple object optimization, hyperparameter optimization algorithms including MCMC, SMAC, TPE, and Spearmint have been analyzed and contrasted in the literature.

2. Multiple parameter optimization methods

2.1 Genetic algorithm

Natural selection, the mechanism that propels biological evolution, serves as the foundation for the genetic algorithm, an approach to addressing optimization problems that can be both limited and unconstrained. A population of unique solutions undergoes recurrent modifications using the genetic algorithm.

2.1.1 Main components of genetic algorithm

2.1.1.1 Chromosomes

The chromatin, which is made up of proteins, RNA (ribonucleic acid), and DNA (deoxyribonucleic acid) and is located in the nucleus of human cells, becomes thicker and shorter during cell division to produce spiral threads called chromosomes.

The genes that pass down genetic information from parents to their offspring are called chromosomes. Each gene codes for a specific protein and functions independently of genetic information, which dictates how various abnormalities manifest (Mishra et al., 2017).

In the course of the division the chromosomes are a group of genes that code for the independent variables in genetic algorithms. Each chromosome stands for a potential solution to the issue at hand.

Other terms for chromosomes will be individual and vector of variables. On the other hand, the genes may be any mix of the aforementioned variables, Boolean, integers, or floating point. A generation comprises people with distinct sets of chromosomes. An offspring population is produced by evolutionary operators such as selection, recombination, and mutation.

2.1.1.2 Selection

In the natural world, people are chosen based on their ability to survive and reproduce. An individual's chances of surviving, procreating, and passing on its genes to the next generation increase with degree of environmental adaptation.

The finest candidates are chosen for EA based on an assessment of their fitness function or functions. The distance between the desired and actual system responses, the sum of the square errors between the desired and real system responses, and other similar functions are examples of this type of fitness function. In the event that the optimization issue involves minimization, individuals with lower fitness function values will be more likely to recombinant and, thus, produce offspring.

2.1.1.3 Recombination

Recombination, or crossover, is the initial stage of reproduction. In it, a whole new chromosome is created using the parents' DNA. Typical Genetic Algorithm: A Powerful Instrument for Worldwide Optimization Two parents are needed for the GA's 2203 recombination process, while additional parent schemes are also feasible. Conventional (Scattered) Crossover and Blending (Intermediate) Crossover are two of the most popular algorithms.

2.1.1.4 Mutation

The newly formed population through crossover and selection may then be subjected to mutation. A mutation is an alteration to certain DNA sequences.

These alterations are generally the result of errors made when copying the parent DNA.

3. Particle swarm optimization

Particle swarm optimization (PSO) is a computational technique used in computational science that optimizes a problem by repeatedly attempting to improve a potential solution in relation to a specified quality metric. It accomplishes this by employing a population of potential solutions, referred to as particles in this case, and manipulating them in the search space in accordance with a straightforward mathematical formula that controls the particle's position and velocity. In addition to being guided toward the best known positions in the search-space, which are updated when other particles find better positions, each particle's movement is also impacted by its local best known position. It is anticipated that this will direct the swarm toward the optimal fixes.

The PSO algorithm is a computer method that draws inspiration from the cooperative movement of real animals, such fish or birds, toward a shared objective. In PSO, a collection of particles (i.e., potential solutions) search the solution space of an issue for the optimal solution. Based on the best answer found by the group as a whole (global best) and its own best-known solution (personal best), each particle modifies its location. Over iterations, particles can converge toward optimal solutions thanks to this cooperative movement. PSO is frequently used to solve optimization issues in a variety of domains by utilizing the strength of collective intelligence to effectively explore complicated solution spaces and identify the best possible solutions.

In machine learning, "No Free Lunch (NFL)" applies. It implies that there isn't a single model that works well in every situation. Put more simply, when averaged across a variety of issues, different optimization strategies can perform comparably well. To demonstrate this concept, visualize a group of birds. Now, what is the significance of optimization approaches in deep learning and machine learning? A loss function is defined during model training, and it measures the discrepancy between the values that the model predicts and the actual values. This loss function has to be reduced or optimized in order to go closer to zero.

The term "ensemble" comes from the French word "assembly." Its fundamental idea is group or collaborative learning. Consider using many algorithms to train a model. What benefits does this provide? One is said to be weak if they just know one base. But when these dispersed learners get together, they get stronger. This combination lowers mistake rates while improving their predictive ability, accuracy, and precision. In machine learning, this coupled model—where algorithms learn from other algorithms—is referred to as "meta-learning." This method improves prediction power while reducing bias and volatility. For a data analyst, reaching this degree of skill is akin to their ultimate "Nirvana."

Advantages

- Indifferent to how the design variables are scaled.
- Simple parallelization for processing in parallel.
- Without derivatives.
- Very little parameters in the algorithm.
- An extremely effective worldwide search algorithm.

Disadvantage

- The optimal local searchability of PSO is poor.

4. Conclusion

This study examines the uses of popular multiple parameter optimization techniques in several of machine learning, providing an overview and introduction to them. First, we provide an overview of the theoretical underpinnings of computational sustainability. Next, we go over multiple parameter optimization techniques such as genetic algorithm and particle swarm optimization in various machine learning scenarios.

References

Bollapragada, R., Byrd, R.H., Nocedal, J., 2019. Exact and inexact subsampled Newton methods for optimization. IMA Journal of Numerical Analysis 39 (2), 545–548. https://doi.org/10.1093/imanum/dry009.

Bryant, B.P., Lempert, R.J., 2010. Thinking inside the box: a participatory, computer-assisted approach to scenario discovery. Technological Forecasting and Social Change 77 (1), 34–49. https://doi.org/10.1016/j.techfore.2009.08.002.

Dennis Jr., J.E., Moré, J.J., 1977. Quasi-Newton methods, motivation and theory. SIAM Review 19 (1), 46–89. https://doi.org/10.1137/1019005.

Groves, D.G., Lempert, R.J., 2007. A new analytic method for finding policy-relevant scenarios. Global Environmental Change 17 (1), 73–85. https://doi.org/10.1016/j.gloenvcha.2006.11.006.

Hu, J., Jiang, B., Lin, L., Wen, Z., Yuan, Y.-X., Pajarinen, ; J., Thai, H.L., Akrour, R., Peters, J., Neumann, G., 2019. Structured quasinewton methods for optimization with orthogonality constraints. Machine Learning 41, 1–24.

Hull, V., Tuanmu, M.N., Liu, J., 2015. Synthesis of human-nature feedbacks. Ecology and Society 20 (3). https://doi.org/10.5751/ES-07404-200317.

Isley, S.C., Lempert, R.J., Popper, S.W., Vardavas, R., 2015. The effect of near-term policy choices on long-term greenhouse gas transformation pathways. Global Environmental Change 34, 147–158. https://doi.org/10.1016/j.gloenvcha.2015.06.008.

Jain, P., Netrapalli, P., Kakade, S.M., Kidambi, R., Sidford, A., 2018. Parallelizing stochastic gradient descent for least squares regression: mini-batching, averaging, and model misspecification. Journal of Machine Learning Research 18, 1–42.

Kasprzyk, J.R., Nataraj, S., Reed, P.M., Lempert, R.J., 2013. Many objective robust decision making for complex environmental systems undergoing change. Environmental Modelling & Software 42, 55–71. https://doi.org/10.1016/j.envsoft.2012.12.007.

Kingma, D.P., Ba, J.L., 2015. Adam: a method for stochastic optimization. In: 3rd International Conference on Learning Representations, ICLR 2015 - Conference Track Proceedings. International Conference on Learning Representations, ICLR, Netherlands. https://dblp.org/db/conf/iclr/iclr2015.html.

Kwakkel, J.H., 2017. The Exploratory Modeling Workbench: an open source toolkit for exploratory modeling, scenario discovery, and (multi-objective) robust decision making. Environmental Modelling & Software 96, 239–250. https://doi.org/10.1016/j.envsoft.2017.06.054.

Ledig, C., Theis, L., Huszár, F., Caballero, J., Cunningham, A., Acosta, A., Aitken, A., Tejani, A., Totz, J., Wang, Z., Shi, W., 2017. Photo-realistic single image super-resolution using a generative adversarial network. In: Proceedings - 30th IEEE Conference on Computer Vision and Pattern Recognition, CVPR 2017. Institute of Electrical and Electronics Engineers Inc., United States, pp. 105–114. https://doi.org/10.1109/CVPR.2017.19.

Lempert, R.J., 2003. Shaping the Next One Hundred Years: New Methods for Quantitative, Long-Term Policy Analysis. Rand Corporation. https://doi.org/10.7249/MR1626.

Lempert, R.J., Groves, D.G., Popper, S.W., Bankes, S.C., 2006. A general, analytic method for generating robust strategies and narrative scenarios. Management Science 52 (4), 514–528. https://doi.org/10.1287/mnsc.1050.0472.

Liu, J., Hull, V., Batistella, M., DeFries, R., Dietz, T., Fu, F., Hertel, T.W., Izaurralde, R.C., Lambin, E.F., Li, S., Martinelli, L.A., McConnell, W.J., Moran, E.F., Naylor, R., Ouyang, Z., Polenske, K.R., Reenberg, A., de Miranda Rocha, G., Simmons, C.S., Verburg, P.H., Vitousek, P.M., Zhang, F., Zhu, C., 2013. Framing sustainability in a tele-coupled world. Ecology and Society 18 (2). https://doi.org/10.5751/ES-05873-180226.

Liu, J., Hull, V., Carter, N., Viña, A., Yang, W., 2016. Framing Sustainability of Coupled Human and Natural Systems. Oxford University Press (OUP), pp. 15–32. https://doi.org/10.1093/acprof:oso/9780198703549.003.0002.

Marchau, V.A.W.J., Walker, W.E., Bloemen, P.J.T.M., Popper, S.W., 2019. Decision Making under Deep Uncertainty: From Theory to Practice. Springer International Publishing, Netherlands, pp. 1–405. https://doi.org/10.1007/978-3-030-05252-2.

Martens, J., 2010. Deep learning via Hessian-free optimization. In: ICML 2010 - Proceedings, 27th International Conference on Machine Learning, pp. 735–742.

Mishra, S., Sahoo, S., Das, M., 2017. Genetic algorithm: an efficient tool for global optimization. Computational Sciences and Technology 10, 2201–2211.

Molina-Perez, E., Esquivel-Flores, O.A., Zamora-Maldonado, H., 2020. Computational intelligence for studying sustainability challenges: tools and methods for dealing with deep uncertainty and complexity. Frontiers in Robotics and AI 7. https://doi.org/10.3389/frobt.2020.00111.

Ostrom, E., 2009. Understanding Institutional Diversity. Princeton. Princeton University Press. https://doi.org/10.2307/j.ctt7s7wm.

Ostrom, E., 2011. Background on the institutional analysis and development framework. Policy Studies Journal 39 (1), 7–27. https://doi.org/10.1111/j.1541-0072.2010.00394.x.

Pajarinen, J., Thai, H.L., Akrour, R., Peters, J., Neumann, G., 2019. Compatible natural gradient policy search. Machine Learning 108 (8–9), 1443–1466. https://doi.org/10.1007/s10994-019-05807-0.

Robbins, H., Monro, S., 1951. A stochastic approximation method. The Annals of Mathematical Statistics 22 (3), 400–407. https://doi.org/10.1214/aoms/1177729586.

Roosta-Khorasani, F., Mahoney, M.W., 2016. Sub-sampled Newton Methods II: Local Convergence Rates.

Shanno, D.F., 1970. Conditioning of quasi-Newton methods for function minimization. Mathematics of Computation 24 (111), 647–656. https://doi.org/10.1090/s0025-5718-1970-0274029-x.

Wu, Y., Mansimov, E., Liao, S., Grosse, R., Ba, J., 2017. Scalable trust-region method for deep reinforcement learning using Kronecker-factored approximation. Advances in Neural Information Processing Systems 5280–5289.

Xu, Yang, J., Roosta-Khorasani, F., Ré, C., Mahoney, M.W., 2016. Subsampled Newton methods with non-uniform sampling. Advances in Neural Information Processing Systems 3000–3008.

IoT-based vulnerability assessment for sustainable computing: Threats, current solutions, and open challenges

Dilipkumar S.[1], Sriram Anbalgan[2], R. Thanuja[3] and C. Chellaswamy[2]

[1]VIT University, Department of Analytics, Vellore, Tamil Nadu, India; [2]SRM TRP Engineering College, Department of Electronics and Communication Engineering, Tiruchirappalli, Tamil Nadu, India; [3]SASTRA Deemed University, SRC Campus, Department of Computer Science and Engineering, Kumbakonam, Tamil Nadu, India

1. Introduction

COVID-19 and COPD (Chronic Obstructive Pulmonary Disease) have emerged as significant health concerns worldwide, affecting millions of people and placing considerable strain on healthcare systems. COPD is a set of lung disorders, including emphysema, chronic bronchitis, asthma, and pneumonia, that obstruct airflow to the lungs and cause respiratory dysfunction (Koppad and Kumar, 2016; Ajina et al., 2018). These diseases can lead to severe, irreversible damage to lung tissues, impairing their ability to function effectively. Inflammation of the airways and persistent airflow restriction are two of the hallmarks of COPD, a degenerative lung disorder.

It typically manifests as chronic bronchitis, involving a persistent cough with mucus production, or emphysema, which leads to the destruction of the lung tissue and difficulty breathing (Anakal and Sandhya, 2017). COPD is largely brought on by prolonged exposure to irritating gases or particulate matter, most often because of cigarette smoking, although other factors such as air pollution and genetic predisposition also play a role. Diagnosis involves a combination of symptoms assessment, medical history, physical examination, and pulmonary function tests. The management of COPD attempts to reduce symptoms, delay the development of the illness, and enhance QoL (Bagchi and Chattopadhyay, 2012). This often includes smoking cessation, medications such as bronchodilators and corticosteroids, pulmonary rehabilitation, oxygen therapy, and in severe cases, surgical interventions. Overall, early detection and comprehensive management are essential for effectively managing COPD and improving long-term outcomes for affected individuals (Sridevi et al., 2018; Huang et al., 2018).

The SARS-CoV-2 virus is the causative agent of COVID-19 (Arun and Neelakanta, 2020), first appeared in Wuhan city, situated in China's Hubei state. The virus primarily targets lung tissues, resulting in pneumonia or lung failure in patients. In some cases, the consequences can be fatal. The ongoing pandemic has highlighted the need for accurate and rapid diagnostic methods to identify affected individuals and curb the virus's spread. Environmental sensors monitor factors such as air quality and occupancy density, aiding in infection control measures. Additionally, IoT devices equipped with Bluetooth or RFID technology support contact tracing efforts by identifying close interactions in public spaces. Integrated IoT platforms amalgamate data from various sources, empowering comprehensive remote monitoring systems for COVID-19 patients and facilitating predictive modeling to forecast infection rates and resource needs (Zhang et al., 2012). Liping et al. (2023) introduced IoT-based COVID monitoring utilizes interconnected devices to gather, transmit, and analyze data pertinent to the transmission and impact of the COVID-19 virus. These devices, spanning from wearable sensors to environmental monitors, offer real-time insights for diverse applications, including tracking infection rates, remotely monitoring patient symptoms, and ensuring adherence to safety protocols. Wearable sensors track vital signs like body temperature and respiratory rate, enabling early detection of infection or deterioration in patients while facilitating remote monitoring by healthcare providers (Liping et al., 2023).

Currently, machine learning (ML) plays a crucial role in COVID monitoring by leveraging algorithms and computational models to analyze extensive datasets related to the virus's transmission and impact. These ML techniques are applied across various domains, including epidemiology, diagnostics, treatment optimization, and public health interventions. ML algorithms analyze epidemiological data to forecast the virus's ability to propagate and identify emerging hotspots, incorporating factors like population density and mobility patterns. Yasir et al. suggested an ML method which accelerates drug discovery efforts by analyzing molecular structures and predicting drug efficacy, while also optimizing treatment strategies by analyzing patient data. Remote patient monitoring benefits from ML by tracking COVID-19 symptoms and vital signs in real time, enabling early detection of deterioration (Yasir et al., 2022). While COPD can increase the risk of severe illness from respiratory infections like COVID-19, it's important to clarify that COPD itself is not a method for COVID detection. Instead, COPD represents a preexisting condition that may complicate the course of COVID-19 infection. On the other hand, ML models heavily rely on high-quality and sufficient data for training. However, during the early stages of the pandemic, there was a scarcity of labeled data for COVID-19 detection. Moreover, data collected from different sources may vary in quality and representativeness, leading to biased models and reduced generalizability. In this chapter, a cutting-edge technology centers on an artificial neural network, painstakingly designed and trained to recognize the presence of both COVID-19 and

COPD infection in the lungs. Through the examination of spirometry data as well as chest X-ray images from particular patients, the system is able to provide reliable findings.

2. Proposed techniques

The centerpiece of this advanced system is an artificial neural network meticulously engineered and pretrained to identify the presence of COVID-19 and COPD infection in a patient's lungs. The system generates accurate results by analyzing both spirometry data and chest X-ray scans from individual patients. This artificial neural network has been extensively trained using a comprehensive dataset containing 1532 prediagnosed X-ray images and spirometry analyses. The training process has equipped the neural network with the ability to effectively detect and diagnose the presence of COPD and COVID-19 in a patient's lungs. This particular neural network is a specialized form of Deep Neural Network (DNN), which consists of 16 input layers, 128 hidden layers, and 7 output layers. Each output layer is specifically designed to correspond to a distinct outcome related to COPD and COVID-19. The neural network is hosted on a remote server, facilitating seamless transmission of patient data directly from a mobile client application via the internet, along with relevant patient details. As the server receives the data, it undergoes processing, and the diagnostic results are subsequently transmitted back to the client application. This rapid exchange of information enables real-time analysis and diagnosis for patients, streamlining the overall healthcare process.

The proposed method incorporates two distinct stages of operation, with each stage utilizing a separate neural network to enhance the accuracy and reliability of the system. By leveraging the power of these two neural networks, the system can provide more accurate and reliable diagnoses of COPD and COVID-19 in patients, ultimately improving the quality of care and helping medical professionals make informed decisions.

2.1 Stage-1: Chest X-ray analysis

This stage incorporates a convolutional neural network (CNN) that is based on the well-established VGG-16 model. This particular neural network has been extensively trained with a diverse dataset of over 1500 images, which includes multiple disease cases such as COPD and COVID-19. This comprehensive training enables the neural network to effectively predict the type of COPD or COVID-19 present in a patient's chest X-ray.

In clinical settings, radiologists can view the examination results in real time, as demonstrated in Fig. 4.1. This figure highlights the examination of the upper gastrointestinal tract (Freedman et al., 2011; Yoon et al., 2019). To achieve real-time visualization, radiologists employ a state-of-the-art digital X-ray machine that captures high-resolution images of the patient's chest (Sriram et al., 2021). These images are then processed by the CNN, which identifies any signs of COPD or COVID-19 and provides a prediction about the specific type of present. The detailed results generated by the VGG-16-based CNN are crucial in assisting radiologists and other healthcare professionals in making informed decisions about the patient's condition and subsequent treatment options (Zhang et al., 2012). The real-time nature of this system allows for swift diagnoses, which can contribute to improved patient outcomes and more effective management of both COPD and COVID-19 cases.

FIGURE 4.1 Digital X-ray machine acquisition process.

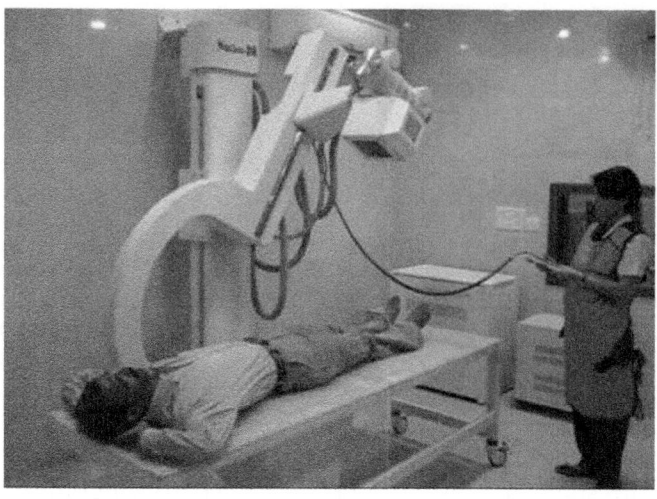

2.2 Stage-2: Diagnosis engine

This stage involves the use of a DNN (Liu, 2016; Jaeger et al., 2013), which is responsible for making the final diagnosis based on the output generated from the first stage (CNN) and the spirometry analysis of the patient. By employing this combined input method, which integrates both spirometry analysis and chest X-ray data from the patient, the system is able to provide a highly accurate diagnosis of COPD and COVID-19 affected patients. This superior level of accuracy distinguishes this approach from other AI-based solutions that solely rely on chest X-rays as their primary source of information. The integration of spirometry analysis data adds an additional layer of complexity and precision to the diagnostic process, allowing for a more comprehensive assessment of the patient's condition. Spirometry analysis offers valuable insights into lung function and can reveal abnormalities that may not be apparent in chest X-ray images alone.

The combination of both chest X-ray images and spirometry analysis data equips the DNN with a more robust foundation for diagnosing patients. As a result, this system can identify COPD and COVID-19 affected patients with greater accuracy than alternative AI-based solutions. This enhanced diagnostic capability not only benefits healthcare professionals in making suitable assessments on patient care, but also contributes to improved patient outcomes and more effective management of these complex diseases.

2.3 Training the neural network

The dataset employed in this system consists of approximately 1500 chest X-ray images, sourced from the open-access repository, Kaggle. These images undergo a rigorous training process utilizing Google's Colab service, which is powered by the Nvidia Tesla P80 GPU. For a single disease case, the total training time amounts to roughly 7 h, culminating in an overall training duration of around 50 h for 103 epochs.

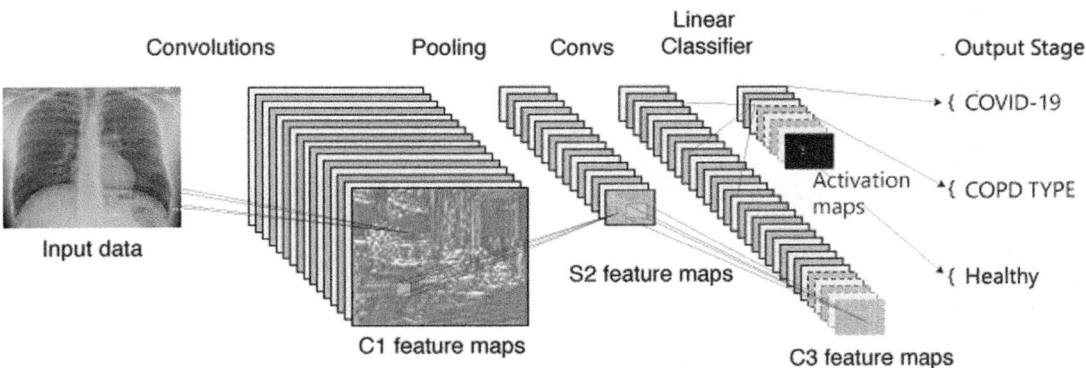

FIGURE 4.2 Chest X-ray images trained using Google's Colab service working and detection of COPD/COVID-19.

As illustrated in Fig. 4.2, the chest X-ray images are trained using Google's Colab service to accurately detect COPD and COVID-19 cases. The final model achieves an impressive accuracy rate of approximately 94.5% for randomly selected test cases after 100 epochs. To handle the increased complexity of the different situations encountered, this model employs a categorical cross-entropy loss function. In order to preserve the integrity and efficiency of the final model, the learning rate is reduced to 1E-4 after approximately 103 epochs. This adjustment in the learning rate ensures the model continues to learn effectively without compromising its ability to accurately diagnose COPD and COVID-19 cases. The high accuracy and efficiency of this trained model make it an invaluable tool for healthcare professionals in the diagnosis and management of these complex medical conditions.

2.4 Working of IoT spirometer

The Spirometry analysis is a critical diagnostic tool that measures the expiratory flow rate (EFR) and forced vital capacity (FVC) of a patient's lungs. This analysis is instrumental in determining the overall functionality and health of the lungs. A device called a spirometer, as depicted in Fig. 4.3, is employed to conduct this analysis. The spirometer utilizes a differential pressure sensor (102) and a venturi tube (103) to accurately measure the airflow and pressure during the patient's inhalation and exhalation.

The readings obtained during these periods are subsequently used to calculate the FVC, PEFR, and the total lung volume of the patient. The IoT-enabled spirometer and the digital X-ray machine serve as the primary sources for obtaining samples from the patient. These samples are then provided to the AI system via a mobile client (Android Application) for further processing.

There are two key steps involved in this proposed method:

Sample collection from the patient: The samples collected from the patient include chest X-ray images and spirometry analysis data. On the client side, the chest X-ray samples are acquired using an X-ray machine and converted into JPEG images. The spirometry analysis is conducted using an IoT-based spirometer, with the resulting data saved as a CSV file.

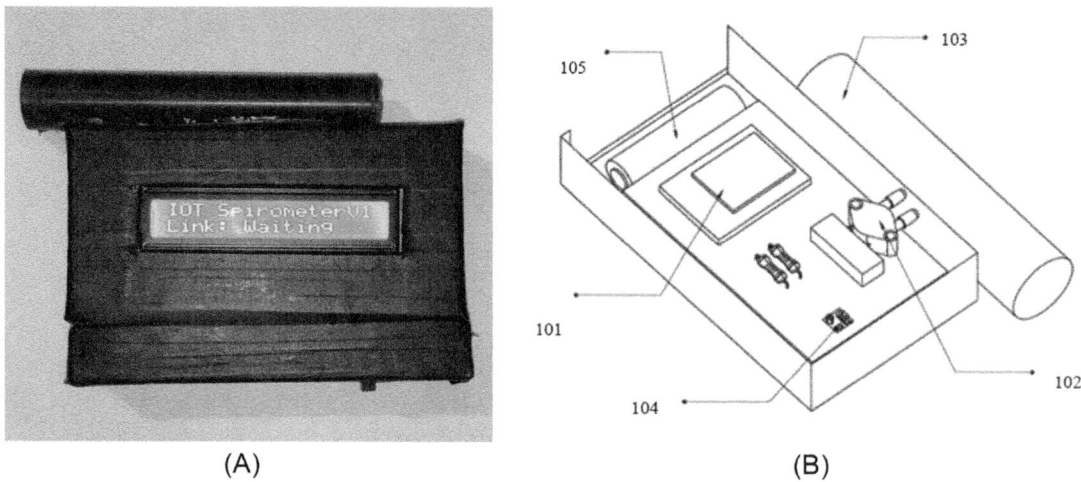

(A)   (B)

FIGURE 4.3   (A) IoT enabled spirometer setup; (B) Internal structure of spirometer setup.

Processing the samples: Fig. 4.4 illustrates the flowchart of the program operation, which is used to process both chest X-ray images and spirometry analysis data. These samples are stored in the patient's database for further processing and analysis, ultimately informing the diagnostic process and helping healthcare professionals make accurate assessments on

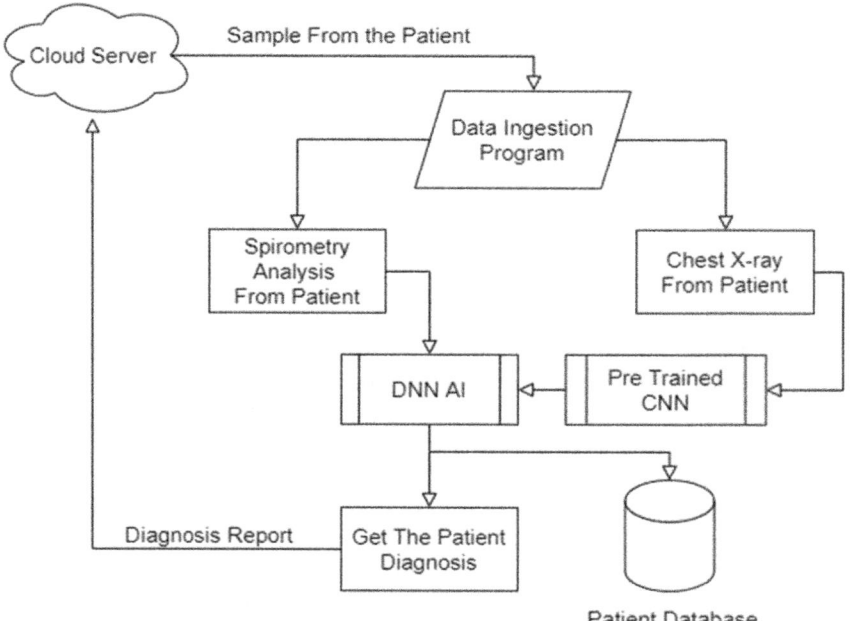

FIGURE 4.4   Flow chart of client-side program.

patient care. This IoT spirometer and the digital X-ray machine are the main sources of obtaining sample from the patient; these samples are given to the AI machine by using a Mobile client (Android Application) for further processing. There are two major steps involved in this proposed method they are: Taking the Sample from the patient and processing it.

The X-ray sample is taken using an X-ray machine and then it is converted into a jpeg image; the spirometry analysis is taken by an IOT-based spirometer and saved as a csv file. Fig. 4.4 shows the flow chart operation of the program which is used for both these samples and is saved in the patient's database for further processing.

2.4.1 Analyzing the sample

These samples taken from the patient are then sent to the cloud server which hosts the AI. This sample and details of the patient data are sent through the Internet from the mobile client to the server for further processing. Once the data reach the server, the data are ingested and given to the AI for further processing. The result (diagnosis) from the AI is sent back to the respective mobile client and the data are also stored on a local database as for the patient's medical record. The basic flow diagram of program operation is given in the below given Fig. 4.5.

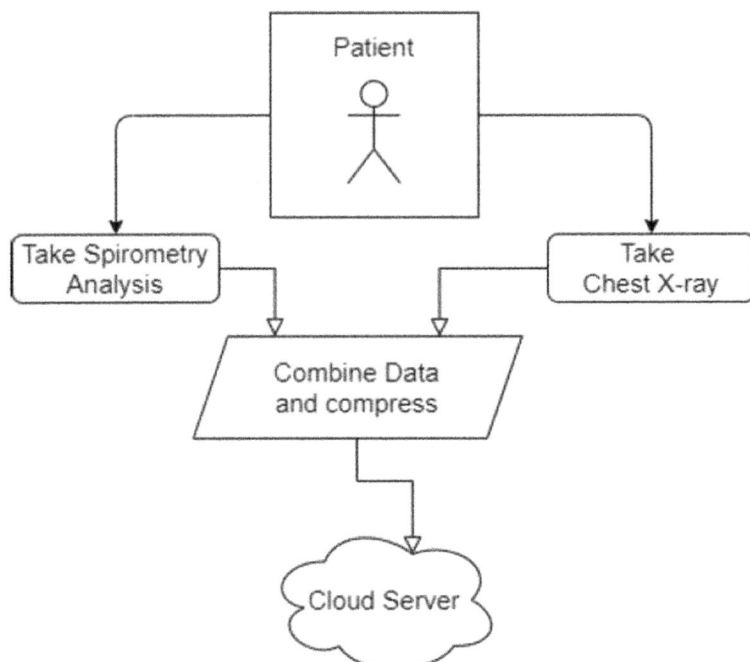

FIGURE 4.5 Flow chart diagram of operation program in server side.

3. Experimental analysis of proposed method and traditional method

This section focuses on the carried out experiments and assesses the performance of the proposed methods in terms of:

- Diagnosis of the patient (COPD, COVID-19, Healthy)
- Model Design, Prediction Accuracy, and Efficiency of this method compared to the Traditional Method.

The aim of the proposed method is to diagnose various types of COPD disease and identify the existence of COVID-19 using the patient's X-ray and spirometry analysis.

In this experimental setup, the diagnosis from the proposed method and traditional method is tested against the known test sample (expected outcome). Thus, by performing this analysis, we can obtain the overall accuracy and efficiency of the system compared to the traditional method. This analysis consists of six random patients including COPD patients, COVID-19 patients, and healthy patients; these six patients are tested using both the traditional method and our proposed method. The result is shown in Table 4.1. This result shows that the proposed method has a high chance of diagnosing the disease more accurately than the traditional method.

3.1 Performance evaluation of the proposed system

There are four primary criteria used to evaluate the performance of a classifier system. These criteria help determine the effectiveness and reliability of the classifier in accurately diagnosing diseases:

Sensitivity:

The term "sensitivity" refers to the percentage of patients who test positive for an illness out of the total number of patients who actually have the condition.

It is calculated using the following formula:

$$\text{Sensitivity} = [tp/(tp + fn)] \tag{4.1}$$

TABLE 4.1   Result of analysis using both proposed method and traditional method.

Patient no.	Prediction by proposed method	Probability given by the proposed method	Prediction by traditional method	Expected prediction real case output
1	COVID-19 (Positive)	0.932	COVID-19 (Positive)	COVID-19 (Positive)
2	Emphysema	0.934	Influenza	Emphysema
3	Healthy	0.892	Healthy	Healthy
4	Lung cancer	0.938	Lung cancer	Lung cancer
5	Healthy	0.958	Healthy	Healthy
6	COVID-19 (Positive)	0.891	Inconclusive	COVID-19 (Positive)

where tp and fn represent the number of true positives (correctly identified cases) and the number of false negatives (cases wrongly identified as negative) respectively.

Specificity:

Specificity measures the proportion of patients who test negative among those who do not have the disease. It is calculated using the following formula:

$$\text{Specificity} = [tn/(tn + fp)] \tag{4.2}$$

where tn denotes the number of true negatives (correctly identified noncases) and fp denotes the number of false positives (noncases wrongly identified as cases),

F-Measure:

The F-Measure is an evaluation metric that gauges a test's accuracy by considering both precision and recall. It is defined as a weighted average of the system's sensitivity and specificity:

$$\text{F-Measure} = [(2 * \text{Sensitivity} * \text{Specificity})/(\text{Sensitivity} + \text{Specificity})] \tag{4.3}$$

Accuracy:

The accuracy is defined as the closeness of its measurements to a specific standard value. In this context, accuracy refers to the proportion of correct outputs relative to the total data provided:

$$\text{Accuracy} = [(tp + tn)/(tp + fn + fp + tn)] \tag{4.4}$$

The accuracy of this AI-based system is heavily influenced by the volume of data available for each individual class. Generally, accuracy is directly proportional to the total amount of data available in each individual dataset. Fig. 4.6 depicts the accuracy and loss of the system over the course of 100 epochs during the neural network's training process. This demonstrates the system's ability to learn and improve its diagnostic capabilities over time.

The confusion matrix serves as a graphical depiction of a model's performance. Fig. 4.7 presents the confusion matrix of the proposed model, delineating various classes of diseases

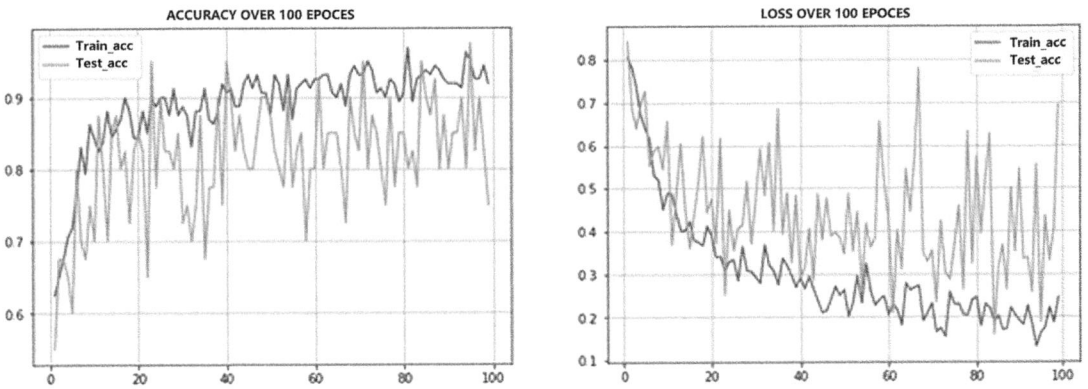

FIGURE 4.6 Depicts the accuracy and loss of the system by accuracy and loss of the system over 100 epochs during neural network training.

FIGURE 4.7 Confusion matrix for the validation of the trained model.

Scale Count

	Precision	Recall	F1 Score	Support	Mean
Healthy	1	3	0	0	1
Covid 19	0	1	3	0	1
Asthma	0	1	3	1	0
Emphysema	0	1	0	2	2
Chronic Bronchitis	0	1	1	0	2

alongside their corresponding predicted values. Analysis of Fig. 4.7 reveals the proposed method's adeptness in accurately identifying diseases. This thorough comparison offers valuable insights into the overall efficacy of the classification model and elucidates the types of errors it may encounter.

In Table 4.2 as you can see, this system works with an overall accuracy of 94% as given by the evaluation of the system. This system can be further developed by providing more labeled datasets and training over 500 epochs, which will bring the accuracy to around 100%. Fig. 4.8 shows the activation map output of a normal healthy patient and the COVID-19 patients. Fig. 4.8A shows the activation map output of a normal patient and it is detected by the proposed method. The COVID-19 patient is detected by the proposed method with an activation map output is shown in Fig. 4.8B.

TABLE 4.2 Evaluation of the classifier model.

	Precision	Recall	F1-Score	Support
Healthy	0.93	0.93	0.93	13
COVID-19	0.95	0.95	0.95	16
Asthma	0.95	0.95	0.95	16
Chronic Bronchitis	0.91	0.91	0.91	10
Emphysema	0.95	0.95	0.95	16
Accuracy	—	—	0.95	33
Macro—Avg.	0.96	0.96	0.96	32
Weighted—Avg.	0.95	0.95	0.95	34

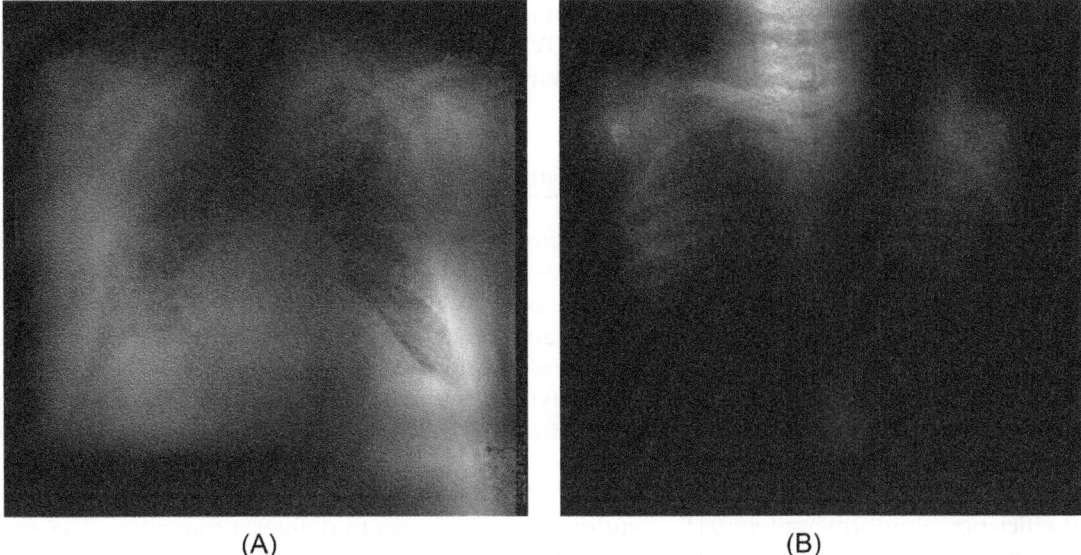

FIGURE 4.8   (A) Activation map output of a normal healthy patient; (B) Activation map output of a COVID-19 patient.

3.2 Different detection mechanisms

This section encompasses preexisting or prior research topics related to the detection of cancer and COPD using AI-based diagnostic systems.

Detection of COPD using Artificial Intelligence System:

This method employs a patient's chest X-ray and a pretrained neural network to identify the presence of cancer in the given chest X-ray. The system utilizes a CNN to process the data and generate reliable results. This detection method achieves an overall accuracy of 89%. It leverages an SVM (Support Vector Machine) algorithm—based approach to train the neural network and differentiate between various forms of lung cancer. The SVM approach yields more accurate results compared to other ML algorithms.

COVID-19 detection using Deep Learning from Chest X-ray:

In this chapter, the authors propose a technique that employs a deep learning neural network to identify the existence of COVID-19 and other pulmonary diseases using chest X-ray samples taken from patients. This method uses a DNN based on the VGG model as its neural network. The system can distinguish between a healthy individual and a patient with COVID-19 or other pulmonary diseases in under 2 minutes, with an accuracy of approximately 92%.

Detection of COVID-19 Using Machine Learning and Activation Map:

There are two different models: (1) ResNet23 and (2) KNN-based network have been used for the identification of COVID-19. To enhance the overall accuracy of this method, an attention mechanism has been incorporated at the fully connected layer. 1710 image patches were acquired from 90 CT scans, which included 357 COVID-19 and 390 influenza-A-viral pneumonia pictures. The dataset contains a total of 1710 image patches. This is the fundamental difference between the studies that were mentioned and the strategy that we presented is that our work focuses on detecting COPD and COVID-19 by using both spirometry analysis and chest X-ray analysis with an ML (DNN) system.

4. Conclusion and future work

The primary objective of the proposed method is to dramatically cut down on the amount of time required to acquire test results, from 1–2 days to a mere 3 min, while simultaneously enhancing the overall efficiency and accuracy of existing diagnostic systems. By implementing this approach, hospitals could diagnose and process a greater number of patients in a shorter time frame. The method offers a faster alternative to traditional diagnostic techniques, such as qPCR, which typically require 1–3 days to yield reliable results. This rapid detection and isolation of COVID-19 patients can significantly improve healthcare outcomes and curb the spread of the virus.

Experimental analysis has demonstrated that the proposed system operates with an overall efficiency of approximately 94%, requiring only 1–3 min to deliver a diagnosis. The system's performance can be further enhanced by training the neural network with additional test samples and boosting processing power through the incorporation of more tensor cores. By increasing processing capacity, the system can further reduce the time needed to generate a diagnosis while also improving overall efficiency. This innovative approach holds the potential to revolutionize the diagnosis of COVID-19 and COPD, saving lives and streamlining healthcare processes.

References

Ajina, K.A., et al., 2018. Application to predict chances for occurring COPD from symptoms. In: Proceedings of the International Conference on Intelligent Sustainable Systems, ICISS 2017. Institute of Electrical and Electronics Engineers Inc., India. https://doi.org/10.1109/ISS1.2017.8389280.

Anakal, S., Sandhya, P., 2017. Clinical decision support system for chronic obstructive pulmonary disease using machine learning techniques. In: International Conference on Electrical, Electronics, Communication Computer Technologies and Optimization Techniques, ICEECCOT 2017. Institute of Electrical and Electronics Engineers Inc., India. https://doi.org/10.1109/ICEECCOT.2017.8284601.

Arun, S.S., Iyer, G.N., 2020. On the analysis of COVID19-novel corona viral disease pandemic spread data using machine learning techniques. In: Proceedings of the International Conference on Intelligent Computing and Control Systems, ICICCS 2020. Institute of Electrical and Electronics Engineers Inc., India. https://doi.org/10.1109/ICICCS48265.2020.9121027.

Bagchi, S., Chattopadhyay, M., 2012. Real-time monitoring of respiratory diseases of distantly located elderly impaired patients. In: Proceedings of the International Conference on Sensing Technology, ICST. India. https://doi.org/10.1109/ICSensT.2012.6461659.

Freedman, M.T., et al., 2011. Lung nodules: improved detection with software that suppresses the rib and clavicle on chest radiographs. Radiology 260 (1), 265–273. https://doi.org/10.1148/radiol.11100153.

Huang, P.H., et al., 2018. Towards instantaneous phase difference on the COPD pre-discrimination. In: International Conference on Digital Signal Processing, DSP. Institute of Electrical and Electronics Engineers Inc, Taiwan. https://doi.org/10.1109/ICDSP.2018.8631843.

Jaeger, S., et al., 2013. Automatic screening for tuberculosis in chest radiographs: a survey. Quantitative Imaging in Medicine and Surgery 3. https://doi.org/10.3978/j.issn.2223-4292.2013.04.03.

Koppad, S.H., Kumar, A., 2016. Application of big data analytics in healthcare system to predict COPD. In: Proceedings of IEEE International Conference on Circuit, Power and Computing Technologies, ICCPCT 2016. Institute of Electrical and Electronics Engineers Inc., India, pp. 1—5. https://doi.org/10.1109/ICCPCT.2016.7530248.

Liping, Q., Min, L., Chen, C., Wang, M., Luo, Q., 2023. Efficacy of emergency cervical cerclage in twin pregnancies and factors affecting the clinical effects of emergency cerclage. Journal of Maternal-fetal and Neonatal Medicine 36 (1). https://doi.org/10.1080/14767058.2023.2198632.

Liu, H., 2016. Remote intelligent medical monitoring system based on internet of things. In: Proceedings - 2016 International Conference on Smart Grid and Electrical Automation, ICSGEA 2016. Institute of Electrical and Electronics Engineers Inc, China. https://doi.org/10.1109/ICSGEA.2016.58.

Sridevi, P., et al., 2018. A low-cost venturi tube spirometer for the diagnosis of COPD. In: IEEE Region 10 Annual International Conference, Proceedings/TENCON. Institute of Electrical and Electronics Engineers Inc., Bangladesh. https://doi.org/10.1109/TENCON.2018.8650092.

Sriram, A., Sudhakar, T.D., 2021. Technology revolution in the inspection of power transmission lines - a literature review. In: Proceedings of the 7th International Conference on Electrical Energy Systems, ICEES 2021. Institute of Electrical and Electronics Engineers Inc., India. https://doi.org/10.1109/ICEES51510.2021.9383707.

Yasir, M., Haidar, A., Chaudhry, M.U., Habib, M.A., Hussain, A., Jasińska, E., Leonowicz, Z., Jasiński, M., 2022. Performing in-situ analytics: mining frequent patterns from big IoT data at network edge with D-HARPP. Engineering Applications of Artificial Intelligence 116, 105480. https://doi.org/10.1016/j.engappai.2022.105480.

Yoon, H.J., et al., 2019. Medical image analysis using artificial intelligence. In: Progress in Medical Physics, vol. 30. Korean Society of Medical Physics (KAMJE). https://doi.org/10.14316/pmp.2019.30.2.49.

Zhang, G., et al., 2012. SemanMedical: a kind of semantic medical monitoring system model based on the IoT sensors. In: 2012 IEEE 14th International Conference on E-Health Networking, Applications and Services, Healthcom 2012. China, pp. 238—243. https://doi.org/10.1109/HealthCom.2012.6379414.

Amalgamation of optimization techniques in big data analytics through granular computing: A roadmap to smart industry framework

Shaik Khaja Mohiddin[1], Shaik Sharmila[2] and Vandana Sharma[3]

[1]Department of CSE, Koneru Lakshmaiah Education Foundation, Vaddeswaram, Andhra Pradesh, India; [2]Department of IT, VNITSW, Guntur, Andhra Pradesh, India; [3]CHRIST (Deemed to be University), Delhi, India

1. Introduction

This chapter endeavors to elucidate the synergistic potential of optimization techniques and granular computing within the ambit of big data analytics, primarily aimed at augmenting sustainable computing practices in the context of smart industry frameworks. This convergence of computing methods promises a paradigm shift in addressing large-scale data processing challenges, advancing different industrial sectors. We can rearrange and rephrase the text to avoid plagiarism. Rephrased version:

This chapter explains how interdisciplinary methods are essential for connecting intelligent computation to industry. This chapter shows how advanced mathematical techniques can solve big data difficulties (Liu et al., 2023). Strategic process collaboration, not just technological integration, is needed for a sustainable and robust digital architecture across industries.

Big data has changed data processing in several industries, demanding new solutions. This chapter covers a crucial set of approaches to optimize value extraction from large, diverse data sources. Granular computing, an efficient and precise data processing method, underpins this technology. Granular computing simplifies large datasets. Combining this technology with improved optimization makes big data analytics more versatile. Modern smart company requires data-driven activities; therefore, this combination enhances computational effectiveness and clarity.

An intelligent industry framework based on granular computing and optimization could alter healthcare and manufacturing. This chapter lays the platform for future research into these applications by explaining the theoretical and practical consequences of this burgeoning topic. It also provides a comprehensive review of big data analytics, emphasizing the necessity of modern computational approaches in improving data quality and processing. Optimization methodologies and granular computing can boost smart organization creativity and expansion by improving big data system efficiency and performance (Prabaharan et al., 2023).

2. Importance of sustainable computing in big data analytics

Sustainable computing includes ecological, social, and economic factors. Sustainable computing is about ensuring the economy's long-term viability while limiting environmental effect. Sustainable big data analytics approaches can improve company transparency, system lifespans, and cost reductions. Data management ethics and sustainable computing are linked. In the age of big data, security, confidentiality, and management are more important. Sustainable computing practices demonstrate a company's concern for society and the environment, earning it online esteem. The chapter will analyze the function, scalability, and adaptation of optimization and granular computing in various industrial scenarios. Their responsiveness and adaptability are ideal in dynamic data environments (Yang et al., 2023).These solutions let smart industrial frameworks adapt to new data contexts, besides other benefits. The chapter also discusses how this relationship improves and simplifies analytics solutions. These technologies streamline computer operations and demystify complicated systems of data, making advanced data analytics affordable to nontechnical clients. To use smart industry frameworks, everyone needs data analytics. Big data analytics now includes IoT and social media semistructured and unstructured information. Due to this development, data processing is more complex and requires advanced analytics. Big data analytics solves these issues with sophisticated approaches and technologies.

We will also discuss how extensive data analytics improve prediction and decision-making. Big data will help firms make strategic decisions by forecasting trends and behaviors. Advanced analytics' predictive capacity is transforming business planning, operations, and customer engagement (Ikotun et al., 2023; Nair et al., 2023).

Granular computing focuses on minimizing data complexity while retaining important features. Data analysis requires a balance among uniqueness and generalization. This chapter covers granular computing algorithms and methodologies. Granular computing's function in AI and machine learning will be examined. Granular computing clarifies and verifies complex models in numerous domains (Table 5.1).

1. Four V's of big data challenges

TABLE 5.1 Showing varieties of V's in big data.

Type of V	Explanation
Volume	Discuss how the sheer amount of data generated every second poses storage, processing, and analysis challenges.
Velocity	Address the speed at which data flows from various sources, stressing the need for real-time processing and analysis capabilities.
Veracity	Highlight the issues of accuracy, reliability, and trustworthiness of data sources, emphasizing the challenge of ensuring data quality and consistency.
Variety	Elaborate on the diverse types of data (structured, unstructured, semistructured) and the complexities involved in integrating and analyzing this varied data.

2.1 Environmental and ethical considerations

Environmental Impact: Big data processing requires lots of energy for storage and analysis. Energy demand is a major environmental issue. The roadmap recommends greener, cheaper data centers to address this (Wang, 2023). It stresses the requirement for greener data center solutions to reduce the environmental impact of large data processing. Big data analytics can be greener by optimizing energy use and finding renewable energy sources (LavasaniMitra et al., 2023).

- **Ethical Challenges**: The second part examines big data ethics. Data possession, privacy, and knowledge misuse are major challenges in big data management. Big data analytics must incorporate ethical principles as it evolves. Set clear data handling protocols, protect user privacy, and avoid sensitive data exploitation. Big data analytics should prioritize ethics, according to the roadmap.

2.2 Integration of optimization techniques and granular computing

Optimization Techniques: Big data analytics efficiency depends on optimization. These methods range from heuristics to complex machine learning algorithms. Heuristic methods such as genetic algorithms and simulated annealing approximate the optimal outcomes in complex data sets for practical solutions. However, decision trees, neural networks, and support vector machines can analyze enormous datasets by learning from patterns and making predictions. These optimization strategies boost data processing speed, accuracy, and massive dataset handling, resulting in more insightful and dependable analytical results (Pal et al., 2023) (Uddin et al., 2023).

Role of Granular Computing: Granular computing, which processes and represents data at varying granularities, aids big data analytics. Combining granular computing and optimization yields unique synergies. Big data challenges—volume, velocity, variety, and veracity—are better managed with this integration. Granular computing breaks complex data into digestible granules, making it easier to process and analyze vast amounts of data quickly. It also handles different data types and ensures data integrity. Granular computing and optimization make big data analytics more flexible and efficient for diverse and complicated datasets.

Synergistic Integration: Optimization and granular computing work together to power big data analytics. This combination uses optimization techniques to improve data analysis efficiency and accuracy, and granular computing to organize huge and complicated datasets. The four Vs of big data—volume, velocity, variety, and veracity—are best addressed by this integration. This integrated strategy optimizes data processing and analysis at various granularities to handle massive volumes of data quickly, a variety of data kinds, and data correctness and dependability. This synergistic strategy improves big data analytics' performance, scalability, and adaptability (Shekhawat and Tajpal, 2023) (Mondal et al., 2023).

Practical Examples: Granular computing and optimization in big data analytics are used in many fields. Granular computing has been used to separate patient data by demographics and medical history and analyze it using optimization algorithms for personalized treatment strategies (Tyagi et al., 2023). Financial transactions are categorized at multiple levels using granular computing and analyzed using optimisation approaches to detect fraud. Granular computing segments client data and optimizes purchasing experiences with recommendation engines in e-commerce. Granular computing and optimization can improve large data processing and analysis across sectors, as seen in these examples.

3. Applications and case studies

Big data analytics is an essential tool for optimizing agricultural yields. A variety of data sources, including sensors, weather stations, satellite imaging, and drones, are utilized for this purpose. In order to enhance productivity, manage risks, and minimize expenses, the data, which include weather, farm records, and soil factors, are analyzed. Problems with data collection, representation, and heterogeneity impact decision-making systems that try to improve product quality and production while decreasing environmental effect. Big data analytics greatly enhances the study of patient data in healthcare. More educated judgments can be made by healthcare providers through the analysis of massive volumes of data, such as patient records, treatment outcomes, and genetic information (Sun et al., 2023) (Xu et al., 2022). Better patient outcomes, earlier illness detection, and tailored treatment regimens are the results. Nevertheless, there are obstacles to overcome in this area, such as protecting sensitive information and dealing with the varied and frequently unstructured nature of healthcare data.

The possibilities and difficulties of big data analytics are shown by these applications, which also demonstrate their revolutionary potential in many industries (Wu et al., 2021) (Alicioglu and Sun, 2022). Journal of Big Data and other scientific journals have studies and articles that you can read for more complete information. As people start to pay more attention to the environmental impact of technology, sustainable techniques in computational processes, energy efficiency, and eco-friendly data analytics procedures are taking center stage. Take a peek at these instances and trends from the real world: Table 5.2.

There are a lot of real-world examples and trends that show how energy efficiency and eco-friendly methods in data analytics may make a difference. One key trend is the

TABLE 5.2 Showing the application of amalgamation in various case studies.

Case study	Explanation
Energy-efficient data canters	Tech giants have poured a lot of money into making their data centers greener. Renewable energy sources, better cooling systems, and energy-efficient data center design are all part of this. For example, Google has pioneered the use of machine learning to forecast and decrease energy usage in its data centers, leading to exceptionally high levels of energy efficiency.
Green algorithms	An increasing number of "green algorithms" are being developed. Computing tasks can have their energy usage minimized with the use of these techniques. As an example, in order to promote more energy-conscious programming, the green algorithms project offers scientists tools to quantify the carbon impact of their computational operations.
Sustainable practices in big data analytics	An increasing number of "green algorithms" are being developed. Computing tasks can have their energy usage minimized with the use of these techniques. As an example, in order to promote more energy-conscious programming, the green algorithms project offers scientists tools to quantify the carbon impact of their computational operations.
Carbon footprint tracking and reduction	New platforms and tools are appearing to assist businesses in monitoring and lowering the environmental impact of their computational operations. With these resources, companies may track their energy usage in real-time and make changes to become more environmentally friendly.
AI for environmental sustainability	Environmental problems are being addressed through the application of AI and ML. Climate modeling, energy consumption pattern prediction, and renewable energy source optimization are a few examples of areas that make use of AI. To further cut down on waste and increase efficiency, businesses are also analyzing environmental data with AI (Zheng et al., 2023).
Recycling and reducing electronic waste	Recycling programs and proper disposal of obsolete devices are gaining more and more attention as ways to decrease electrical waste. To further reduce waste, businesses should think about using second-hand hardware and extend the lifespan of their gadgets.
Sustainable smart cities	For efficient usage of utilities like water and electricity, "smart cities" leverage big data and internet of things (IoT) tracking devices. Technologies that control traffic, smart grids, and intelligent lighting all work together to lessen our impact on the environment.

development of Green Data Centers, which focus on energy-efficient design and operation to reduce environmental impact. Cloud computing services and data centers, like those operated by Fujitsu, have the potential to be more environmentally friendly than traditional onsite IT infrastructure. By optimizing airflow and reconfiguring cooling units, significant reductions in energy consumption can be achieved, which in Fujitsu's case, led to a 48% reduction in energy use in their Australian data centers. This optimization not only lowers operational costs but also significantly reduces the carbon footprint (Gupta and Pawar, 2022; Thenmozhi and Chandrakala, 2022).

The Center for Sustainable Systems at the University of Michigan emphasizes several sustainable alternatives in technology, such as virtualization and combined heat and power systems. Virtualization allows for more efficient use of servers, reducing the number of physical servers needed and promoting greater utilization. In data centers, energy efficiency can be improved by utilizing systems where heat recovered from electricity generation is used for cooling. Additionally, office equipment energy consumption could be significantly reduced by using devices with low-power modes and ensuring they are turned off when not in use. The emphasis on buying Energy Star–certified products, consolidating devices, and recycling unused electronics is also highlighted as essential actions for sustainability. Real-world examples and trends in energy-efficient and eco-friendly data analytics practices highlight how companies and organizations are adapting to sustainable models. Here are some notable examples and trends (Wang and Zhao, 2021; Ahmed and Babu, 2024).

4. Green data centers

Companies such as Fujitsu are leading the way in creating sustainable data centers. These centers are designed for energy efficiency, such as optimizing cooling technology, leading to significant reductions in energy consumption. For instance, by reconfiguring cooling units and optimizing airflow, Fujitsu achieved a 48% reduction in energy usage in two of its Australian data centers, equating to the annual energy consumption of 350 average households

- **Virtualization and Efficient Server Use**: Virtualization technology, where one physical server runs many independent programs or operating systems, is becoming a popular method to reduce the number of physical servers needed. This approach increases the utilization of each server, reducing material waste, electricity use, and cooling loads, contributing to significant environmental benefits (https://css.umich.edu/publications/factsheets/built-environment/green-it-factsheet).
- **Energy Star–Certified Products**: The use of Energy Star certified products in offices and data centers is another trend. These products are, on average, 30% more energy-efficient than standard servers. Adopting such certified products can lead to substantial savings in energy and greenhouse gas emissions (Aswanandini and Deepa, 2021a).
- **AI in Climate Modeling and Renewable Energy**: AI and machine learning are increasingly being used for environmental sustainability. They are applied in areas like climate modeling, predicting energy consumption patterns, and optimizing renewable energy sources. This application of AI helps in analyzing environmental data to reduce waste and improve efficiency across various industries.
- **E-Waste Management and Recycling**: It is becoming more important to deal with electronic garbage. Recycling electronic waste and finding better ways to dispose of obsolete devices are becoming more popular causes. To lessen the world's electronic trash, several businesses have teamed up with accredited recycling groups, and efforts are underway to improve hardware lifecycle management and encourage more responsible recycling practices (Aswanandini and Deepa, 2021a; Sundarakumar et al., 2023).

TABLE 5.3 Showing some real word scenarios.

Real world scenario	Explanation
Hospitality and retail	Businesses that have made good use of big data include Marriott Hotels and Amazon with their dynamic pricing and product suggestion systems. Marriott utilises data to change hotel rates in response to things like local events and weather, whereas Amazon analyses things like purchasing trends and competition prices to set its prices. Both revenue and the quality of the client experience have improved as a result of this strategy.
Entertainment and media	One company that has successfully used big data in the entertainment industry is Netflix. Netflix improves the user experience and keeps customers engaged by analyzing data on their viewing periods, search terms, and browsing tendencies to provide personalized movie selections.
Healthcare	Businesses in the healthcare sector are utilizing big data to offer individualized treatment suggestions; examples include Tempus and SOPHiA GENETICS. For screening and diagnosis, Tempus makes use of electronic medical records, genomic information, and imaging scans, while SOPHiA GENETICS provides insights derived from biological, clinical, genomic, and radiomics datasets.
Transportation	In order to optimize fuel efficiency and maintain safety, airlines like Southwest have begun using GE's Flight efficiency services, which analyze large amounts of data produced by airplanes. Because of this, airline operations are now more efficient and have a smaller impact on the environment.
Food service	By analyzing preparation times and making logistical adjustments based on that information, Uber Eats is able to optimize meal delivery. To further improve the efficiency and dependability of their service, they also hire meteorologists to foretell how the weather would affect deliveries.

4.1 Real-world scenario showing the effectiveness of the integration

Many sectors have benefited greatly from the introduction of new ideas and advancements brought forth by big data analytics. Let's have a look at a few examples of real-life situations where this integration has worked really well: Table 5.3.

Many industries will look different in the future due to new developments in smart industrial frameworks and big data analytics. Some important currents and potential avenues for further study are as follows: Tables 5.4 and 5.5.

5. Potential future developments in computational intelligence

1. Advancements in Learning Algorithms: which means that machine learning algorithms will get a lot better, which means that models will be more efficient and accurate. That encompasses advancements in RL and DL as well.
2. Quantum Computing: Quantum computing has the potential to completely alter the field of computational intelligence by facilitating the lightning-fast resolution of extremely complicated issues.
3. Improving Human—AI Collaboration: Developing better algorithms and user interfaces to facilitate greater human—AI collaboration, which in turn will lead to more effective and intuitive AI applications across industries.

TABLE 5.4 Showing the amalgamation application in various fields.

Type of application	Explanation
Integration of AI and machine learning	One notable development is the growing use of AI and ML algorithms in big data analytics. Improved predictive analytics and decision-making made possible by this connection let companies delve farther into their data.
Edge computing	Processing data closer to its point of generation (edge computing) is gaining importance as data volumes increase. More efficient and quicker data processing is possible with this method since it decreases latency and bandwidth consumption.
Automation and real-time analytics	More and more, people are opting for automated systems that can give them insights in real time. Companies in the transportation and industrial sectors, for example, where real-time data analysis can greatly enhance efficiency, must make this transition immediately.
Enhanced data security and privacy	Security and privacy issues are becoming more noticeable due to the ever-increasing data processing volumes. Stronger security protocols and analytics methods that protect user privacy are the subject of current research.
Sustainable big data practices	Sustainable and ecologically friendly data practices are gaining prominence. Reducing energy use and carbon impact involves optimizing data storage and processing.
IoT and smart devices proliferation	The exponential growth of the internet of things and smart devices: Data is being generated at an exponential rate due to the widespread use of smart sensors and the internet of things. Because of this shift, sophisticated big data frameworks are required to process massive volumes of data, both structured and unstructured, from a wide variety of sources.
Predictive maintenance in manufacturing	More and more factories are implementing predictive maintenance systems that use big data analytics. Businesses can cut down on maintenance expenses and downtime by analyzing data from machinery and equipment to identify faults before they happen.
Healthcare personalization	The healthcare industry is leveraging big data to enhance patient outcomes through personalized treatment regimens. Accurate diagnoses and individualized treatment plans are made possible by the analysis of massive datasets gathered from many sources.
Blockchain for data security and integrity:	Investigations exploring the potential integration of big data and blockchain technologies are continuing. Industries like healthcare and finance can greatly benefit from blockchain technology because of the increased security, transparency, and integrity it can bring to their data.
Data democratization	There's been an effort to make data more accessible and democratise it so that more individuals in an organization may access and analyze it without specific expertise. Creating platforms and tools that are easy for users to navigate is a current trend.
Quantum computing in big data	An exciting new field of study is the investigation of quantum computing as a potential solution to the problems posed by big data. New approaches to solving difficult issues may be possible with the advent of quantum computers, which might handle massive datasets at a fraction of the time it takes classical computers.
Data privacy and security	It is critical to do research into stronger data security and privacy safeguards due to the increasing frequency of data breaches. Innovative methods of data encryption and safe procedures for exchanging sensitive information are part of this effort.

TABLE 5.4 Showing the amalgamation application in various fields.—cont'd

Type of application	Explanation
Advanced AI and machine learning integration	Improving AI and ML algorithms is anticipated to be the primary focus of future studies. To improve data processing automation and predictive analytics, this incorporates neural networks and deep learning.
Edge computing enhancements:	An ever-growing number of internet of things (IoT) devices is elevating the significance of edge computing. Improving data processing at network edges to decrease latency and improve real-time data analysis is the main focus of research in this area.
Green computing	Research on green computing practices is crucial in light of the growing concern for the environment caused by data centers and other forms of IT infrastructure. Among these, you can find renewable energy sources and energy-efficient computers.
Blockchain for enhanced data integrity	An emerging field of research is the use of blockchain technology to guarantee the security and integrity of data. The storage and sharing of data across industries could be completely transformed by this technology.
Ethical AI and big data governance:	Research on AI's ethical implications and the creation of big data governance frameworks are crucial given AI's growing involvement in decision-making.
Personalized healthcare analytics	Personalized medicine utilizing big data is becoming more and more of an emphasis in the healthcare industry. More personalized treatment strategies could be possible in the future if genomic data and electronic health records could be integrated.
Human-AI collaboration	One important area of study is how to make AI easier to use and understand by studying how people and AI systems work together.
Smart cities and urban data analytics	There is a rising interest in studying how smart cities may use big data analytics to enhance public services, traffic control, and urban planning.

6. Ethical and social implications of these technologies

1. **Privacy Concerns**: Concerns about the security and privacy of personal information are growing in importance as AI finds more and more uses in everyday life.
2. **Job Displacement**: There needs to be an emphasis on retraining and education because AI automation of jobs could cause people to lose their jobs.
3. **Bias and Fairness**: More and more people are worried about whether or not AI systems are impartial and free of bias (Sundarakumar et al., 2023).

6.1 Directions for future research in sustainable computing and smart industries

1. **Energy-Efficient Computing**: Research into developing more energy-efficient computing methods to reduce the environmental impact of technology.
2. **Smart Industry Applications**: Exploring the integration of AI in industries for enhanced efficiency, productivity, and sustainability.
3. **Ethical AI**: Developing frameworks and guidelines to ensure AI is used ethically in industrial applications.

TABLE 5.5 Showing a list of optimization techniques used in various big data techniques.

Authors	Contribution	Methodology used	Outcome	Application	Key findings
Y. Liu, H. Xu, W. C. Lau (Liu et al., 2023)	Optimization of cloud configurations for recurring batch-processing applications.	Big data analysis in cloud computing.	Enhanced efficiency in cloud processing tasks.	Cloud computing.	Identified optimal configurations for specific cloud processing scenarios.
G. Prabaharan et al. (Prabaharan et al., 2023)	Introduced a group teaching optimization algorithm with ML for big data analytics.	Machine learning in big data analytics.	Improved data analysis efficiency.	Big data analytics.	Demonstrated effectiveness of the new algorithm in handling large datasets.
Yang, Chao-Tung et al. (Yang et al., 2023)	Analysis of big data and ML in bioprocessing.	ML-driven bioprocessing techniques.	Insights into recent trends and critical analysis in bioprocessing.	Bioprocessing industry.	Highlighted the impact of ML on bioprocessing efficiency and accuracy.
Ikotun, Abiodun M. et al. (Ikotun et al., 2023)	Comprehensive review of K-means clustering algorithms in big data.	Variants analysis of K-means clustering.	Advanced understanding of clustering algorithms.	Data clustering in big data.	Provided a detailed comparison of various K-means algorithms.
Nair, Akarsh K. et al. (Nair et al., 2023)	Privacy preservation in IoMT using federated learning and edge computing.	Federated learning framework for big data analysis.	Enhanced data privacy in IoMT.	IoMT and big data analysis.	Showcased how federated learning can secure IoMT data.
Sadat Lavasani, Mitra et al. (LavasaniMitra et al., 2023)	Opportunities of big data analytics in process engineering.	Review of applications in process engineering.	Identified applications in various engineering processes.	Process engineering.	Explored diverse applications of big data in process engineering.
Wang, Si. et al. (Wang, 2023)	Optimization of health service management using big data.	Big data knowledge management in healthcare.	Improved health service management.	Healthcare management.	Focused on how big data can revolutionize health service management.

Reference	Description	Focus area	Outcome	Application	Contribution
Pal, Souvik et al. (Pal et al., 2023)	Developed a hybrid edge-cloud system for networking service optimization.	IoT and big data in networking services.	Enhanced networking service efficiency.	Networking and IoT.	Demonstrated a novel approach for optimizing networking services.
Uddin, Md Galal et al. (Uddin et al., 2023)	Assessing optimization techniques for water quality models.	Optimization in environmental models.	Improved water quality assessment.	Environmental science.	Analyzed effective optimization techniques for environmental modeling.
J. Shekhawat, Tajpal et al. (Shekhawat and Tajpal, 2023)	Improved configuration of SVM using particle swarm optimization for cybersecurity.	Particle swarm optimization in cybersecurity.	Enhanced cyber security measures.	Cybersecurity in big data.	Showed improved results in cybersecurity using SVM and particle swarm optimization.
Tyagi, Tirbhuwan et al. (Tyagi et al., 2023)	Neuro-optimization techniques for inventory models in manufacturing.	Neuro-optimization in manufacturing.	Improved inventory management.	Manufacturing sector.	Explored the application of neuro-optimization in inventory management.
Mondal, Partha Pratim et al. (Mondal et al., 2023)	Review on ML-based bioprocess optimization and monitoring.	ML in bioprocess systems.	Advanced bioprocess monitoring and control.	Bioprocessing industry.	Discussed advancements in ML for bioprocess optimization.
G. Sun et al. (Sun et al., 2023)	Survey on mathematical optimization in data visualization.	Mathematical techniques in big data visualization.	Enhanced data visualization methods.	Data visualization in big data.	Surveyed various mathematical optimization methods for data visualization.

7. Conclusion

With quantum computing and learning algorithms projected to make significant strides in the future, computational intelligence is set for a game-changing expansion. Nevertheless, there are social and ethical concerns that must be addressed in relation to these technologies. Researchers in the fields of sustainable computing and smart industries need to think about how to progress technology while also taking into account ethical concerns, sustainability, and the societal impact. Finding a middle ground between technical advancement and ethical accountability is essential as we move forward with AI. While advanced AI's widespread adoption holds great potential for improved productivity and life quality, it also begs the question of how this technology will affect people and the planet in the long run. Researchers, lawmakers, and business moguls must work together to solve these problems and make sure that technology advances benefit society. The path forward in AI goes beyond achieving technical benchmarks; it also involves molding a future that is congruent with our moral principles and societal objectives (Aswanandini and Deepa, 2021b).

References

Ahmed, M., Babu, G.R.M., 2024. Hyper-heuristic multi-objective online optimization for cyber security in big data. International Journal of System Assurance Engineering and Management 15 (1), 314–323.

Alicioglu, G., Sun, B., 2022. A survey of visual analytics for explainable artificial intelligence methods. Computers & Graphics 102, 502–520.

Aswanandini, R., Deepa, C., 2021a. Hyper-heuristic firefly algorithm based convolutional neural networks for big data cyber security. Indian Journal of Science and Technology 14, 2934–2945.

Aswanandini, R., Deepa, C., 2021b. Network intrusion classification using configuration optimized support vector machines. In: 2021 International Conference on Advancements in Electrical, Electronics, Communication, Computing and Automation (ICAECA). IEEE, pp. 1–6.

Gupta, V., Pawar, S., 2022. An effective structure of multi-modal deep convolutional neural network with adaptive group teaching optimization. Soft Computing 26 (15), 7211–7232.

Ikotun, A.M., et al., 2023. K-means clustering algorithms: a comprehensive review, variants analysis, and advances in the era of big data. Information Sciences 622, 178–210.

Lavasani, S., Mitra, et al., 2023. Big data analytics opportunities for applications in process engineering. Reviews in Chemical Engineering 39 (3), 479–511.

Liu, Y., Xu, H., Lau, W.C., May 2023. Cloud configuration optimization for recurring batch-processing applications. IEEE Transactions on Parallel and Distributed Systems 34 (5), 1495–1507. https://doi.org/10.1109/TPDS.2023.3246086.

Mondal, P.P., et al., 2023. Review on machine learning-based bioprocess optimization, monitoring, and control systems. Bioresource Technology 370, 128523.

Nair, A.K., Sahoo, J., Deni Raj, E., 2023. Privacy preserving Federated Learning framework for IoMT based big data analysis using edge computing. Computer Standards & Interfaces 86, 103720.

Pal, S., et al., 2023. A hybrid edge-cloud system for networking service components optimization using the internet of things. Electronics 12.3, 649.

Prabaharan, G., Venkatesh, K., Mary Sundararajan, S.C., et al., 2023. Group teaching optimization algorithm with machine learning model for big data analytics. In: 2023 International Conference on Sustainable Communication Networks and Application (ICSCNA), Theni, India, pp. 147–152. https://doi.org/10.1109/ICSCNA58489.2023.10370497.

Shekhawat, J., Tajpal, 2023. A multi-objective hyper-heuristic improved configuration of Svm based on particle swarm optimization for big data cyber security. In: 2023 3rd International Conference on Advance Computing and Innovative Technologies in Engineering (ICACITE), Greater Noida, India, pp. 2223–2225. https://doi.org/10.1109/ICACITE57410.2023.10182425.

Sun, G., et al., 2023. Application of mathematical optimization in data visualization and visual analytics: a survey. IEEE Transactions on Big Data 9 (4), 1018–1037. https://doi.org/10.1109/TBDATA.2023.3262151.

Sundarakumar, M.R., Salanagai Nayagi, D., Vinodhini, S., Vinayaga Priya, S., Marimuthu, M., Basheer, S., Johny Renoald, A., 2023. A heuristic approach to improve the data processing in big data using enhanced Salp Swarm algorithm (ESSA) and MK-means algorithm. Journal of Intelligent and Fuzzy Systems 1–16.

Thenmozhi, L., Chandrakala, N., 2022. Developed modified particle swarm optimization for feature selection on learning based big data in cloud computing. Journal of Algebraic Statistics 13 (1), 310–320.

Tyagi, T., et al., 2023. A novel neuro-optimization techniques for inventory models in manufacturing sectors. Journal of Computational and Cognitive Engineering 2 (3), 204–209.

Uddin, Md G., et al., 2023. Assessing optimization techniques for improving water quality model. Journal of Cleaner Production 385, 135671.

Wang, S., 2023. Optimization health service management platform based on big data knowledge management. Optik 273, 170412.

Wang, J., Zhao, B., 2021. Intelligent system for interactive online education based on cloud big data analytics. Journal of Intelligent and Fuzzy Systems 40 (2), 2839–2849.

Wu, A., Wang, Y., Shu, X., Moritz, D., Cui, W., Zhang, H., et al., 2021. Ai4vis: survey on artificial intelligence approaches for data visualization. IEEE Transactions on Visualization and Computer Graphics 28 (12), 5049–5070.

Xu, H., Berres, A., Liu, Y., Allen-Dumas, M.R., Sanyal, J., 2022. An overview of visualization and visual analytics applications in water resources management. Environmental Modelling & Software 153, 105396.

Yang, C.-T., et al., 2023. Big data and machine learning driven bioprocessing-recent trends and critical analysis. Bioresource Technology 128625.

Zheng, L., Wang, C., Chen, X., Song, Y., Meng, Z., Zhang, R., 2023. Evolutionary machine learning builds smart education big data platform: data-driven higher education. Applied Soft Computing 136, 110114.

6

Computational intelligence for data analysis in pattern recognition and biomedical fields

Sudha Senthilkumar and Vaishnavi Munusamy

School of Computer Science and Engineering, Vellore Institute of Technology, Vellore, Tamil Nadu, India

1. Introduction

Prior to the 20th century, doctors made diagnoses mostly by using their manual and visual pattern recognition skills while observing physical symptoms and relying on their knowledge and intuition. Medical imaging, particularly X-rays, which introduced objective pattern recognition to medicine and allowed doctors to detect illnesses through images, brought about a significant change in the 20th century. Early rule-based computer-based expert systems made an effort in the 1970 and 1980s to mimic the diagnostic reasoning of medical professionals. For the diagnosis and prognosis of diseases, machine learning techniques such as decision trees and neural networks were widely used in the late 20th century. The examination of medical images such as CT scans and MRIs has become more automated owing to the developments in digital technology and computer vision. The Human Genome Project and genomic developments in the 21st century introduced pattern recognition to genetic and genomic data processing, assisting in the discovery of genetic patterns related to diseases. Convolutional and recurrent neural networks (CNNs) and other deep learning approaches such as deep neural networks have revolutionized medical image processing since the decade of the 2010s. Predictive modeling and customized medicine were made possible by the simultaneous growth of electronic health records (EHRs) and big data analytics, which made it possible to extract patterns from vast amounts of patient data. In addition, Natural Language Processing (NLP) has grown to be crucial for gathering information from clinical notes and medical literature to aid in clinical decision-making, disease surveillance, and research. In telemedicine and remote monitoring, pattern recognition has become increasingly significant in the 2010s, making it possible to examine patient data gathered outside of conventional clinical settings. The development and application of pattern recognition techniques in medicine have been further accelerated in the current era by interdisciplinary collaborations between computer scientists, data analysts, and medical professionals, promising more precise diagnosis, individualized treatment, and improved patient care.

Pattern recognition is a process in which a system or an observer identifies and interprets recurring, meaningful patterns or structures within data. These patterns can manifest in various forms, including visual patterns in images, audio patterns in sound signals, text patterns in written or spoken language, or patterns in numerical data. Pattern recognition involves the extraction of relevant features or characteristics from data and the subsequent classification or interpretation of those patterns. It is used in a wide range of fields, including machine learning, computer vision, natural language processing, and the biomedical sciences, to recognize and make sense of complex information and to support decision-making and automation.

Pattern recognition is invaluable in the biomedical field due to its ability to identify complex and subtle patterns within extensive datasets, encompassing medical images, genetic sequences, and patient records. These patterns play a pivotal role in disease diagnosis, patient outcomes, biomarker identification, drug discovery, and treatment optimization. By automating the analysis of medical data and deriving meaningful insights, pattern recognition significantly enhances the precision and efficiency of medical decision-making. This, in turn, leads to improvements in patient care, advancements in research, and the acceleration of medical breakthroughs. Therefore, the importance of pattern recognition in the biomedical field cannot be overstated. Its significance is underscored by its diverse applications (Fig. 6.1):

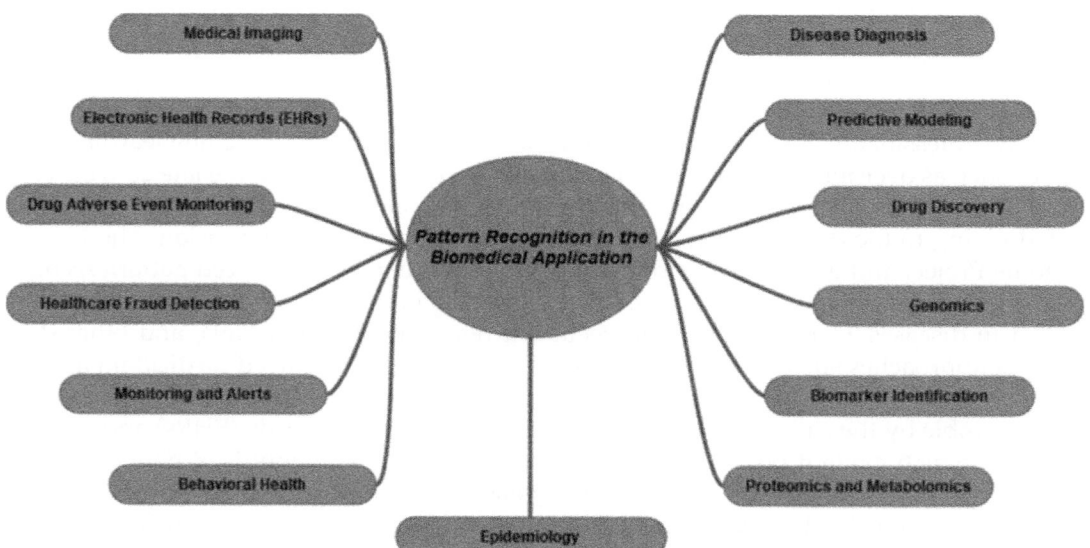

FIGURE 6.1   Pattern recognition in the bio-medical applications.

1. **Disease Diagnosis:** Pattern recognition algorithms scrutinize medical data, including images, genetic sequences, and patient records, to uncover disease-associated patterns, aiding in early and accurate diagnosis. For example, it can facilitate the detection of anomalies in medical images such as X-rays, MRIs, or CT scans.
2. **Predictive Modeling:** Through pattern recognition, predictive models can be constructed to forecast disease outcomes, patient responses to treatments, and the likelihood of disease recurrence, fostering personalized medicine and tailored treatment plans.
3. **Drug Discovery:** Pattern recognition techniques analyze chemical structures and biological data to expedite the discovery of new drug compounds and predict their potential efficacy in treating diseases.
4. **Genomics:** In genomics, pattern recognition is leveraged to analyze DNA and RNA sequences, uncovering genes, regulatory elements, and genetic variations linked to diseases, thereby advancing our understanding of the genetic underpinnings of ailments and the design of targeted therapies.
5. **Biomarker Identification:** The identification of specific biomarkers or patterns in biological samples, such as blood or urine, facilitates early disease detection, continuous disease monitoring, and the assessment of treatment effectiveness.
6. **Proteomics and Metabolomics:** Pattern recognition is instrumental in deciphering complex data from proteomics and metabolomics studies, enabling the recognition of patterns associated with diseases at the protein and metabolite level.
7. **Medical Imaging:** In radiology and pathology, pattern recognition streamlines the interpretation of medical images, making it more efficient for healthcare professionals to identify anomalies, tumors, and abnormalities.
8. **Electronic Health Records (EHRs):** Analyzing EHR data with pattern recognition can identify trends and risk factors for various diseases, which in turn enhances patient care and healthcare management.
9. **Drug Adverse Event Monitoring:** By analyzing patient records and reporting systems, pattern recognition can pinpoint adverse events associated with medications, thereby bolstering drug safety monitoring.
10. **Healthcare Fraud Detection:** Pattern recognition techniques are applied to detect fraudulent activities in healthcare billing, insurance claims, and prescription drug utilization, safeguarding the healthcare system's integrity.
11. **Monitoring and Alerts:** Pattern recognition algorithms are employed to continuously monitor patient vital signs, alerting healthcare providers to any deviations from normal patterns, thus enabling timely interventions.
12. **Behavioral Health:** Pattern recognition delves into behavioral data to uncover patterns associated with mental health conditions and addiction, thereby supporting diagnosis and treatment strategies.
13. **Epidemiology:** Pattern recognition contributes to epidemiological studies by identifying disease outbreaks and tracing their propagation through the analysis of epidemiological data, facilitating effective public health responses.

Pattern recognition in the biomedical field involves various techniques and algorithms for analyzing complex data to identify patterns. Some of the commonly used techniques and algorithms include.

1.1 Machine learning algorithms

- **Supervised Learning**: Algorithms such as Support Vector Machines (SVM), Random Forest, and Neural Networks are trained on labeled data to recognize patterns, such as disease classification based on patient records.
- **Unsupervised Learning**: Clustering methods such as k-Means and hierarchical clustering can identify patterns within data without prior labeling.
- **Dimensionality Reduction**: Techniques such as Principal Component Analysis (PCA) reduce the dimensionality of data while preserving important patterns.

1.2 Deep learning

- **Convolutional Neural Networks (CNN)**: Used in medical image analysis, including CT scans, MRIs, and X-rays.
- **Recurrent Neural Networks (RNN)**: Suitable for sequential data, such as time-series patient records and electrocardiograms (ECG).
- **Long Short-Term Memory (LSTM)**: Useful for analyzing sequential data and time series, as it can capture long-term dependencies.

1.3 Image processing techniques

- **Image Segmentation**: Dividing medical images into meaningful regions, such as segmenting tumors in radiological images.
- **Feature Extraction**: Techniques such as texture analysis, edge detection, and wavelet transform help in extracting relevant information from images.

1.4 Signal processing

- **Fourier Transform**: Analyzes frequency components in signals, useful in electroencephalogram (EEG) and electrocardiogram (ECG) analysis.
- **Wavelet Transform**: Reveals signal details at different scales and is beneficial in analyzing time-frequency patterns in biomedical signals.

1.5 Natural language processing (NLP)

- NLP algorithms such as Named Entity Recognition and Sentiment Analysis process clinical notes, research articles, and electronic health records to extract information for clinical decision support.

1.6 Genomic and sequencing data analysis

- **Alignment Algorithms**: Tools such as BLAST and Bowtie align DNA or RNA sequences to a reference genome.
- **Variant Calling Algorithms**: Identify genetic variations and mutations in sequencing data, e.g., GATK (Genome Analysis Toolkit).

1.7 Pattern matching and classification

- Algorithms for sequence alignment, such as Needleman-Wunsch or Smith-Waterman, are used in genetics and proteomics.
- k-Nearest Neighbors (k-NN) and Decision Trees can be applied in classification tasks.

1.8 Time series analysis

- Methods such as Autoregressive Integrated Moving Average (ARIMA) or Exponential Smoothing analyze time-dependent data, such as patient vitals or disease progression.

1.9 Data mining and feature selection

- Techniques such as Apriori and association rule mining can discover relevant patterns in large datasets, while feature selection methods help identify the most important variables.

1.10 Ensemble learning

- Combining the predictions of multiple models, such as Random Forests and Gradient Boosting, to enhance pattern recognition accuracy.

1.11 Anomaly detection

Algorithms such as Isolation Forest and One-Class SVM are useful for identifying rare or abnormal patterns in datasets, which can be crucial in early disease detection.

2. GAIT-based pattern recognition in biomedical field

2.1 Gait recognition and its applications in the medical field

Gait recognition is a biometric technology that identifies individuals based on their unique walking patterns or gait, similar to how fingerprints, facial features, or iris patterns serve as distinct biometric identifiers. Gait recognition systems typically use data from sensors or cameras, such as video cameras, depth sensors, accelerometers, or pressure-sensitive mats, to capture an individual's walking pattern. The system then analyzes various gait features, including step length, stride duration, walking speed, and the motion of body parts such as the head, arms, and legs. These gait features are unique to each person and can be used for identification and authentication. Gait recognition finds applications in various fields, including security, healthcare, and assistive technologies. In security, it can be employed for access control, surveillance, and identity verification. In healthcare, gait analysis aids in diagnosing and monitoring medical conditions related to mobility and balance. Additionally, gait recognition is used in developing personalized human-computer interfaces and assistive devices for individuals with mobility impairments. Pattern recognition is highly valuable in abnormal gait recognition within the biomedical field, where abnormal gait patterns can indicate various neurological, musculoskeletal, or neuromuscular disorders, making early detection and assessment critical for diagnosis and treatment planning. Pattern recognition algorithms can analyze motion data from sensors, such as accelerometers or video recordings, to identify deviations from a normal gait pattern. These algorithms quantify parameters such as gait speed, stride length, step symmetry, and joint angles, enabling healthcare professionals to detect abnormalities and assess their severity. Abnormal gait recognition assists in diagnosing conditions such as Parkinson's disease, multiple sclerosis, or injuries, tracking disease progression, evaluating treatment effectiveness, and providing objective measures for rehabilitation programs. It plays a vital role in improving patient care and enhancing our understanding of various medical conditions affecting mobility.

2.2 Pattern recognition is exceptionally helpful in abnormal gait recognition in the biomedical field for several key reasons

1. Early Detection: Pattern recognition algorithms can identify subtle changes in gait patterns even before visible symptoms manifest. Early detection is crucial for timely intervention and treatment, which can significantly improve patient outcomes.
2. Objective Assessment: Gait abnormalities are often assessed subjectively by clinicians, which can be prone to bias and variation. Pattern recognition provides an objective and quantitative means of evaluating gait, making assessments more reliable and consistent.
3. Monitoring Progression: Abnormal gait patterns can change over time, especially in progressive diseases. Pattern recognition allows for continuous monitoring, enabling healthcare providers to track disease progression and adjust treatment plans accordingly.
4. Treatment Evaluation: When patients undergo interventions such as physical therapy or surgery, pattern recognition can quantify changes in gait parameters, helping healthcare professionals assess the effectiveness of treatments and make informed decisions.
5. Rehabilitation: Pattern recognition can be used in rehabilitation settings to design personalized exercise programs and track a patient's progress during recovery, ultimately facilitating a faster and more tailored rehabilitation process.
6. Research and Insights: Pattern recognition enables researchers to analyze large datasets of gait data, uncover hidden patterns or correlations, and gain insights into the biomechanics and pathophysiology of various conditions affecting gait.
7. Telemedicine and Remote Monitoring: In the era of telemedicine and remote patient monitoring, pattern recognition can play a pivotal role in assessing gait remotely, allowing for continuous care and reducing the need for frequent in-person visits.
8. Enhancing Mobility Aids: By recognizing abnormal gait patterns, pattern recognition can be integrated into assistive devices such as prosthetics or exoskeletons, improving their functionality and adaptability to the individual's gait.
9. Fall Prevention: Abnormal gait recognition can help identify individuals at risk of falls due to their gait abnormalities. This information can be used to implement fall prevention strategies and reduce injury risks, especially in the elderly population.

2.3 Hidden Markov model and deep neural network based model

The research introduces (Wang et al., 2022) a comprehensive approach for recognizing a wide range of abnormal behaviors, particularly in elderly individuals in their homes, by combining deep neural network algorithms and a hidden Markov model (HMM). The methodology involves collecting data from multiple sensors in a smartphone, including accelerometers, gyroscopes, and orientation sensors, followed by noise reduction and addressing imbalances in data categories through techniques such as SMOTE (Chawla et al., 2002). The data is then segmented and normalized using min-max scaling. An attention–CNN–LSTM model is introduced to identify various behavioral actions, such as falls and daily activities, by extracting time-dependent features through 1D convolutional layers and LSTM networks (Tan et al., 2019; Chen et al., 2021) equipped with an attention mechanism. Additionally, the HMM (Asghari et al., 2020; Choudhury et al., 2019; Liu et al., 2021; Forney, 1973) is utilized to construct daily behavior sequences based on long-term data, associating hidden and observed states, with sequence identification carried out using the Viterbi algorithm (Forkan et al., 2015). Abnormal behaviors are detected in two stages: the initial phase employs a behavior classification model, while the subsequent stage utilizes the HMM to detect deviations from regular daily activities. This approach provides a comprehensive method for identifying multiple anomalies in specific behaviors and behavioral states using data from a variety of sensors, with a primary focus on elderly individuals in a home environment (Fig. 6.2).

2.4 LSTM-CNN based model

In (Gao et al., 2019), the authors utilize a gait information localization algorithm to process data and present a deep neural network that includes CNN and LSTM layers for the

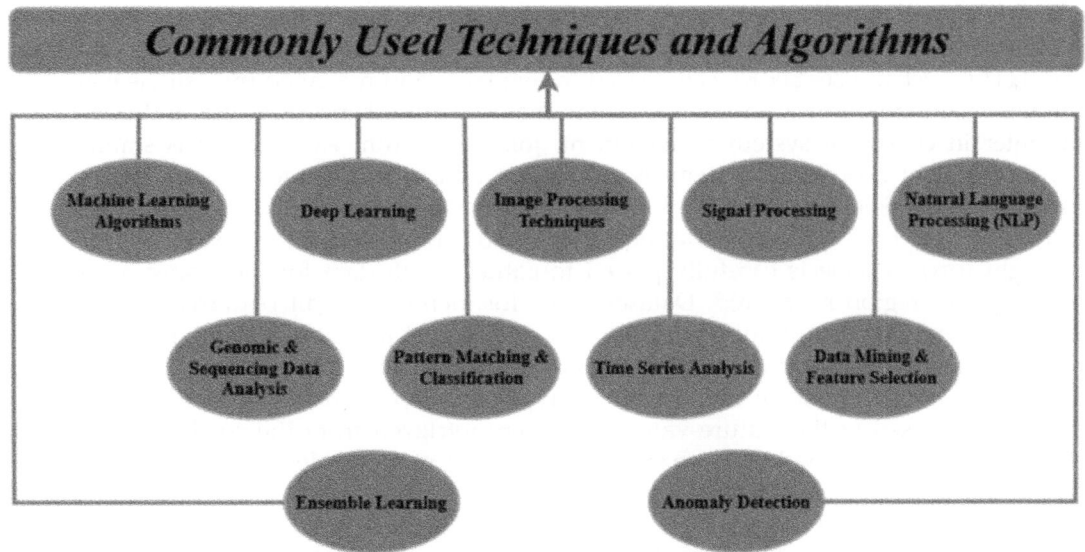

FIGURE 6.2 Some commonly used techniques and algorithm in the pattern recognition.

classification of abnormal gait patterns. Furthermore, they introduce a WhSPP model aimed at optimizing the LSTM-CNN pooling layer, leading to improved accuracy in classification. To collect gait data and identify abnormal gait patterns based on attitude angles after data fusion, the Dynamic Step Conjugate Gradient Algorithm is applied using inertial sensors. This involves the simulation of a smartphone's IMU by two sensors (Asuncion et al., 2018), specifically focusing on capturing rolling, pitching, and deflection angles postfusion. The authors demonstrate the potential of CNN-based gait recognition algorithms for biometrics and tackle challenges related to sensor noise and data fusion by employing the conjugate gradient algorithm, customized for enhanced precision. This method amalgamates conjugation with the steepest descent approach to iteratively refine attitude estimation for the recognition of abnormal gait, effectively addressing concerns related to step size through the introduction of a dynamic step adjustment that correlates with the physical angular velocity of human motion. The chapter also covers data preprocessing, encompassing the collection of both normal and abnormal gait data through experimental means. It introduces a sliding window segmentation method and calculates data segment similarity to selectively retain feature data with a similarity threshold of at least 80%. This chapter introduces the LCSWnet framework, which combines LSTM and CNN layers for the classification of abnormal gait. The LSTM layer captures temporal features, while the CNN layer extracts spatial features. The addition of a WhSPP layer optimizes the final pooling layer, thereby enhancing feature extraction and classification accuracy. During training, the primary goal is to minimize cross-entropy error. This chapter presents a comprehensive approach to abnormal gait recognition, covering various aspects such as data preprocessing, dynamic step adjustment, data segmentation, and the architecture of the LCSWnet model.

2.5 RNN-LSTM based model

In order to improve the accuracy of depth data and jointly acquire information, a system using several Kinects is introduced in (Lee et al., 2019), solving the shortcomings of a single Kinect. This system uses six Kinects placed in two rows and managed by a single computer, creating a configuration with six computers managing the Kinect devices and one central computer in charge of system management. Joint data from each Kinect is standardized into a consistent coordinate system by calibration and synchronized in terms of timing to assure data consistency. On a 10-meter walkway (Geerse et al., 2015), data is gathered over the center 6 m with special attention paid to gait-related information. With the left and right rows of Kinects carefully placed to gather depth data for the respective sides of the body, this region is covered. Datasets used for identifying gait patterns are created by combining the joint information from the left and right Kinects and averaging the results. In order to collect data for this chapter's main goal of identifying six distinct gait patterns, four people had to replicate each of these atypical gait patterns. An RNN-LSTM model's input is comprised of the feature values that were retrieved from the combined joint data. The joint data is replicated and separated into distinct sequence lengths to increase dataset size, resulting in dataset sizes that vary between 18,000 and 50,000 sets. These datasets are divided into testing and training portions, with testing taking up a fraction of each. Additionally, the chapter examines 14 feature types produced from 3D joint data for feature extraction and evaluates how they affect test validity. Additionally, the impact of sequence duration on test performance is examined in an effort to perhaps streamline the data gathering system by lowering the number of Kinects needed.

2.6 RNN-based autoencoder model

The chapter (Jun et al., 2020) provides an in-depth analysis of the key elements found in recurrent neural network (RNN) designs, with a focus on the fundamental RNN, LSTM, and GRU models, which are crucial to the chapter's methodology. The RNN DM (Deep Model), RNN AE (Autoencoder), and the ground-breaking hybrid RNN AE-DM model are then introduced. The RNN DM's main goal is to identify abnormal gait patterns, which includes evaluating a variety of features and helps to build the hybrid RNN AE-DM model. The RNN AE, on the other hand, is designed to extract features from skeletal data, improving data alignment and boosting the hybrid RNN AE-DM model. Features from the RNN AE are fed into the RNN DM in the hybrid RNN AE-DM framework in order to identify gait patterns. This section goes into the fundamental ideas behind RNN designs, highlighting their ability to handle sequential input and the crucial part hidden states play in these structures. It elaborates on the mathematical formulas used to calculate hidden states and output values within recurrent layers, illuminating the impact that previous knowledge has on the data now being processed. The vanishing gradient problem is skilfully addressed by LSTM (Chung et al., 2014) and GRU (Hochreiter, 1998) architectures, which are given special attention because of their gate designs and equations. In order to evaluate the efficacy of various variables generated from skeletal gait data, the RNN DM is introduced as a crucial tool for the identification of aberrant gaits using skeleton data. For identifying gait patterns, the RNN DM consists of an input layer, a 4-layer RNN structure, and a softmax classifier. Notably, cross-entropy cost functions, L2 regularization, and the Adam optimization technique are all used throughout the training procedure. The RNN AE, envisaged as a sequence-to-sequence model using either LSTM or GRU architectures, is also provided as a mechanism for feature extraction. The encoder in this approach decreases dimensions and extracts features, while the decoder raises dimensions and reconstructs data. The mean squared error between the input and the reconstructed data is the definition of the loss function. The hybrid RNN AE-DM model, which uses the RNN DM to recognize gait patterns using characteristics extracted from the RNN AE, is finally presented. The RNN AE and RNN DM are trained separately, with the former focusing on feature extraction and the latter on gait pattern identification. The reduction of discrepancies between input and reconstructed data is essential for feature extraction to be effective.

2.7 Gait-structural graph convolutional network (AGS-GCN) based model

The chapter (Tian et al., 2022) introduces the concept of a "gait graph" for gait recognition, which involves constructing spatial-temporal skeleton graphs using skeleton joints as vertices and bones as edges. For abnormal gait recognition, the gait graph is based on abnormal gait characteristics and body skeletal structure, with a focus on lower limb skeleton data often collected through devices such as Kinect sensors (Eltoukhy et al., 2017; Tanaka et al.,

2018). Notably, the accuracy of knee and hip angle measurements from Kinect sensors is consistent with Vicon system results (McGinley et al., 2009; Hassan et al., 2007). However, foot information is less accurate. To address this, the chapter constructs the gait graph using seven lower limb joints and defines the relationships between corresponding joints on the right and left sides as "gait-links." The spatial gait graph comprises these seven joints as vertices with bones and gait-links as edges. The chapter also introduces Graph Convolutional Networks (GCNs) (Kipf and Welling, 2016) for analyzing graph-structured data, where GCNs use matrices X and A as inputs and propagate features through layers. Furthermore, the chapter outlines the architecture of the Spatio-Temporal Attention (Hu et al., 2018; Wu et al., 2020) Enhanced Gait-Structural Graph Convolutional Network (AGS-GCN) designed for abnormal gait recognition, involving three AGS-GCN blocks, temporal convolution, and global average pooling. Finally, the chapter discusses the use of spatio-temporal attention to enhance fine-grained learning, mitigate oversmoothing, and generate an enhanced gait feature map through layer-wise attention-weighted feature combination.

2.8 Sparse representation classifier based model

In (Sithi Shameem Fathima et al., 2023), the gait recognition process begins with the collection and preprocessing of gait data from various individuals. Preprocessing includes the removal of unwanted pixels and the merging of disjoint lines in silhouettes to ensure continuity. This cleaned data is then used to extract crucial features such as gait cycles, height, and width of the silhouettes, which are stored in a database during the training phase. During the testing phase, random gait patterns are acquired, and their features are extracted and preprocessed similarly to the training phase. These features are then compared and matched with the existing dataset using a Sparse Representation Classifier, which classifies individuals as either having a normal or abnormal gait based on these features. The feature extraction process involves recording individuals' walking motions via a camera and converting them into JPEG format frames, resulting in a substantial dataset for analysis. Images are transformed into silhouettes and further into skeletons using morphological operations, with angle values representing various body segments, including gait cycles, serving as feature vectors for the experiment. Data analysis is carried out to classify gait patterns as normal or abnormal, leveraging the powerful Sparse Representation Classifier (SRC) (Sithi Shameem Fathima and Wahida Banu, 2015), known for its effectiveness in pixel-wise classification, particularly in handling noisy data. Height and width measurements are crucial components of gait analysis. The maximum height and width values during a gait cycle provide valuable information for distinguishing between normal and abnormal gait patterns. Additionally, the width of the outer contour of binarized silhouettes is used as a feature vector. To calculate the centroid of human silhouettes, pixel positions are averaged, and the resulting data are analyzed using Matlab, yielding classification results for normal and various abnormal gait patterns. It presents a classification rate table, demonstrating the accuracy of classification for different gait patterns, with the Sparse Representation Classifier proving effective in managing noisy data. In summary, this comprehensive process encompasses gait data collection, preprocessing, feature extraction, and classifier-based classification of gait patterns as normal or abnormal, with height, width measurements, and elliptical features playing pivotal roles in the process (Fig. 6.3).

FIGURE 6.3   Key reasons—how the pattern recognition helpful in identifying abnormal gait recognition.

2.9 DNN based model

The work (Chakraborty and Nandy, 2020) describes a sequence of crucial procedures for preparing and analyzing gait signal data within the context of using Deep Neural Networks (DNN) for gait classification. These preprocessing methods involve adjusting for errors through calibration, eliminating stationary data by identifying peaks and valleys (clipping), and removing noise and unwanted frequencies via Fourier spectrum analysis (Chakraborty and Nandy, 2019). Segmentation is carried out to divide gait signals into cycles using auto-correlation functions, with the segment length determined based on periodicity values (Dehzangi et al., 2017). Wavelet decomposition is utilized to extract pertinent information from signals that exhibit nonstationary behavior. The classification process with DNN comprises a highly optimized architecture featuring multiple convolutional layers, pooling, dropout mechanisms, and a dense layer. This architecture utilizes ReLU activation, "rmsprop" optimization, and "binary cross-entropy" loss for binary classification. In summary, the proposed DNN framework effectively processes decomposed sensor data, fine-tunes various parameters, and generates feature maps to facilitate gait classification.

2.10 RNN based model

The proposed approach (Jinnovart et al., 2020) makes use of OpenPose (Cao et al., 2018) for estimating human poses and employs a deep neural network model to categorize various gait abnormalities in both walking and standing postures. These abnormalities, which encompass spastic gait, scissors gait, steppage gait, waddling gait, and propulsive gait, are identified based on an individual's walking behavior and how their body weight is distributed. To train and validate the model, the Institute of New Imaging Technologies (INIT) (Ortells et al., 2018) Gait Database, which consists of high-quality binary silhouettes, is used. This dataset includes both normal and abnormal gait patterns. Human pose estimation is carried out using Open-Pose in conjunction with TensorFlow, where 135 key points are extracted from video frames, and Convolutional Neural Networks (CNNs) are used to validate the detected postures. Feature extraction primarily involves calculating body weight distribution along the x and y axes, and these features are used as input for the predictive model that classifies gait abnormalities. The system incorporates three variants of Recurrent Neural Networks (RNNs)—Simple RNN, LSTM, and GRU—to address issues related to convergence that are often observed in conventional RNNs. The model architecture comprises five key layers, including recurrent layers for the upper/lower body and left/right body, a fully connected layer, and an output layer that estimates confidence levels for the identified gait abnormalities. Gait is considered normal when all five confidence levels are below 0.2. The evaluation process entails splitting the INIT Gait dataset into training and testing sets and evaluating the model's performance using confusion matrices. Additionally, the system supports training and inference on two GPUs, utilizing parallel computing to reduce processing times and improve responsiveness.

2.11 Combination of CNN, LSTM, SVM and KNN based model

In this (Chen et al., 2022), a group of seven healthy individuals, aged between 23 and 29, with no history of neurological or musculoskeletal ailments, participated in an experiment that made use of an Azure Kinect DK integrated camera, an h/p/cosmos treadmill, and a laptop for the gathering and analysis of data. The primary focus of the study was on three distinct gait patterns: normal gait (NG), pelvic obliquity (PO) gait (Stanhope et al., 2014; Akbas et al., 2019), and knee hyperextension (HK) gait, with potential applications in detecting falls. Participants underwent practice sessions for these gait patterns on the treadmill, with their movements being recorded in real-time via the Azure Kinect camera, capturing data on 32 joint point angles. These joint angle measurements served as the fundamental features for gait recognition. The gathered data was subsequently divided into 3-second time segments, resulting in a total of 15 sets of valid experimental datasets that were later subject to offline analysis. To classify gait patterns, four machine learning classifiers were utilized, including Convolutional Neural Network (CNN) (Krizhevsky et al., 2017; Santos et al., 2019; Zheng et al., 2018; Zhao et al., 2020), Long Short-Term Memory (LSTM) neural network (Sarshar et al., 2021; Khokhlova et al., 2019), Support Vector Machines (SVM) (Smola and Schölkopf, 1998), and K-Nearest Neighbors (KNN) (Altman, 1992; Saçlı et al., 2019; You et al., 2022), each being optimized using distinct approaches. The study's statistical analysis encompassed one-way ANOVA, Pearson's chi-square, and Bonferroni correction to evaluate the accuracy of gait pattern recognition and disparities in the range of motion (ROM) across various gait patterns.

2.12 CNN based model

A 3D Convolutional Neural Network (CNN) is employed (Shoryabi et al., 2021), to perform the classification of gait abnormalities. The decision to use this network is guided by three key factors: its capacity to effectively leverage the spatial structure of the problem, its ability to establish hierarchies of intricate features, and its potential for enhanced computational efficiency when utilizing GPU acceleration. The network's architectural components consist of input layers for processing two 3D images, convolution layers featuring ReLU activation functions, max pooling layers to simplify data, and a fully connected layer. For classification, the output layer incorporates softmax activation, with the number of neurons corresponding to the number of classes, which includes a sixth class representing normal gait. These layers within the neural network are meticulously designed to accurately process and classify gait patterns.

3. Gait-based recognition for Parkinson's disease

Pattern recognition is a valuable tool in the biomedical field for understanding and diagnosing diseases such as Parkinson's disease. Parkinson's disease is a complex neurodegenerative disorder that primarily affects motor function, and it can be challenging to diagnose and track its progression. Here's how pattern recognition can be helpful in the context of Parkinson's disease.

1. Early Diagnosis: Pattern recognition techniques can be used to identify subtle patterns and biomarkers in medical data, such as brain imaging scans, genetic information, or even voice and gait patterns, that may indicate the presence of Parkinson's disease before clinical symptoms become apparent. Early diagnosis can enable timely intervention and treatment, potentially slowing the progression of the disease.
2. Disease Progression Monitoring: Parkinson's disease is a progressive disorder, and monitoring its progression is crucial for optimizing treatment and care. Pattern recognition algorithms can analyze longitudinal patient data to detect changes in symptoms and motor function over time, helping healthcare providers adjust treatment plans accordingly.
3. Differential Diagnosis: Differential diagnosis is essential in distinguishing Parkinson's disease from other neurodegenerative disorders that may have similar symptoms. Pattern recognition can assist in making more accurate distinctions by identifying unique patterns associated with each condition.
4. Predictive Modeling: Machine learning and pattern recognition algorithms can be used to create predictive models that estimate a patient's risk of developing Parkinson's disease based on factors such as genetic predisposition, lifestyle, and exposure to environmental risk factors. These models can help with preventive measures and early intervention.
5. Treatment Personalization: Parkinson's disease management often involves medication adjustments and therapies tailored to each patient's specific needs. Pattern recognition can help identify how individual patients respond to different treatments and their unique disease progression patterns, allowing for more personalized and effective treatment plans.

6. Remote Monitoring: Telemedicine and wearable devices can collect continuous data on a patient's motor function and overall health. Pattern recognition techniques can analyze this data in real-time, providing insights into a patient's condition, enabling remote monitoring, and potentially preventing hospitalization.
7. Speech and Gait Analysis: Changes in speech and gait are common symptoms of Parkinson's disease. Pattern recognition can analyze voice recordings and gait patterns to identify subtle alterations that may be indicative of the disease, even before other symptoms become apparent.
8. Drug Discovery: In the development of new medications for Parkinson's disease, pattern recognition can help identify potential drug candidates by analyzing their impact on disease-related molecular and cellular patterns.
9. Biomarker Discovery: Identifying reliable biomarkers associated with Parkinson's disease can aid in early detection, tracking progression, and developing new treatments. Pattern recognition can be used to uncover such biomarkers in various types of data, including genomics, proteomics, and neuroimaging (Fig. 6.4).

3.1 PCA and LDA based model

In their research (Cho et al., 2009), the authors prepared silhouette images of both normal subjects and individuals with Parkinson's disease, resizing them to a uniform 64×64 pixel format to optimize computational efficiency. They followed standard machine learning practices by dividing the dataset into separate training and testing sets. To extract relevant features, they applied Principal Component Analysis (PCA), a technique that involved transforming silhouettes into vectors, calculating mean vectors, and selecting key eigenvectors to reduce dimensionality. Additionally, they employed Linear Discriminant Analysis (LDA) to improve recognition capabilities, with a specific focus on categorizing subjects into two groups: those without Parkinson's disease and those with the condition. Their primary objective was to generate feature vectors suitable for machine learning-based classification, with a particular emphasis on effectively distinguishing between these two categories.

3.2 CNN based model

The authors in (Xia et al., 2018), offer a comprehensive overview of their research methodology concerning the detection of Freezing of Gait (FOG) (Bachlin et al., 2010) in individuals with Parkinson's disease (PD). They begin by discussing the dataset, which was gathered through on-body accelerometers and involved the participation of 10 PD patients, providing specific details about sensor placements and the patients' demographics. Signal preprocessing procedures included the utilization of the "three-sigma-rule" (Hausdorff et al., 2000; Wu and Krishnan, 2009; Pukelsheim, 1994) to eliminate outliers from the acceleration data. Data segmentation was conducted using a sliding window approach, with labels assigned based on the most frequently occurring class. The authors then introduce the architecture of Convolutional Neural Networks (CNNs) (Hubel and Wiesel, 1968; Sainath et al., 2013), with an emphasis on convolutional layers, max pooling, ReLU activation, and fully connected layers (Glorot et al., 2011). They delve into the CNN learning

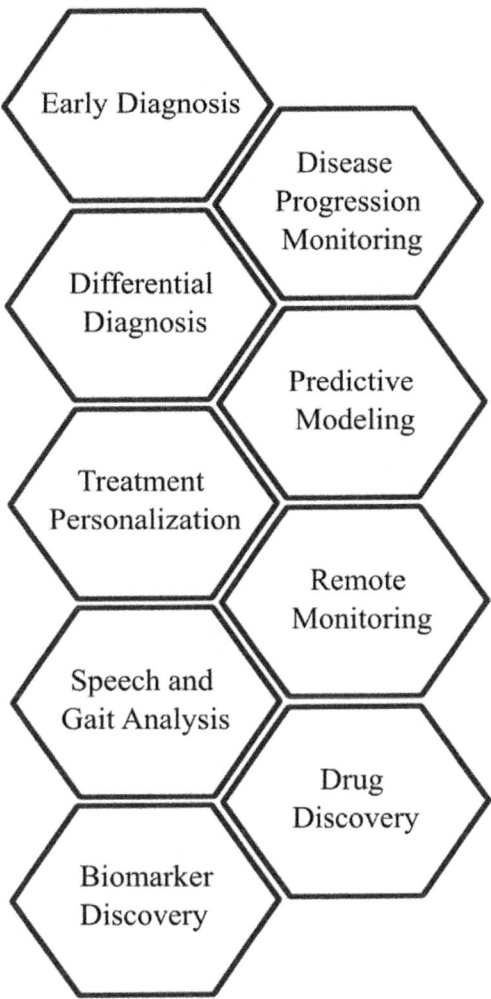

FIGURE 6.4 Unlocking the potential of pattern recognition in Parkinson's disease.

mechanism, describing loss functions and optimization parameters. Furthermore, the proposed CNN architecture is elaborated upon, highlighting feature extraction and fusion strategies. The assessment of performance covered both patient-dependent and patient-independent experiments, with specificity, sensitivity, and classification accuracy serving as the key evaluation metrics.

3.3 RNN-LSTM based model

The study (Wang et al., 2021), aims to develop a real-time gait detection model for the identification of Heel Strike (HS) events, which are clinically significant in Neurodevelopmental Treatment (NDT) rehabilitation (Wang et al., 2018; Wang et al., 2020). Data was gathered from five healthy participants, utilizing Inertial Measurement Units (IMUs) attached to their legs. To restrict joint movement, a gaiter with a 2 kg mass was employed. During walking in a corridor, three axial accelerations and three axial angular velocities were recorded for each leg. Gait events, including Maximum Swing (MS) (Aminian et al., 2002; Rebula et al., 2013; Yun et al., 2012), HS, and Toe Off (TO), were discerned based on angular velocity data, with a specific focus on abnormalities in the constrained leg. The development of an RNN (Tieleman and Hinton, 2012) model, featuring Long Short-Term Memory (LSTM), facilitated real-time HS event detection. The model underwent training using normalized six-axial IMU data, and validation was conducted through Leave-One-Out Cross Validation (LOOCV). Performance assessment revealed a notably high Positive Predictive Value (PPV) and exceptionally low Mean Absolute Error (MAE), underscoring the model's effectiveness in real-time HS event detection for both constrained and unconstrained legs.

3.4 LSTM-DNN based model

The suggested hybrid neural network consists of two fundamental layers introduced in (Paragliola and Coronato, 2018) is the Classification Layer (CL) and the Reduction Layer (RL). With the CL, sequential data is processed in an all-encompassing manner, capturing information throughout the sequence, by utilizing a combination of a Long Short-Term Memory (LSTM) network (Goodfellow et al., 2016) and a Deep Neural Network (DNN). By estimating functions to classify input data, the DNN—which operates as a multilayer perceptron (MLP) (Orbach, 1962)—is able to discriminate between regular and aberrant gait patterns. Crucially, the output of the LSTM simplifies the classification procedure by acting as an input for the DNN, which is designed especially for identifying walking patterns. On the other hand, high-dimensional multivariate time series (MTS) (Coronato et al., 2014; Park et al., 2016; Paragliola and Coronato, 2013, 2017) data management is the focus of the RL. The implementation of a dimensionality reduction procedure maximizes classification efficiency. Investigations are conducted into four distinct methods of dimensionality reduction: Under-sampling, Fourier Transformation, Autoencoder, and Convolutional Neural Network (CNN). The goal of each of these methods is to produce a more condensed representation of the MTS while maintaining the vital details required for precise classification. It is important to emphasize that the choice of a particular approach and the parameters that go along with it have a big impact on how well the classification task performs. Determining the compression rate, window length, and aggregation functions are crucial parameters for under-sampling. When it comes to Fourier transformation, the number of filtered coefficients (F) that are utilized to describe the data is what matters most. The implementation of an autoencoder involves taking into account many parameters, such as reconstruction error, window size, and compression rate. Convolutional neural networks, on the other hand, depend on variables such as pooling, filter, and kernel sizes. The process of dimensionality reduction and the parameters associated with it are key factors that influence how successful a hybrid neural network is. The study strategically uses grid search algorithms to carefully test different combinations of parameters. These reduction strategies are carefully designed to preserve classification accuracy while reducing the computing load caused by high-dimensional MTS data.

3.5 SVM-ANN based model

The chapter (Goh et al., 2022) involved data preprocessing steps, including data cleaning, normalization, and feature derivation. Data cleaning addressed issues such as missing values (which were not found) and the removal of outliers marked by zero values in specific features. Data normalization, vital for the Support Vector Machines (SVM) algorithm, was achieved using Z-score normalization. New features, namely stride count and stride speed, were created to enhance dataset richness. The SVM (Osisanwo et al., 2017), selected for its efficiency and adaptability in various data spaces, used the RBF kernel, ideal for multiclass and imbalanced datasets. Parameter fine-tuning covered C (affecting the decision margin) and gamma (controlling the boundary's curvature) through gridsearch. Additionally, Artificial Neural Networks (ANNs) with two hidden layers and ReLU activation functions were employed, alongside Softmax for multi-class problems. Performance assessment involved a 10-fold cross-validation, focusing on accuracy and F1 score to gauge predictive accuracy and manage imbalanced class distributions.

3.6 Different machine learning based techniques

The study investigates (Ricciardi et al., 2020), gait analysis was conducted using the BTS Bioengineering system, specifically the SMART DX optical system equipped with an array of sensors and markers. The Davis protocol (Davis et al., 1991) entailed a series of activities, including gathering anthropometric data, positioning markers on subjects, a stationary phase, and a 10-meter walking task under different conditions. The study focused on distinguishing between Parkinson's disease (PD) patients with and without mild cognitive impairment (MCI) through statistical analyses. To detect MCI, machine learning techniques, including Decision Trees (DT), Random Forest (RF), and k-Nearest Neighbors (KNN), were employed. The dataset was balanced using the Synthetic Minority Oversampling technique (SMOTE), and robust assessment was achieved through leave-one-out cross-validation, with performance metrics encompassing accuracy, sensitivity, specificity, and the Area Under the Curve Receiver Operating Characteristic (AUCROC).

Initial data analysis, appraisal of gait pattern, and the extraction of pertinent features using statistical and kinematic analyses are all steps in the suggested framework (Balaji et al., 2020) for determining the severity of Parkinson's disease (PD). It uses a rank correlation method to pinpoint important gait-related signals, which are subsequently used by a classifier model to forecast the stages of PD (Mu et al., 2018). Six indicators are used to assess how well the model is performing. Gait patterns between healthy individuals and people with Parkinson's disease differ noticeably, especially when the mean values from VGRF sensors are compared. A 10-fold cross-validation method and the Shapiro—Wilk test are used to make sure that the model evaluation is reliable and confirm the normal distribution of the data attributes. The kinematic study focuses on gait cycle metrics and offers crucial biomarkers for comprehending PD-related aberrant gait patterns. Variations across different PD stages are highlighted by a close examination of both temporal and spatial aspects. Different machine learning algorithms, such as Decision Trees (DT), Support Vector Machines (SVM), Ensemble Classifier (EC), and Bayes Classifier (BC), are used for PD classification. Based on the Hoehn and Yahr (H & Y) scale, the DT classifier creates a hierarchical tree structure to classify PD episodes. By maximizing the margin between classes, SVM manages jobs including both linear and nonlinear classification. The Bayes classifier uses probabilistic predictions to ensure accurate classification based on previous class-related data, whereas the ensemble classifier integrates decision tree models to improve classification performance. These modeling and analytical methods are intended to provide an accurate assessment of the severity of PD.

3.7 Different machine learning and ANN based techniques

In (Trabassi et al., 2022), the researchers initially utilized an independent sample t-test to identify specific gait parameters capable of distinguishing individuals with Parkinson's disease (referred to as pwPD) from a control group of healthy subjects (referred to as HSmatched). They analyzed a dataset comprising 22 gait parameters collected through inertial measurement units (IMUs) and ensured that their analysis adhered to statistical assumptions using the Shapiro—Wilk test. To handle multicollinearity issues, they computed partial Pearson correlation coefficients while accounting for age and gait speed, ultimately selecting variables with modest to low correlations ($r < 0.5$) with other parameters and retaining those with clinical relevance. Subsequently, they employed a sequential backward selection (SBS) method to assess the significance of these chosen gait variables, with a focus on reducing feature dimensionality while preserving classifier performance. They further refined feature importance using the random forest technique and employed five supervised machine learning algorithms to classify gait patterns in pwPD versus HSmatched individuals. The data was preprocessed through z-score normalization, missing values were addressed via "mean imputation," and class labels were encoded as integers. The dataset was split into distinct training and test sets. They evaluated model performance and optimized hyperparameters using k-fold cross-validation and grid search, respectively. The study also explored the application of tree-based algorithms, K-nearest neighbors (KNN), support vector machines (SVM), and artificial neural networks (ANN) in the classification task, all fine-tuned with specific hyperparameters to enhance model accuracy. In summary, the research aimed to identify effective gait parameters for distinguishing pwPD from HSmatched individuals and to construct machine learning models for this classification task.

4. Gait-based recognition for cerebral palsy

Cerebral palsy (CP) is a group of lifelong neurological disorders affecting movement, muscle coordination, and posture, with causes rooted in damage to the developing brain, often occurring before or during infancy. This condition varies widely in severity and presentation. It primarily manifests as motor function impairment, encompassing muscle stiffness, weakness, lack of coordination, or involuntary movements, categorized into types such as spastic, dyskinetic, ataxic, or mixed CP. While primarily motor-centric, it can also involve cognitive and communication challenges. CP is a lifelong condition, but early intervention, therapies, and medical treatments can help improve quality of life by managing symptoms and adapting to evolving needs. A multidisciplinary approach to care, including healthcare professionals, therapists, and family support, is essential for individuals with CP.

Cerebral palsy gait recognition employs technology, including sensors and pattern recognition algorithms, to analyze and identify distinctive patterns and abnormalities in the walking patterns of individuals with cerebral palsy. This neurological disorder often results in various motor impairments, and gait recognition plays a vital role in assessing, diagnosing, and treating individuals with cerebral palsy. It involves data collection through sensors such as motion-capture systems and accelerometers, capturing parameters such as joint angles and step length, and data analysis using pattern recognition algorithms to detect irregularities in gait. These systems can detect specific abnormalities, offer quantitative assessments, and support clinical applications, enabling personalized interventions and aiding research and development. Furthermore, it allows for remote monitoring, reducing the need for frequent in-person assessments, and thus improving the overall management and care of individuals with cerebral palsy. Pattern recognition is of paramount importance within the biomedical domain, particularly in the sphere of cerebral palsy gait analysis, for a multitude of significant reasons.

1. Objective Assessment: Cerebral palsy affects individuals in varying degrees, and gait abnormalities are a common symptom. Pattern recognition allows for objective assessment of these abnormalities by analyzing data from sensors and video recordings, reducing the subjectivity associated with traditional clinical evaluations.
2. Early Diagnosis: Pattern recognition algorithms can identify subtle gait abnormalities that might not be apparent through visual observation alone. Early diagnosis of cerebral palsy and monitoring of its progression are essential for timely intervention and improved patient outcomes.
3. Quantitative Data: Biomechanical and kinematic data from gait recognition provide quantitative measurements of gait parameters such as joint angles, step length, and gait speed. This quantitative data is essential for tracking changes over time and assessing the effectiveness of interventions.
4. Customized Treatment Plans: Gait recognition can aid in the development of personalized treatment plans. By analyzing the specific gait patterns of individuals with cerebral palsy, healthcare providers can tailor interventions, such as physical therapy and orthotic devices, to address the unique needs of each patient.
5. Monitoring Progress: Regular gait assessments through pattern recognition allow healthcare professionals to monitor a patient's progress over time. This helps in adjusting treatment strategies as needed and ensuring that improvements are being made.
6. Research and Development: Gait recognition data can be invaluable for researchers studying cerebral palsy and its impact on gait. It provides a wealth of information for developing new treatments, interventions, and assistive devices.
7. Objective Outcome Measures: In clinical trials and research studies, pattern recognition can provide objective outcome measures. This is crucial for determining the efficacy of new therapies and interventions in a standardized and quantifiable manner.
8. Rehabilitation Support: Gait recognition technology can be integrated into rehabilitation programs to provide real-time feedback to patients. This interactive approach can motivate patients and improve their gait through biofeedback mechanisms.

9. Remote Monitoring: Pattern recognition can enable remote monitoring of a patient's gait. This is particularly important in situations where patients have limited access to healthcare facilities or during tele-health consultations, ensuring that healthcare providers can assess progress without the need for physical presence.
10. Data-driven Insights: Pattern recognition can analyze vast amounts of gait data, uncovering trends and patterns that might not be apparent through manual analysis. These insights can help in refining treatment approaches and understanding the condition better.

4.1 CNN based model

AbnormNet (Bajpai et al., 2022), a neural network, comprises three segments: the first segment with two sets of convolutional layers, the second segment with fully connected layers, and the third segment with additional fully connected layers. The first segment's convolutional layers maintain a constant filter size of 3, but the number of filters in each layer varies (ranging from 64 to 1). LeakyReLU activation functions introduce nonlinearity after convolution. The outcomes from the first segment are merged and flattened into 102 nodes, establishing connections with the second segment, which consists of three fully connected layers. The third segment incorporates four fully connected layers, all employing LeakyReLU activation except for the output layer, which uses sigmoid activation to convert values into discrete ones. The training process for AbnormNet unfolds in three stages. In the initial stage, the first and second modules predict the abnormality index (AIIkneeS) using the Automated Gait Assessment Score (A-GAS) (Bajpai and Joshi, 2021) for the knee joint in the sagittal plane. The second stage classifies AIIkneeS into 10 categories representing distinct gait patterns, while the third phase fine-tunes all three modules. The dataset consists of 1719 data points, partitioned into 80% training and validation sets, and a 20% testing set. A five-fold cross-validation is applied while maintaining consistent initial weights. The three-phase training approach aims to boost classification performance, compare training approaches, and facilitate the acquisition of gait-related features to achieve optimal performance. Each stage involves different numbers of epochs and batch sizes, with the initial phase employing 240 epochs, the second phase using 50, and the third phase utilizing 20 epochs, each with its specified batch size.

4.2 Different machine learning based techniques

In (Zhang and Ma, 2019), Gait data from a cohort of 200 children diagnosed with spastic diplegia were analyzed under specific inclusion criteria, including age range, motor function classification, independent walking capability, and the absence of recent lower extremity treatments. Their lower limb kinematics were recorded using a motion analysis system and force platforms, synchronized with ground reaction forces. The gait patterns were categorized into four groups: true equinus, jump gait, apparent equinus, and crouch gait. To establish a classification system, essential features related to ankle, knee, and hip movements were extracted. Seven supervised machine learning algorithms, namely artificial neural network (ANN), discriminant analysis, naive Bayes, decision tree, support vector machines

FIGURE 6.5 Enhancing cerebral palsy management with pattern recognition.

(SVM), random forest, and k-nearest neighbors (KNN), were employed (Wu et al., 2008). These models were trained on the extracted features and assessed for their proficiency in classifying the four sagittal gait patterns. Performance metrics such as sensitivity, specificity, and the area under the receiver operating characteristic curve (AUC) were used to gauge their classification capabilities. The models underwent a rigorous 10-fold cross-validation process to ensure their predictive accuracy and effectiveness in distinguishing between different gait patterns (Fig. 6.5).

4.3 Few deep learning (FCN, LSTM, CNN and transformer) based model

To ensure the models' (Kolaghassi et al., 2023), general applicability, data from typically developing children were randomly split into three subsets. These subsets underwent a preprocessing step that involved dividing gait trials into samples. Each sample included an input matrix consisting of 100 time-step joint angle values and a target vector representing the subsequent time-step, corresponding to a full gait cycle. Using a sliding window approach, this resulted in 41,120 training samples, 7316 testing samples, and 4832 validation samples. Additionally, two sets for long-term predictions, with 200 time-step outputs, were created to evaluate the models' forecasting capabilities. Data from children with cerebral palsy were exclusively reserved for testing, covering both one-step-ahead and long-term predictions. To account for variations in joint angle ranges between CP and typically developing children, all datasets were normalized through min-max normalization. The study introduced four deep learning models (FCN, LSTM, CNN, and Transformer) designed to predict gait trajectories, with potential applications in assisting children with mobility challenges.

5. Conclusion

In conclusion, this study has outlined the historical development of gait and pattern recognition in the biomedical domain, demonstrating the noteworthy advancements made possible by the combination of many deep learning and machine learning methodologies. It has been demonstrated how important computer vision is to biomedical applications, and how it may be used to automate the analysis of complicated data—such as medical images—and improve the precision of diagnostic procedures. The research also provided an overview of the various methods that may be used to identify irregularities in gait, emphasizing how these methods can be used to diagnose disorders such as cerebral palsy and Parkinson's disease. These methods have proven their ability to deliver accurate and rapid assessments of gait patterns by utilizing state-of-the-art technologies, such as deep neural networks and conventional machine learning techniques. This has significant advantages for early diagnosis and treatment planning. Biomedical pattern recognition is heading for a new era as interdisciplinary cooperation between data analysts, computer scientists, and medical experts continue to thrive. More precise and customized diagnosis and treatment are anticipated as a result of this synergy, which will eventually improve patient care and improve health outcomes. The methods and strategies covered in this chapter point to a bright future for healthcare, one in which early identification and treatment are crucial to improving the quality of life for people with a range of medical disorders. The biomedical profession is poised to make great advancements in the next years by utilizing the creative potential of deep learning, machine learning, and computer vision along with the power of pattern recognition.

References

Akbas, T., Prajapati, S., Ziemnicki, D., Tamma, P., Gross, S., Sulzer, J., 2019. Hip circumduction is not a compensation for reduced knee flexion angle during gait. Journal of Biomechanics 87, 150–156. https://doi.org/10.1016/j.jbiomech.2019.02.026.

Altman, N.S., 1992. An introduction to kernel and nearest-neighbor nonparametric regression. The American Statistician 46 (3), 175–185. https://doi.org/10.1080/00031305.1992.10475879.

Aminian, K., Najafi, B., Büla, C., Leyvraz, P.-F., Robert, P., 2002. Spatio-temporal parameters of gait measured by an ambulatory system using miniature gyroscopes. Journal of Biomechanics 35 (5), 689–699. https://doi.org/10.1016/s0021-9290(02)00008-8.

Asghari, P., Soleimani, E., Nazerfard, E., 2020. Online human activity recognition employing hierarchical hidden Markov models. Journal of Ambient Intelligence and Humanized Computing 11 (3), 1141–1152. https://doi.org/10.1007/s12652-019-01380-5.

Asuncion, L.V.R., Mesa, J. X. P. de, Juan, P.K.H., Sayson, N.T., Cruz, A.R.D., 2018. 'Thigh motion-based gait analysis for human identification using inertial measurement units (IMUs. In: Proc. IEEE 10th Int. Conf. Hum, pp. 1–6.

Bachlin, M., Plotnik, M., Roggen, D., Maidan, I., Hausdorff, J.M., Giladi, N., Troster, G., 2010. Wearable assistant for Parkinson's disease patients with the freezing of gait symptom. IEEE Transactions on Information Technology in Biomedicine 14 (2), 436–446. https://doi.org/10.1109/titb.2009.2036165.

Bajpai, R., Joshi, D., 2021. A-GAS: a probabilistic approach for generating automated gait assessment score for cerebral palsy children. IEEE Transactions on Neural Systems and Rehabilitation Engineering 29, 2530–2539. https://doi.org/10.1109/TNSRE.2021.3131466.

Bajpai, R., Tiwari, A., Joshi, D., Khatavkar, R., 2022. AbnormNet: a neural network based suggestive tool for identifying gait abnormalities in cerebral palsy children. In: 2022 International Conference for Advancement in Technology, ICONAT 2022. Institute of Electrical and Electronics Engineers Inc., India https://doi.org/10.1109/ICONAT53423.2022.9725832.

Balaji, E., Brindha, D., Balakrishnan, R., 2020. Supervised machine learning based gait classification system for early detection and stage classification of Parkinson's disease. Applied Soft Computing 94. https://doi.org/10.1016/j.asoc.2020.106494.

Cao, Z., Hidalgo, G., Simon, T., Wei, S., Sheikh, Y., 2018. Openpose:Realtime Multi-Person 2d Pose Estimation Using Part Affinity Fields.

Chakraborty, J., Nandy, A., 2019. Periodicity detection of quasi-periodic slow-speed gait signal using IMU sensor. Lecture Notes in Computer Science 11581, 140–152. https://doi.org/10.1007/978-3-030-22216-1_11.

Chakraborty, J., Nandy, A., 2020. Discrete wavelet transform based data representation in deep neural network for gait abnormality detection. Biomedical Signal Processing and Control 62. https://doi.org/10.1016/j.bspc.2020.102076.

Chawla, N.V., Bowyer, K.W., Hall, L.O., Kegelmeyer, W.P., 2002. SMOTE: synthetic minority over-sampling technique. Journal of Artificial Intelligence Research 16, 321–357. https://doi.org/10.1613/jair.953.

Chen, Z., Wu, M., Cui, W., Liu, C., Li, X., 2021. An attention based CNN-LSTM approach for sleep-wake detection with heterogeneous sensors. IEEE Journal of Biomedical and Health Informatics 25 (9), 3270–3277. https://doi.org/10.1109/JBHI.2020.3006145.

Chen, B., Chen, C., Hu, J., Sayeed, Z., Qi, J., Darwiche, H.F., Little, B.E., Lou, S., Darwish, M., Foote, C., Palacio-Lascano, C., 2022. Computer vision and machine learning-based gait pattern recognition for flat fall prediction. Sensors 22 (20). https://doi.org/10.3390/s22207960.

Cho, C.W., Chao, W.H., Lin, S.H., Chen, Y.Y., 2009. A vision-based analysis system for gait recognition in patients with Parkinson's disease. Expert Systems with Applications 36 (3), 7033–7039. https://doi.org/10.1016/j.eswa.2008.08.076.

Choudhury, H., Mandal, S., Prasanna, S.R.M., 2019. Exploiting forced alignment of time-reversed data for improving HMM-based handwriting segmentation. Expert Systems with Applications 121, 158–169. https://doi.org/10.1016/j.eswa.2018.12.012.

Chung, J., Gulcehre, C., Cho, K., Bengio, Y., 2014. Empirical Evaluation of Gated Recurrent Neural Networks on Sequence Modeling.

Coronato, A., De Pietro, G., Paragliola, G., 2014. A situation-aware system for the detection of motion disorders of patients with Autism Spectrum Disorders. Expert Systems with Applications 41 (17), 7868–7877. https://doi.org/10.1016/j.eswa.2014.05.011.

Davis, R.B., Õunpuu, S., Tyburski, D., Gage, J.R., 1991. A gait analysis data collection and reduction technique. Human Movement Science 10 (5), 575–587. https://doi.org/10.1016/0167-9457(91)90046-Z.

Dehzangi, O., Taherisadr, M., ChangalVala, R., 2017. IMU-based gait recognition using convolutional neural networks and multi-sensor fusion. Sensors 17 (12). https://doi.org/10.3390/s17122735.

Eltoukhy, M., Oh, J., Kuenze, C., Signorile, J., 2017. Improved kinect-based spatiotemporal and kinematic treadmill gait assessment. Gait & Posture 51, 77–83. https://doi.org/10.1016/j.gaitpost.2016.10.001.

Forkan, A.R.M., Khalil, I., Tari, Z., Foufou, S., Bouras, A., 2015. A context-aware approach for long-term behavioural change detection and abnormality prediction in ambient assisted living. Pattern Recognition 48 (3), 628–641. https://doi.org/10.1016/j.patcog.2014.07.007.

Forney, G.D., 1973. The viterbi algorithm. Proceedings of the IEEE 61 (3), 268–278. https://doi.org/10.1109/PROC.1973.9030.

Gao, J., Gu, P., Ren, Q., Zhang, J., Song, X., 2019. Abnormal gait recognition algorithm based on LSTM-CNN fusion network. IEEE Access 7, 163180–163190. https://doi.org/10.1109/access.2019.2950254.

Geerse, D.J., Coolen, B.H., Roerdink, M., 2015. Kinematic validation of a multi-Kinect v2 instrumented 10-meter walkway for quantitative gait assessments. PLoS One 10 (10). https://doi.org/10.1371/journal.pone.0139913.

Glorot, X., Bordes, A., Bengio, Y., 2011. Deep sparse rectifier neural networks. Journal of Machine Learning Research 15, 315–323.

Goh, C.H., Koh, C.H., Chong, Y.Z., Lim, W.Y., 2022. Gait classification of Parkinson's disease with supervised machine learning approach. In: 7th IEEE-EMBS Conference on Biomedical Engineering and Sciences, IECBES 2022—Proceedings. Institute of Electrical and Electronics Engineers Inc., Malaysia, pp. 112–116. https://doi.org/10.1109/IECBES54088.2022.10079640.

Goodfellow, I., Bengio, Y., Courville, A., Bach, F., Deep Learning. 2016. 2016. MIT Press.

Hassan, E.A., Jenkyn, T.R., Dunning, C.E., 2007. Direct comparison of kinematic data collected using an electromagnetic tracking system versus a digital optical system. Journal of Biomechanics 40 (4), 930–935. https://doi.org/10.1016/j.jbiomech.2006.03.019.

Hausdorff, J.M., Lertratanakul, A., Cudkowicz, M.E., Peterson, A.L., Kaliton, D., Goldberger, A.L., 2000. Dynamic markers of altered gait rhythm in amyotrophic lateral sclerosis. Journal of Applied Physiology 88 (6), 2045–2053. https://doi.org/10.1152/jappl.2000.88.6.2045.

Hochreiter, S., 1998. The vanishing gradient problem during learning recurrent neural nets and problem solutions. International Journal of Uncertainty, Fuzziness and Knowledge-Based Systems 6 (2), 107–116. https://doi.org/10.1142/S0218488598000094.

Hu, J., Shen, L., Sun, G., 2018. Squeeze-and-Excitation networks. In: Proceedings of the IEEE Computer Society Conference on Computer Vision and Pattern Recognition. IEEE Computer Society, United Kingdom, pp. 7132–7141. https://doi.org/10.1109/CVPR.2018.00745.

Hubel, D.H., Wiesel, T.N., 1968. Receptive fields and functional architecture of monkey striate cortex. The Journal of Physiology 195 (1), 215–243. https://doi.org/10.1113/jphysiol.1968.sp008455.

Jinnovart, T., Cai, X., Thonglek, K., 2020. Abnormal gait recognition in real-time using recurrent neural networks. In: Proceedings of the IEEE Conference on Decision and Control. Institute of Electrical and Electronics Engineers Inc., Australia, pp. 972–977. https://doi.org/10.1109/CDC42340.2020.9304106.

Jun, K., Lee, D.W., Lee, K., Lee, S., Kim, M.S., 2020. Feature extraction using an RNN autoencoder for skeleton-based abnormal gait recognition. IEEE Access 8, 19196–19207. https://doi.org/10.1109/ACCESS.2020.2967845.

Khokhlova, M., Migniot, C., Morozov, A., Sushkova, O., Dipanda, A., 2019. Normal and pathological gait classification LSTM model. Artificial Intelligence in Medicine 94, 54–66. https://doi.org/10.1016/j.artmed.2018.12.007.

Kipf, T.N., Welling, M., 2016. 'Semi-supervised Classification with Graph Convolutional Networks.

Kolaghassi, R., Marcelli, G., Sirlantzis, K., 2023. Deep learning models for stable gait prediction applied to exoskeleton reference trajectories for children with cerebral palsy. IEEE Access 11, 31962–31976. https://doi.org/10.1109/access.2023.3252916.

Krizhevsky, A., Sutskever, I., Hinton, G.E., 2017. ImageNet classification with deep convolutional neural networks. Communications of the ACM 60 (6), 84–90. https://doi.org/10.1145/3065386.

Lee, D.W., Jun, K., Lee, S., Ko, J.K., Kim, M.S., 2019. Abnormal gait recognition using 3D joint information of multiple Kinects system and RNN-LSTM. In: Proceedings of the Annual International Conference of the IEEE Engineering in Medicine and Biology Society, EMBS. Institute of Electrical and Electronics Engineers Inc., South Korea, pp. 542–545. https://doi.org/10.1109/EMBC.2019.8857607.

Liu, J., Gao, L., Guo, S., Ding, R., Huang, X., Ye, L., Meng, Q., Nazari, A., Thiruvady, D., 2021. A hybrid deep-learning approach for complex biochemical named entity recognition. Knowledge-Based Systems 221. https://doi.org/10.1016/j.knosys.2021.106958.

McGinley, J.L., Baker, R., Wolfe, R., Morris, M.E., 2009. The reliability of three-dimensional kinematic gait measurements: a systematic review. Gait & Posture 29 (3), 360–369. https://doi.org/10.1016/j.gaitpost.2008.09.003.

Mu, Y., Liu, X., Wang, L., 2018. A Pearson's correlation coefficient based decision tree and its parallel implementation. Information Sciences 435, 40–58. https://doi.org/10.1016/j.ins.2017.12.059.

Orbach, J., 1962. Principles of neurodynamics. Perceptrons and the theory of brain mechanisms. Archives of General Psychiatry 7 (3). https://doi.org/10.1001/archpsyc.1962.01720030064010.

Ortells, J., Herrero-Ezquerro, M.T., Mollineda, R.A., 2018. Vision-based gait impairment analysis for aided diagnosis. Medical, & Biological Engineering & Computing 56 (9), 1553–1564. https://doi.org/10.1007/s11517-018-1795-2.

Osisanwo, F.Y., Akinsola, J.E.T., Awodele, O., Hinmikaiye, J.O., Olakanmi, O., Akinjobi, J., 2017. Supervised machine learning algorithms: classification and comparison. International Journal of Computer Trends and Technology 48 (3), 128–138. https://doi.org/10.14445/22312803/IJCTT-V48P126.

Paragliola, G., Coronato, A., 2013. Intelligent monitoring of stereotyped motion disorders in case of children with autism. In: Proceedings—9th International Conference on Intelligent Environments, IE 2013, pp. 258–261. https://doi.org/10.1109/IE.2013.12.

Paragliola, G., Coronato, A., 2017. 'A deep learning-based approach for the recognition of sleep disorders in patients with cognitive diseases: a case study. Proceeding of Federated Conference in Computer Science 12, 43–48. https://doi.org/10.15439/2017F532.

Paragliola, G., Coronato, A., 2018. Gait anomaly detection of subjects with Parkinson's disease using a deep time series-based approach. IEEE Access 6, 73280–73292. https://doi.org/10.1109/access.2018.2882245.

Park, J., Kim, D., Yang, C., Ko, H., 2016. SVM based dynamic classifier for sleep disorder monitoring wearable device. In: 2016 IEEE International Conference on Consumer Electronics, ICCE 2016. Institute of Electrical and Electronics Engineers Inc., South Korea, pp. 309–310. https://doi.org/10.1109/ICCE.2016.7430624.

Pukelsheim, F., 1994. The three sigma rule. The American Statistician 48 (2), 88–91. https://doi.org/10.1080/00031305.1994.10476030.

Rebula, J.R., Ojeda, L.V., Adamczyk, P.G., Kuo, A.D., 2013. Measurement of foot placement and its variability with inertial sensors. Gait & Posture 38 (4), 974–980. https://doi.org/10.1016/j.gaitpost.2013.05.012.

Ricciardi, C., Amboni, M., De Santis, C., Ricciardelli, G., Improta, G., D'Addio, G., Cuoco, S., Picillo, M., Barone, P., Cesarelli, M., 2020. Machine learning can detect the presence of Mild cognitive impairment in patients affected by Parkinson's Disease. In: IEEE Medical Measurements and Applications, MeMeA 2020—Conference Proceedings. Institute of Electrical and Electronics Engineers Inc., Italy https://doi.org/10.1109/MeMeA49120.2020.9137301.

Saçlı, B., Aydınalp, C., Cansız, G., Joof, S., Yilmaz, T., Çayören, M., Önal, B., Akduman, I., 2019. Microwave dielectric property based classification of renal calculi: application of a kNN algorithm. Computers in Biology and Medicine 112.

Sainath, T.N., Mohamed, A.R., Kingsbury, B., Ramabhadran, B., 2013. Deep convolutional neural networks for LVCSR. In: ICASSP, IEEE International Conference on Acoustics, Speech and Signal Processing—Proceedings, pp. 8614–8618. https://doi.org/10.1109/ICASSP.2013.6639347.

Santos, G.L., Endo, P.T., Monteiro, K.H.d.C., Rocha, E.d.S., Silva, I., Lynn, T., 2019. Accelerometer-based human fall detection using convolutional neural networks. Sensors 19 (7). https://doi.org/10.3390/s19071644.

Sarshar, M., Polturi, S., Schega, L., 2021. Gait phase estimation by using LSTM in IMU-based gait analysis—proof of concept. Sensors 21 (17). https://doi.org/10.3390/s21175749.

Shoryabi, M., Foroutannia, A., Rowhanimanesh, A., 2021. A 3D deep learning approach for classification of gait abnormalities using microsoft kinect V2 sensor. In: 26th International Computer Conference, Computer Society of Iran, CSICC 2021. Institute of Electrical and Electronics Engineers Inc., Iran https://doi.org/10.1109/CSICC52343.2021.9420611.

Sithi Shameem Fathima, S.M.H., Wahida Banu, R.S.D., 2015. Elliptical model for normal and abnormal gait classification. Research Journal of Applied Sciences, Engineering and Technology 11 (11), 1238–1244. https://doi.org/10.19026/rjaset.11.2231.

Sithi Shameem Fathima, S.M.H., Jyotsna, K.A., Srinivasulu, T., Archana, K., Tulasi rama, M., Ravichand, S., 2023. Walking pattern analysis using GAIT cycles and silhouettes for clinical applications. Measurement: Sensors 30. https://doi.org/10.1016/j.measen.2023.100893.

Smola, A.J., Schölkopf, B., 1998. On a kernel-based method for pattern recognition, regression, approximation, and operator inversion. Algorithmica 22 (1–2), 211–231. https://doi.org/10.1007/pl00013831.

Stanhope, V.A., Knarr, B.A., Reisman, D.S., Higginson, J.S., 2014. Frontal plane compensatory strategies associated with self-selected walking speed in individuals post-stroke. Clinical Biomechanics 29 (5), 518–522. https://doi.org/10.1016/j.clinbiomech.2014.03.013.

Tan, H.X., Aung, N.N., Tian, J., Chua, M.C.H., Yang, Y.O., 2019. Time series classification using a modified LSTM approach from accelerometer-based data: a comparative study for gait cycle detection. Gait & Posture 74, 128–134. https://doi.org/10.1016/j.gaitpost.2019.09.007.

Tanaka, R., Takimoto, H., Yamasaki, T., Higashi, A., 2018. Validity of time series kinematical data as measured by a markerless motion capture system on a flatland for gait assessment. Journal of Biomechanics 71, 281–285. https://doi.org/10.1016/j.jbiomech.2018.01.035.

Tian, H., Ma, X., Wu, H., Li, Y., 2022. Skeleton-based abnormal gait recognition with spatio-temporal attention enhanced gait-structural graph convolutional networks. Neurocomputing 473, 116–126. https://doi.org/10.1016/j.neucom.2021.12.004.

Tieleman, T., Hinton, G., 2012. 'Lecture 6.5-RMSPROP: divide the gradient by a running average of its recent magnitude. Neural Network in Machine Learning 4, 26–31. https://doi.org/10.4135/9781412983907.n1717.

Trabassi, D., Serrao, M., Varrecchia, T., Ranavolo, A., Coppola, G., De Icco, R., Tassorelli, C., Castiglia, S.F., 2022. Machine learning approach to support the detection of Parkinson's disease in IMU-based gait analysis. Sensors 22 (10). https://doi.org/10.3390/s22103700.

Wang, F.C., Lin, Y.Y., Yu, C.H., 2018. Robust loop-shaping control design and implementation for an automatic neuro-developmental-treatment device. In: 2018 European Control Conference, ECC 2018. Institute of Electrical and Electronics Engineers Inc., Taiwan, pp. 997–1002. https://doi.org/10.23919/ECC.2018.8550387.

Wang, F.C., Lin, Y.Y., Li, Y.C., Chen, P.Y., Yu, C.H., 2020. Development of an automated assistive trainer inspired by neuro-developmental treatment. Sensors and Materials 32 (9), 3019–3037. https://doi.org/10.18494/SAM.2020.2708.

Wang, F.C., Li, Y.C., Kuo, T.Y., Chen, S.F., Lin, C.H., 2021. Real-time detection of gait events by recurrent neural networks. IEEE Access 9, 134849–134857. https://doi.org/10.1109/ACCESS.2021.3116047.

Wang, L., Zhou, Y., Li, R., Ding, L., 2022. A fusion of a deep neural network and a hidden Markov model to recognize the multiclass abnormal behavior of elderly people. Knowledge-Based Systems 252. https://doi.org/10.1016/j.knosys.2022.109351.

Wu, Y., Krishnan, S., 2009. Computer-aided analysis of gait rhythm fluctuations in amyotrophic lateral sclerosis. Medical, & Biological Engineering & Computing 47 (11), 1165–1171. https://doi.org/10.1007/s11517-009-0527-z.

Wu, X., Kumar, V., Ross, Q.J., Ghosh, J., Yang, Q., Motoda, H., McLachlan, G.J., Ng, A., Liu, B., Yu, P.S., Zhou, Z.H., Steinbach, M., Hand, D.J., Steinberg, D., 2008. Top 10 algorithms in data mining. Knowledge and Information Systems 14 (1), 1–37. https://doi.org/10.1007/s10115-007-0114-2.

Wu, H., Ma, X., Li, Y., 2020. Convolutional networks with channel and STIPs attention model for action recognition in videos. IEEE Transactions on Multimedia 22 (9), 2293–2306. https://doi.org/10.1109/TMM.2019.2953814.

Xia, Y., Zhang, J., Ye, Q., Cheng, N., Lu, Y., Zhang, D., 2018. Evaluation of deep convolutional neural networks for detection of freezing of gait in Parkinson's disease patients. Biomedical Signal Processing and Control 46, 221–230. https://doi.org/10.1016/j.bspc.2018.07.015.

You, L., Jiang, H., Hu, J., Chang, C.H., Chen, L., Cui, X., Zhao, M., 2022. GPU-accelerated Faster Mean Shift with euclidean distance metrics. In: Proceedings—2022 IEEE 46th Annual Computers, Software, and Applications Conference, COMPSAC 2022. Institute of Electrical and Electronics Engineers Inc., United States, pp. 211–216. https://doi.org/10.1109/COMPSAC54236.2022.00037.

Yun, X., Calusdian, J., Bachmann, E.R., McGhee, R.B., 2012. Estimation of human foot motion during normal walking using inertial and magnetic sensor measurements. IEEE Transactions on Instrumentation and Measurement 61 (7), 2059–2072. https://doi.org/10.1109/TIM.2011.2179830.

Zhang, Y., Ma, Y., 2019. Application of supervised machine learning algorithms in the classification of sagittal gait patterns of cerebral palsy children with spastic diplegia. Computers in Biology and Medicine 106, 33–39. https://doi.org/10.1016/j.compbiomed.2019.01.009.

Zhao, M., Chang, C.H., Xie, W., Xie, Z., Hu, J., 2020. Cloud shape classification system based on multi-channel CNN and improved FDM. IEEE Access 8, 44111–44124. https://doi.org/10.1109/ACCESS.2020.2978090.

Zheng, Q., Yang, M., Yang, J., Zhang, Q., Zhang, X., 2018. Improvement of generalization ability of deep CNN via implicit regularization in two-stage training process. IEEE Access 6, 15844–15869. https://doi.org/10.1109/access.2018.2810849.

A block chain and artificial intelligence—enabled smart IoT framework for the development of sustainable city

Jayashree J. Reddy[1], Aaditri Mittal[2], Vijayashree Reddy[1] and Harshita Das[3]

[1]Department of Database Systems, Vellore Institute of Technology, Vellore, Tamil Nadu, India; [2]Vellore Institute of Technology, Vellore, Tamil Nadu, India; [3]Modi University of Science and Technology, Lakshmangarh, Rajasthan, India

1. Introduction

A smart city may be described as a city that is characterized by its interconnectedness and interlinkage across several domains, including smart government, smart individuals, smart healthcare, smart communication, and smart transportation (Bhushan et al., 2020). Each nation has embarked on the pursuit of smart city initiatives within the context of its own frameworks, with the aim of enhancing the overall quality of life for their citizens. At the moment, scholars from at home as well as abroad have conducted extensive research and collaboration on present safety and confidence issues, guaranteeing safety processes sharing designs, technical personnel methods and models for distributing governance knowledge in smart towns, and suggested an integrated information resource for intelligent cities (Ismagilova et al., 2019). The library approach, the cooperative arrangement of the data resource exchanges a component and the company subsystem, the ERP absorbed governmental data resource merging model, the e-government information resource insertion model, the government data assets based on network of things knowledge integration structure model, and the cloud administration data exchange model are all examples of models. For ensuring protection in the smart city setting, several strategies centered on allocation of resources, processing of signals, and wiretapping coding are investigated.

1.1 Peer-to-peer energy trading system

This section is based on the application of ML and blockchain for smart grid trading in energy. Over the past several decades, nonrenewable energy resources, including natural gas, coal and oil, have emerged as the predominant sources of energy production. However, it is important to note that nonrenewable energy sources are progressively getting more expensive and face challenges in meeting the energy demands of a substantial population (Ikram, 2021). Moreover, it should be noted that nonrenewable energy supplies are not environmentally sustainable. Consequently, there is a notable focus placed on the use of renewable energy sources (RESs), including wind, biomass, solar, tidal, among others. The advancement of renewable energy sources has created opportunities for the implementation of distributed peer-to-peer (P2P) energy trading, especially in residential and commercial buildings. The peer-to-peer energy trading paradigm is often known as the exchange of energy between producers and consumers. Peers have the ability to engage in energy trading between themselves, circumventing the need for conventional energy distributors like the grid (Jamil et al., 2021). The prosumer within the proposed system seeks to engage in the sale of distributed resources of energy to either the utility or the end customer. In the same way, the consumer is the entity that engages in the consumption of energy, while the utility operator assumes the responsibility of overseeing the management of distributed energy resources belonging to the user, in accordance with a mutually agreed-upon settlement.

This section presents a proposal for a predictive system built on blockchain technology. The platform aims to offer real-time assistance, day-ahead control, and production schedule for distributed energy resources. The platform being suggested comprises two distinct components, namely a blockchain-based energy trading segment and the smart contract enabled predictive analytics segment. The blockchain part facilitates real-time monitoring of energy use among peers, enables convenient control over energy trading, implements a reward scheme, and maintains immutable transaction logs for energy trading. The objective of the predictive analytics module, facilitated by the smart contract, is to construct a prediction model using past data of energy consumption in order to forecast energy usage in the near future.

Trading of energies involves the use of real-time control, day-ahead, as well as scheduling strategies to effectively manage distributed energy resources in order to fulfill the load demand of the smart grid. Furthermore, data mining methods are used to conduct time-series analysis, enabling the extraction and examination of hidden trends within the past energy usage data. Time-series analysis is a valuable tool for energy management, as it enables the development of informed strategies and effective resource allocation. By examining historical data and patterns, energy managers may make more informed choices and improve their ability to organize and oversee energy resources in a more efficient manner. Furthermore, an assessment is conducted on the efficacy of the blockchain system using hyperledger caliper, with a focus on metrics such as throughput, latency, and resource consumption (Fan et al., 2020). The experimental findings demonstrate that the suggested approach efficiently facilitates energy crowdsourcing between both producers and consumers, hence achieving a high level of service quality.

The potential for revolutionizing energy production and its consumption exists with the introduction of smart grid technology, which involves the incorporation of distributed energy resources (DERs) as well as microgrids (Kumar et al., 2020). The involvement of prosumers as active participants in the electricity grid enables the provision of energy to central grid storage systems and encourages decentralization of the grid. Additionally, there is a planned modification in the utility offered by power retails to service providers, with the aim of providing prosumers with the opportunity to lease lines for transmission. Furthermore, use of the distributed system also contributes to the improvement of response accuracy to high-precision demand signals, efficiency enhancement, and cost reduction.

This research examines the production of solar energy via photovoltaic systems (PV), the use of dispatchable load such as energy storage systems (ESS), and the incorporation of shippable load such as electric vehicles (EVs) (Ntombela et al., 2023). Various ML modeling and data mining techniques are used to assess the data on energy usage derived from various sources, with the aim of addressing upcoming energy demand. The preprocessed information is used for the purpose of real-time as well as day-ahead management and coordination for distributed energy resources. In addition, the time-series characteristics that have been identified, including hourly, daily, weekly, annual, and seasonal patterns, are used for the purpose of forecasting short-term demand for energy through the utilization of ML models. The establishment of connection between client's applications is facilitated by the RESTful API, which employs the HTTP protocol for the management of user requests.

In order to do experimental study, we used Hyperledger Caliper to assess the efficiency of the system proposed (Abbas et al., 2021). The findings indicate that the suggested predictive platform for trading of energy exhibits superior performance in terms of throughput as well as latency. The identification and use of concealed and revealed patterns have significant value in reducing the expenditure associated with electric consumption of energy for consumers, hence playing a crucial role in the optimization of power engineering. In addition, the objective of the smart contract-enabled forecasting model was to create an advanced model for prediction using XGBoost, LSTM, and RNN random forest algorithms in order to forecast short-term energy consumption. The analysis of the prediction results indicates that the long short-term memory (LSTM) model exhibits the lowest mean absolute percentage error (MAPE) value when compared with other ML models that were applied to time-series data.

1.2 DITrust chain based on IoHT

In this part, we explain the concept of decentralized interoperable trust (DIT), which utilizes smart contracts to ensure the authenticity of budgets (Abou-Nassar et al., 2020). Additionally, we discuss the implementation of the indirect trust inference system (ITIS), which aims to minimize semantic gaps and improve the calculation of the trustworthy factor (TF) by using network nodes and network edges. To maintain security, the DIT IoHT uses a protected blockchain rippled connection in which reliable interaction by validating circuits according to their interoperable structure. This enables controlled communication necessary for resolving fusion and integrating challenges across different areas within the IoHT infrastructure.

There is a well-acknowledged understanding that individuals suffering from chronic diseases, such as high blood pressure, respiratory problems, or type 2 diabetes, need a greater use of healthcare, medical and emergency services compared with those with common illness. Healthcare-oriented Internet of Things (IoHT) encompasses systems that gather data from various sensor devices through the use of middleware. In order to effectively manage the diversity present in IoHT, it is necessary to establish interoperability and address trust-related concerns within the framework of the IoT by using blockchain technology. One potential strategy for ensuring the reliability of IoT data involves the implementation of a distributed transaction system that is trusted by all of its participants. This technique guarantees the immutability of the data, hence enhancing its trustworthiness. Additionally, the implementation of a robust data dependability system would enable government entities to effectively exchange and securely transfer data with citizens.

The preliminary stage is specifically designed for the purpose of gathering and analyzing information, along with implementing any required modifications to the collected data. This layer consists of actuators and sensors that are necessary for various purposes, including the retrieval of the location, blood pressure, body temperature, size, movement, and other related data. These functions are facilitated by the use of conventional ready to use procedures. The second stage includes gateways plus network pathways that are necessary for the transmission of Internet of Things data (Diyan et al., 2022). Gateways play a crucial role in the collection and secure transportation of data from many sources, including devices, distant users, and applications, in order to fulfill specific requirements. The framework's third stage, sometimes referred to as middleware, comprises intermediary sublayers situated between the application tiers and the technology. The Healthedge platform is divided into sublayers, which include blockchain decisions elements, analysis of data, and application supporting layers. Finally, situated at the lowermost level of the architectural framework, we find the application layer, which serves as the interface through which the various capabilities of the system are made accessible to the customers.

The DIT blockchain IoHT architecture is designed as a robust system that facilitates the integration of semantic labels of layers in the Internet of Health Things (IoHT). Cryptographic methods are used to verify, confirm, and ensure the security of various phases involved in data inclusiveness, exchange, and other related processes (Gharaibeh et al., 2017). The suggested model demonstrates superior performance compared with other comparable methods in several aspects, including scaling, interoperability, the availability, mutual authentication, trustworthiness, data integrity, as well as secrecy and privacy. In further research endeavors, researchers want to augment this system by using AI and DL methodologies. The aforementioned technologies are going to be utilized throughout the steps of training to find patterns suggestive of certain ailments by using data obtained from wearable sensors.

IoT has four layers: Device Sense Layer, Network Layer, Middleware Layer and Application Layer (Fig. 7.1).

1.3 Cognitive edge architecture for shared economic service

The use of shared economy services is being facilitated by the incorporation of increasingly powerful IoT devices. One example of contemporary medical technology is the incorporation of IoT devices into healthcare devices, such as the CT Scanners. These IoT embedded gadgets possess the capability to analyze diagnosed data directly at the edge. An organization has the

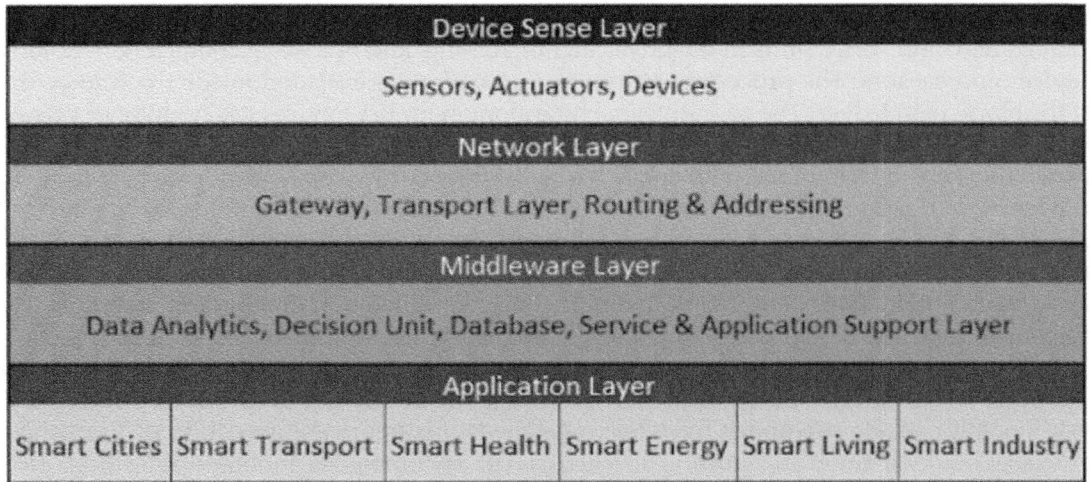

FIGURE 7.1   IoT layers and its components explains the different layers of IOT and its components.

capability to execute various business activities, conduct financial or commercial transactions, and securely store unprocessed multimedia/IoT data minus the need for an intermediary. The implementation of these proposed technologies would effectively facilitate the development of scalable big data, transferring it to a distant cloud, and the subsequent decentralization or centralization of this data within an intelligent city environment (Manimuthu et al., 2021). Nevertheless, the processing of data and the identification of events in various circumstances provide a formidable challenge because of the immense quantity of the data involved.

This section focuses on the task of implementing smart and cognitive processing along the edge. This involves handling the substantial volume of information produced by the IoT and processing it through mobile edge computing (MEC) points (Wang et al., 2019). Additionally, important interactions are concealed and safely stored using blockchain technology. Furthermore, multimedia big data are stored safely using decentralized off-chain remedies to ensure an immutable registry. The use of this suggested framework enables shared economy services to provide safe transactions among both parties without the need for an impartial party. This framework can effectively exploit intelligence stored at the edge to flawlessly coordinate with the collaborate with the IoT processing of data framework. This procedure will facilitate comprehension of the requirements of interaction participants, their past records, plus the information stored in decentralized repositories.

Cognitive computing is predicated upon the cognitive capabilities of human thinking processes, enabling the dissemination of this intellectual capacity to various computerized systems. Cognitive computing encompasses automated ML methodologies, which employ information analytics, pattern detection, and NLP in order to imitate human thought processes (Pramanik et al., 2018). When such systems undergo training, they are capable of functioning autonomously without the need for human intervention. The cognitive engine in question exhibits robust analytical capabilities and data processing prowess, therefore integrating blockchain frameworks with intelligence akin to that of humans. A cognitive machine has various frameworks for analytical and forecasting needs and also is granted permission to document and verify shifts in the physical, functioning, and operating characteristics of IoT devices and sensors. The process of affirming transactions is executed inside the framework of the blockchain ledger. The cognitive machine along with IoT devices use hashing of private keys as a means to authenticate and verify alterations. Several firms have successfully integrated blockchain technology with cognitive computing to develop sophisticated artificial intelligence (AI) systems, exemplified as IBM Watson. Therefore, within the context of blockchain-based automation of shared economic service, the cognitive engine has the responsibility of automating and intelligently overseeing financial transactions and services. The stakeholders engage in reliable activities whereby the cognitive system examines the agreement, and the duties are carried out completely with no human intervention.

For this subsection, it is assumed that the city has implemented 5G cell towers that are equipped with authorized spectrum in order to facilitate device-to-device (D2D) connectivity. Additionally, on-site fog nodes have been installed to support the execution of Blockchain processes and off-chain activities. The client layer encompasses mobile applications, IoT nodes, and an architecture that is closely linked to shared economic services. These client nodes establish connection to the mobile edge layer using several methods of communication, including wireless internet, BLE (Bluetooth low energy), Zigbee, and 5G (Kumhar and Bhatia, 2021). The MEC layer is designed to accommodate many components such as blockchain nodes, decentralized information repository clients, Tor clients, along with other traditional database and cloud apps. The MEC layer assumes the responsibility of coordinating the load synchronization along with the backside cloud layer, contingent upon the load encountered by the MEC layer (Faheem et al., 2018). Given the slim nature of IoT devices, it is reasonable to infer that their ability to establish connections with edge devices operating the complete blockchain nodes is likely to be constrained. The framework facilitates communication between an IoT node alongside adjacent edge nodes that execute smart contracts. It enables the utilization of decentralized communication services, the storage of unprocessed IoT sensory information in a decentralized repository through edge networks, the inclusion of IoT data of relevance in the blockchain, and the establishment of connections with cryptocurrency trading platforms and gateways. The AI system utilizes data obtained from the IoT, Blockchain Technology, and various social network areas to extract emotions, do digital forensics, and identify patterns of interest relevant to different shared economy businesses.

The implementation of our suggested artificial intelligence infrastructure will enable a forthcoming generation of smart cities to provide cyber-physical shared services to the economy by using IoT data. The use of smart contracts enables the framework to provide intricate spatiotemporal solutions on a worldwide basis, eliminating the need for a centralized verification agency.

1.4 Vulnerabilities sustainability of secure edge computing

According to projections from the UN's (United Nations) statistics, it is anticipated that around 75% of the global population would rely exclusively on urban infrastructure and resources for their welfare and livability by year 2050. The integration of smart cities into the Industrial IoT revolution is now known as a potential network that facilitates the distribution of knowledge about the smart resources available inside the city to the general public. The term "IIoT" is sometimes used interchangeably with "Industry 4.0"; nonetheless, they share fundamental similarities as both of them strive to establish a linked industrial realm including corporate strategy in addition to management of supply chains (Awan et al., 2021). Nevertheless, the transmission of wireless data involves significant risks presented by various malevolent activities, such as cyber-attacks and unauthorized access by hackers. These threats are particularly concerning in the context of unprotected IIoT networks, which have been identified as vulnerable entry points for possible breaches. The smart city framework faces significant security challenges that pose threats to the functions of its edge components. These difficulties are characterized by various cyberattacks, which necessitate the implementation of a well-designed security framework. This framework should be capable of effectively detecting and mitigating vulnerabilities related to unauthorized accessibility, clever component hijacking, and privacy breaches within the city engineering.

This research added to the current body of knowledge information by examining the threats to security associated with smart edge technologies in the context of smart city structure by:

- The researchers put forward a reliable framework, named STAPPA (Secure Trust-Aware Philosopher Privacy & Authentication), which utilizes Petri-Net modeling to tackle the vulnerabilities of edge aspects in smart cities. By employing unpredictable distinct event system simulation, the framework aims to mitigate authentication and confidentiality issues within network systems.
- The objective of this study is to explore the development of a decentralized authorization algorithm that is built on the STAPPA model. This algorithm aims to address the issue of unauthorized access by users and identity management inside the social networks of smart cities.
- The Establishment of GARL (Genetic Algorithm-based Reinforcement Learning) for Anomalous Recognition and Routing Path Optimization in Smart City Environments (Malik et al., 2022). This methodology helps the agent in identifying abnormal activity and determining the most efficient routing route using an optimization technique.

The STAPPA concept is used inside the Petri-Net framework to identify instances of anomalous penetration and mitigate unwanted access by users within the network (Ajao and Apeh, 2023). The development and implementation of DAA (Distributed Authorization Algorithm) inside an STAPPA architecture aims to improve identity of user management, ensure safe outflow of information, and mitigate the risk of machine hijacking. The present study introduces a GARL approach, which aims to enhance the process of recognizing anomalies and optimize finding of the shortest path inside an educational setting.

This STAPPA model may be categorized into three distinct sections:

- The defensive layer: functions as a protective barrier that is specifically built to safeguard the city's structure while trying to avoid unapproved individuals from gaining access to the numerous smart components that are there.
- The access control layer: serves the purpose of preventing unapproved access by users and ensuring the confidentiality, integrity, and availability of information provided by users inside the resources layer, hence mitigating the risk of breaches of data.

- The Resource Control Layer: facilitates the oversight and regulation of resource distribution for unapproved users, as well as the management of their priority. Additionally, it facilitates the provision of connections for numerous users to access shared resources in compliance with established regulations and guidelines.

In addition, the implementation of a GARL approach is used to optimize the network's performance during the training phase, to identify anomalies, and to determine the shortest path in the network's setting. The evaluation of the algorithm being used is conducted by assessing its performance in terms of accuracy in detection, which is determined by measuring several metrics including TPR (true positive rate), TNR (true negative rate), sensitivity, FNR (false negative rate), FPR (false positive rate) and specificity. The accuracy of detection achieved in the networked setting during randomized simulation was found to be 98.75%, 99%, 99.50%, 99.75%, and 100%. The acquired rate of detection has an average sensitivity value of 1.000, but the average specificity is 0.867. The evaluation of the rate of learning and the shortest path finding in the GARL model is conducted using the best chromosomal and fitness metrics available throughout a sample of 500 generations. The results obtained from the Genetic Algorithm with Randomized Local Search (GARL) indicate a positive correlation between the total number of chromosomes (representing cluster heads) and the fitness values. Specifically, as the population's size grows, there is a corresponding rise in the fitness values. These methodologies aid in the identification of abnormalities and the mitigation of illegal entry to edge computing elements inside the urban infrastructure.

The next research will focus on examining the identification and reduction of fog computing threats and cloud-based computing vulnerabilities. This will be accomplished via the use of a lightweight authentication method, with the aim of enhancing the security, immutability, and decentralization of large data in smart city environments.

1.5 RON-enhanced blockchain propagation system

Edge computing is a viable solution for addressing the latency requirements of data created by intelligent IoT devices in smart urban areas, since these demands can't be adequately met by cloud-based services. Nevertheless, the deployment of edge-enabled smart city applications is accompanied by a multitude of security issues, including but not limited to extensive centralization, susceptibility to manipulation, and challenges in tracking activities. Blockchain, being a nascent kind of distributed digital ledger technology, has the potential to serve as a viable remedy for these issues. The occurrence of a blockchain fork inside the network for edge-enabled smart city is a possibility, mostly attributed to network propagation latency. Presently, two primary classifications of forks exist: accidental forks and intentional forks. Intentional forks mostly arise as a result of incompatibility of software between older and newer versions, leading to the upgrading of blockchain nodes. The term "accidental fork" pertains to the occurrence of many nodes inside a blockchain network resulting from competitive mining along with the asynchronous characteristics of peer-to-peer (P2P) networks. This leads to double pay scams that may result in financial losses and compromise the integrity of data.

This study presents a novel approach for enhancing blockchain network propagation in edge-enabled smart cities by using the RON (Resilient Overlay Network) (Huang et al.,

2021). The proposed method aims to tackle the aforementioned challenge. This study has three primary contributions:

- It provides an introduction to the idea with the architecture of RON, as well as the RON link Gossip protocol and state route protocol. It also covers some other necessary background information. The framework of the RON-based network on blockchain is specifically developed to include the propagation technique for blockchain network together with the resilient overlaying network.
- It presents a potential technique for route propagation in the context of Blockchain_RON. The use of the shortest route algorithm plus RON policy route is seen in the construction of propagation channels between blockchain nodes inside the Gossip propagation mechanism, as shown by the performance database for RON.
- The purpose of this study is to interrogate the impact of varying the number of nodes upon data transmission simulation. This study aims to simulate the Gossip propagation process using randomized movement models and an Epidemic routing algorithm. Additionally, the study aims to simulate the RON-based blockchain network propagation system using a map-based shortest route movement model and the Focus and Spray routing algorithm. The objective is to contrast the information transmission success rates, delay in time, total no. of route hop, as well as blockchain fork metric between both the propagation mechanisms.

The proposed concept involves the construction of a robust overlay network over the Internet, including several independent domains. This network is structured in a manner similar to blockchain nodes. Subsequently, nodes are chosen from this resilient overlay network to serve as blockchain nodes. Blockchain nodes continue to utilize the Gossip propagation process, albeit with the inclusion of the RON process for message transmission between neighboring nodes. The RON mechanism facilitates the selection of routes, employing a way of routing which optimizes the way for adjacent nodes within the blockchain network. This approach ensures both the smallest possible distance and rapid propagation speed. The main focal areas include two different facets, namely the network's structure of Blockchain_RON as well as block information propagation routing method.

This study focuses on the issue of branching in blockchain networks inside smart cities. This work introduces a novel mechanism for enhancing the pace of block propagation in blockchain networks. Specifically, the suggested method utilizes an RON-based network propagation system, wherein block knowledge is disseminated to adjacent nodes via the Gossip process. However, the data forwarding process employs an efficient overlay network routing tactic at the underlying routing level. The efficiency and feasibility of the RON-based blockchain network's propagation mechanism, in comparison to the Gossip propagation mechanism, have been shown via simulation. This has been evidenced by improvements in message propagation time, systems' overhead, the average routing hops, along with the Message transmission success rate.

1.6 Learning via actions in a smart city context

An observation-based security system is modeled to increase security of communication reliability. Device features and behavior modeling are used to characterize local and global behavioral changes. A neural network-based learning strategy processes the gadget and the service provider's input views to detect problems in accessing resources. The scheme's efficiency balances user safety and resource utilization needs by lowering reaction loss and nonreputable equipment selection.

The expanding population of cities need a sufficient amount of resources as well as ways to communicate based on online mediums. A smart city ecosystem delivers intelligent transportation systems, trash and energy supervisors, smart buildings and houses, and other services via the combination of multiple communication methods and services platforms. IoT sees resources alongside additional information being represented digitally in order to facilitate interactions among many types of individuals and equipment. Linking various technologies improves the dependability of information interchange and transmission in smart city environments utilizing IoT (Jin et al., 2014). Despite the fact that ubiquitous technologies offer smooth support for services as well as application accessibility, cybersecurity remains a barrier in the smart city context. With regard to the varied nature of equipment and linked users, preventing the impacts of information poisoning and fabrication is the primary concern. The kind of information varies depending on the device being used and the people that access it, thus both permission and the authentication process protection must be implemented to maintain the integrity of information and novelty.

Observation-based security system

OSS is intended to enhance users' connectivity and software satisfaction. The gadgets' nature changes depending on their portability and communication capabilities. The goal of OSS construction is to provide autonomous safety features to enable smooth and dependable program and resource use (Nižetić et al., 2020). In this distributed system, security is managed through attribute classification and behavior modeling, which is rendered through recursive training.

- **Attribute classification**

Information is accessible in a smart city setting by exchanging the IoT device identifiers with the app or service operator. Confidential information and storage apps need this form of authorization. By initiating a request, the gadget searches for accessible resources. The components interact with the assets using a multi-hop protocol. As a result, the trustworthiness of the query responder connection is modeled by the characteristics of the standalone devices. Defacing occurs when connectivity is disrupted or users have no way to enter the system. The device classed as join can handle requests for interaction and resource sharing. Unlike traditional learning-based analysis, the condition of the equipment is categorized cooperatively using conclusions obtained from gadget and service supplier variables. The limits in the method of learning are removed by employing behavior modeling, which identifies the device's current state. The mistakes indicate the neighbors' lack of reputation, while the limitation represents the likelihood of a reputed neighbor being available. As a result of the process of education, this analysis is offered.

- **Behavior modeling**

Behavior modeling is used to reduce the adverse impacts of device behavior on distributing resources and accessibility. Behavior modeling takes into account the process of distributing resources and accessibility in line with the limits encountered in attribute categorization. This device prevents other vendors of services from accessing the information/resource shown by it. The device behavior modeling assists suppliers of services in addressing growing concerns in utilization of resources and responsiveness rates. The device's popularity aids in the selection of a trustworthy sharing neighbor for resource utilization and maximum response maximization.

The suggested OSS has a shorter processing time owing to three factors: neighbor categorization, limitation, and analysis of errors (Li et al., 2020). The suggested OSS improves neighbor selection by using popularity. The delay in the other techniques is rather considerable as opposed to OSS since each asset is assigned with a separate source of data and swapped with neighbors. Choosing a group of unreliable neighbors decreases the influence of adversaries on the asset sharing and assignment process. Utilizing the established states, the OSS reduces the effect of such gadgets. The stated design goal of OSS is to maximize resource utilization by lowering adversary influence in the information accessibility ecosystem. The computation and response time are reduced to maximize the utilization of resources.

It presents observation-based safety measures for harmonizing the use of resources and IoT user dependability in smart cities. The information from the device and the internet service supplier is analyzed separately and together using recursive neural learning to detect and limit resource access faults. The suggested approach is proven to be beneficial in enhancing resource utilization and minimizing latency and response losses by boosting reputation-based equipment picking. This situation is ideally suited for smart city security software, and the versatility of these applications will be increased in the future employing distributed computing platforms.

1.7 New smart city development for minimal carbon emissions and environmental sustainability

The blockchain technology is utilized to build a decentralized dispersed using person to person trust service network that is combined using today's PKI/CA safety framework to create an innovative a trust paradigm that allows for multi-CA cohabitation (Sun and Zhang, 2020). From the perspective of big data technology, smart city and block chain integrated, and beneficial impacts associated with pertinent IT summarized on the development of big data network for smart cities. The TOPSIS method's model for smart city project level evaluation was built on this basis.

Through smart reaction, smart cities employ IT to enhance people's lives concerns for example housing, transportation, the government services such as education, and medical care. Smart cities have achieved strides in electronic government, sustainable ecology, and automated transport. The primary purpose of smart city development is to employ IT to detect, send out, incorporate, and analyze essential information from the fundamental framework of operation in the city in order to achieve city intelligence. Big data and blockchain technology are two distinct technologies of Internet. Blockchain interaction is a kind of information less communication. Smart city blockchain ecosystem collaboration of information is expanded to address the confidence territory, privacy of data, and peer management problems that currently exist in smart city information collaboration and interchange (Bagloee et al., 2021). Consistency management and non-real-time transmission are two examples of innovative implementation approaches.

The development of a smart city building model

- **The criteria for a blockchain-based smart city**

The government offers an environment for players in smart cities to play the part in the utilization of big data. Create a model to validate the degree of growth in smart cities using the TOPSIS technique. The computational flow of the structure is shown using the extreme point approach, which is separated into both negative and positive indexes. TOPSIS is a method for making choices with several attributes. It primarily does multiobjective assessment and analysis of choices for a small number of applications. This technique scores the schemes based on how near the assessed scheme is to the optimal solution's favorable and bad aspects.

- **Smart city comprehensive architectural model founded on block architecture**

A blockchain-based government exchange of data system paradigm that is feasible continuously extended, handled, regulated, and open functions is built using different value networks. The mutually beneficial collaboration of smart city fundamental information is broken down segregation and enables efficient collaboration and data exchange across different divisions. There are two key roles in the blockchain foundation for smart cities allocation and transfer model: data generators and data users. The logical structure of the blockchain-based smart city utilization and transfer model is separated into these three layers: network layer, business application layer, and blockchain infrastructure layer.

- **Blockchain-based smart city business collaboration method**

Institutions may develop a network of collaboration in a system centered around Smart city powered by blockchain cooperation in resources and trade. The procedure of assessment, confirmation, and archiving is handled automatically by the blockchain's intelligent contractual technology to assure real-time and immutable alteration of the data transfer process. Structures employ XML standard order rather than alternative techniques because it is significant in handling data interchange in diverse contexts, lowering the burden of developing integration connections between various systems, and generating semantic networking environments.

Increasing the growth of key emerging sectors, boosting business innovation capabilities, enhancing the creation of venues for innovation, and bolstering the development of talented groups, and maximizing the setting for innovation are the six important responsibilities. Use big data to reinvent administration service techniques, develop e-government fiercely, encourage lateral connection and sharing of information of e-government systems across departments and regions, and increase the effectiveness with which government systems operate. Encourage the development of new media for government activities and the enhancement of the digital service to the public system that it incorporates. Coordination of the development developing a cloud-based system for the provincial legislature accomplish extensive production and distribution of the state's facilities as well as heavy administration of data assets.

The technological underpinning of smart cities is big data, and block chain is capable of facilitating data exchange in smart cities. Sharing government information assets necessitates the application of new technology, novel concepts, and novel approaches, as well as the willingness to innovate. Smart cities have evolved from being led by the government and propelled by the government in earlier times to becoming government-regulated, socialized, and participatory by the complete population. Big data in urban areas is a tangible and significant expression of intellect, using cloud services, mobile internet access, AI, big data, IoT, and other next-generation technologies for information are being used to build an information command center.

1.8 Ideal IoT resource management

The IoT is a new smart city technology that connects many digital systems gadgets over the web, hence delivering a wide range of creative solutions, from intellectual to commercial and medical to commerce. It is preferable to have a multimedia system for the huge and brilliant image of every occurrence in smart cities to achieve superior, clear, and highly visible information. From a technical standpoint, IoT has grown into the focal point for realizing and quantifying its incredible impact in the majority of disciplines. Regular inspection of data obtained from IoT-based sensor gadgets provides information exchange that is quick, apparent, transparent, effective, reliable, and right with a powerful omnipresent foundation. On the one hand, the fast multiplication of sensors has made people's lives more delightful and exciting, while on the contrary side, the huge and vivid image of the scene via video communication in smart cities is the basic concept to be realized.

- **Smart cities**

IoT is the foundation of the smart city system, allowing it to serve every part of the globe, from farm through factory, industry to academic. Since the previous few years, the globe has been transformed into an urban community with an extensive online system. As everyone desires to live a comfortable and prosperous life in the city, the rate of migration from rural to urban regions is increasing on a daily basis. With the growing number of cities comes the need over right away and effective knowledge and access to resources in a very short period of time; otherwise, high stress and dissatisfaction will be boosted while other social and environmental variables will affect the daily lives of the people.

- **Architecture of the smart city**

A smart city's architecture is composed of five major levels, numbered one through five (Yin et al., 2015). At first layer, media content is acquired using sensor-enabled technology and a variety of technologies for specific purposes, such as transmission of video. Layer 2 sensor and acquisitions are done, and the power consumption and battery life of the miniaturized electronics are calculated during these phases. encoded information manufacturing, and tracking of media data are done at layer 3 with an obvious evaluation of power waste and battery lifespan. Because the transmission element of the communication system consumes the most power, media information is consolidated, propagated, and transferred at layer 4, and battery consumption and battery lifespan are estimated. Finally, at the smart city's layer 5 implementation phase, evaluation of performance in terms of network parameters such as power (like transmission) disintegration, battery lifespan, and delay is performed. As IoT-enabled smart cities gain popularity, they confront the issue of battery life and energy depletion.

- **Algorithm for Hybrid Adjustable Capacity and Power**

It is critical to note that the authority drain from individual detectors is related to the adjustable component of the sensors rather than the volume of traffic. HABPA reduces visual frame fluctuation even with sufficient buffer capacity. As a result, it exhibits somewhat higher electricity and internet during media consumption in smart cities. As a result, the suggested delay-tolerant streaming algorithm (DSA) solves these issues.

- **Algorithm for delay-tolerant broadcasting**

For broadcasting media in smart cities, we offer the delay-tolerant streaming algorithm (DSA). Our suggested DSA reduces power loss and battery energy utilization by managing the high peak changing rate of media, improving the lifespan and operating period of IoT-based devices with little delay.

Recent developments and methods in miniaturized sensing enabled gadgets have captured everyone's attention, thus efficient and precise delivery of messages, information sharing, and data transfer must be taken into account. As a result, it is preferable to develop environmentally friendly and sustainable transmission of data algorithmic techniques in smart cities that take into account loss of energy.

Because of the way it operates and buffered adjustment during streaming video in smart neighborhoods, HABPA preserves more power than Baseline and DSA, making it a suitable contender for a variety of applications. At various buffer size levels, the energy exhaustion level of Baseline, DSA, and HABPA in intelligent cities is studied. Delay rises with buffer size, which is large for baseline and DSA yet small for HABPA. HAPBA is a better choice than DSA and Baseline for dependable data transfer in smart cities. It is investigated and shown that there is a proportional exchange among buffering capacity and frequency fluctuation, resulting in smoother video transfer in smart cities. It is also found that the distribution of energy drain accounts for the majority of the total amount of energy used in smart cities, making it the decisive factor.

Smart cities have become viable contenders for every nook and cranny. It is critical to consider straightforward and highly apparent information exchange across diverse organizations through multimodal transmission. However, owing to the small and hungry for power nature of sensory devices, as well as their pervasive nature, it is very challenging to develop environmentally friendly and Greener smart cities.

1.9 Applications of artificial intelligence and machine learning in smart cities

ICT (Information and communication technology) plays an important role in policy creation, decision-making, execution, and ultimately profitable services in the smart city idea. Many governments have advocated the notion of smart cities to efficiently oversee resources and optimize energy usage in order to deal with the dramatic rise in urbanization. Smart cities initiatives may specifically address the issue of ensuring a sustainable future by creating and implementing low carbon emission technology. The IoT is the most essential and crucial component aspect of most smart city applications, which generate massive amounts of data. Smart cities, IoT, Blockchain, unmanned aerial vehicles (UAVs), and the use of AI, ML, and DRL-based approaches in various fields are still in the early stages of development and will provide additional possibilities in the future. The most crucial and vital part of a smart city that may realize the ideal notion of a smart city is security online. The nature of our everyday employment is fast changing as a result of AI, which is influencing the conventional approach to human thought and engagement with the natural world.

In order to cope with the tremendous development in development, several governments have championed the notion of smart cities for optimal resource management and utilization of energy. Smart cities efforts may address the challenge of guaranteeing a sustainable future particularly by developing and deploying environmentally friendly technologies. The IoT is the most significant element of most applications for smart cities, which create vast volumes of data. Smart cities, Blockchain, IoT, UAVs, and the application of ML, AI and DRL-based techniques in many industries are still in their early phases of advancement and will give new opportunities in the years to come (Heidari et al., 2022). Internet security is the most crucial and basic aspect of a smart town that may realize the ideal notion of a smart city. Because AI is affecting the standard approach toward human thought processes and involvement with the natural environment, the very essence of our ordinary job is rapidly changing.

- **Intelligent transportation system**

The ITS is a cooperative application of current sensors and control systems, and ICT that produces large amounts of information that has significantly influenced the development of ITS and the notion of cities that are smart. AI, ML, and, in particular, DRL approaches are crucial in observing and calculating current traffic movement in a city context, which is an essential part of a sustainable ITS.

- **Cyber-security**

A smart city is projected to be composed of interconnected actuators, detectors, and relays that gather, analyze, and transfer data in addition to deliver dependable and efficient electronic services. The vast bulk of data is generated by Cloud-based IoT gadgets, which are crucial for numerous applications for smart cities.

- **Smart grids**

The use of big data is transforming the everyday architecture of urban areas and enabling enhanced energy usage. In SGs, heterogeneous data enters from many sources and may be successfully analyzed and utilized to make suitable leadership and operational choices. According to current trends, SGs are effectively using smart meter big data for a variety of applications including as load assessment and forecast, baseline estimate, response to demand, load grouping, and fraudulent information deception assaults.

Applications of DRL-based UAVs in B5G and 5G connectivity

The growing need for fast data transfer rates, reliability, and minimal latency has pushed the current wireless mobile networks toward 5G and B5G communications. DRL-based approaches, ML and AI have been recognized as the most impactful tool for dealing with the numerous complicated communication difficulties utilizing large amounts of network data in order to attain the previously stated objectives. Despite their promising uses, UAVs encounter several unresolved obstacles. Many non-convex optimization issues in sophisticated communication systems are difficult to address effectively. DRL, an ML-based technique, may be the ideal alternative for dealing with many types of such complicated issues.

The shorter wavelength mmWave allows for the efficient organization of small antennas on a single chip to create forming beam station arrays, which are suitable for UAV-assisted communication. The shorter frequency of mmWave allows for the efficient organization of small stations on one microchip to create forming beam arrays of antennas, which are perfect for UAV-assisted connectivity.

Smart city health care and machine learning

Machine learning, AI, and DRL techniques have been extensively applied in better health care procedures known as medical knowledge with the development of upgraded sensors, powerful IoT gadgets, cloud-based computing, and a surge in data rates.

As shown by the current research on smart cities, Machine learning, AI, and DRL techniques have demonstrated promising results. The use of ML-techniques to jointly optimize UAV onboard abilities and trajectory may considerably improve the efficacy of ITS while UAV-vehicle interaction. The primary issue of every SG and electric company is maximizing the effectiveness of power-down scenarios. ML and DRL-based algorithms may be used to develop strategies for switching between 5G technologies for communication while ensuring continuous power supply.

2. Conclusion

In the current context, the advancement of technologies such as the IoT, ML, AI and Blockchain is progressively enhancing many aspects of people's daily lives, making them more efficient and intelligent. The present state of research pertaining to the aforementioned technologies is undergoing fast evolution, accompanied by a wide range of views. It is advisable for researchers to integrate different study domains instead of pursuing their development in isolation. We will develop machine learning well as deep learning applications combining information approaches to serve medical smart towns at cheap medical prices in the not-too-distant future. More sophisticated and massive analytics-based imaginative approaches are necessary to ensure the digital safety and confidentiality of smart city applications. It may assist researchers in achieving great objectives for research and developing a gradual analysis strategy that is valuable to business and academics. Some individuals resist installing specific blockchain systems for any purpose since blockchain is undergoing rapid technical progress after its merger with machine learning; yet, we detailed many uses in smart cities.

References

Abbas, K., Tawalbeh, L.A., Rafiq, A., Muthanna, A., Elgendy, I.A., Abd El-Latif, A.A., 2021. Convergence of blockchain and IoT for secure transportation systems in smart cities. Security and Communication Networks. https://doi.org/10.1155/2021/5597679.

Abou-Nassar, E.M., Iliyasu, A.M., El-Kafrawy, P.M., Song, O.Y., Bashir, A.K., El-Latif, A.A.A., 2020. DITrust chain: towards blockchain-based trust models for sustainable healthcare IoT systems. IEEE Access 8, 111223–111238. https://doi.org/10.1109/ACCESS.2020.2999468.

Ajao, L.A., Apeh, S.T., 2023. Secure edge computing vulnerabilities in smart cities sustainability using petri net and genetic algorithm-based reinforcement learning. Intelligent Systems with Applications 18. https://doi.org/10.1016/j.iswa.2023.200216.

Awan, U., Sroufe, R., Shahbaz, M., 2021. Industry 4.0 and the circular economy: a literature review and recommendations for future research. Business Strategy and the Environment 30 (4), 2038–2060. https://doi.org/10.1002/bse.2731.

Bagloee, S.A., Heshmati, M., Dia, H., Ghaderi, H., Pettit, C., Asadi, M., 2021. Blockchain: the operating system of smart cities. Cities 112. https://doi.org/10.1016/j.cities.2021.103104.

Bhushan, B., Khamparia, A., Sagayam, K.M., Sharma, S.K., Ahad, M.A., Debnath, N.C., 2020. Blockchain for smart cities: a review of architectures, integration trends and future research directions. Sustainable Cities and Society 61. https://doi.org/10.1016/j.scs.2020.102360.

Diyan, M., Nathali Silva, B., Han, J., Cao, Z.B., Han, K., 2022. Intelligent Internet of Things gateway supporting heterogeneous energy data management and processing. Transactions on Emerging Telecommunications Technologies 33 (2). https://doi.org/10.1002/ett.3919.

Faheem, M., Shah, S.B.H., Butt, R.A., Raza, B., Anwar, M., Ashraf, M.W., Ngadi, Md A., Gungor, V.C., 2018. Smart grid communication and information technologies in the perspective of Industry 4.0: opportunities and challenges. Computer Science Review 30, 1–30. https://doi.org/10.1016/j.cosrev.2018.08.001.

Fan, C., Ghaemi, S., Khazaei, H., Musilek, P., 2020. Performance evaluation of blockchain systems: a systematic survey. IEEE Access 8, 126927–126950. https://doi.org/10.1109/access.2020.3006078.

Gharaibeh, A., Salahuddin, M.A., Hussini, S.J., Khreishah, A., Khalil, I., Guizani, M., Al-Fuqaha, A., 2017. Smart cities: a survey on data management, security, and enabling technologies. IEEE Communications Surveys and Tutorials 19 (4), 2456–2501. https://doi.org/10.1109/COMST.2017.2736886.

Heidari, A., Navimipour, N.J., Unal, M., 2022. Applications of ML/DL in the management of smart cities and societies based on new trends in information technologies: a systematic literature review. Sustainable Cities and Society 85. https://doi.org/10.1016/j.scs.2022.104089.

Huang, J., Tan, L., Li, W., Yu, K., 2021. RON-enhanced blockchain propagation mechanism for edge-enabled smart cities. Journal of Information Security and Applications 61. https://doi.org/10.1016/j.jisa.2021.102936.

Ikram, M., 2021. Models for predicting non-renewable energy competing with renewable source for sustainable energy development: case of Asia and Oceania region. Global Journal of Flexible Systems Management 22 (S2), 133–160. https://doi.org/10.1007/s40171-021-00285-7.

Ismagilova, E., Hughes, L., Dwivedi, Y.K., Raman, K.R., 2019. Smart cities: advances in research—an information systems perspective. International Journal of Information Management 47, 88–100. https://doi.org/10.1016/j.ijinfomgt.2019.01.004.

Jamil, F., Iqbal, N., Imran, Ahmad, S., Kim, D., 2021. Peer-to-Peer energy trading mechanism based on blockchain and machine learning for sustainable electrical power supply in smart grid. IEEE Access 9, 39193–39217. https://doi.org/10.1109/access.2021.3060457.

Jin, J., Gubbi, J., Marusic, S., Palaniswami, M., 2014. An information framework for creating a smart city through internet of things. IEEE Internet of Things Journal 1 (2), 112–121. https://doi.org/10.1109/JIOT.2013.2296516.

Kumar, N.M., Chand, A.A., Malvoni, M., Prasad, K.A., Mamun, K.A., Islam, F.R., Chopra, S.S., 2020. Distributed energy resources and the application of ai, iot, and blockchain in smart grids. Energies 13 (21). https://doi.org/10.3390/en13215739.

Kumhar, M., Bhatia, J., 2021. Emerging communication technologies for 5G-enabled internet of things applications. Blockchain for 5G-Enabled IoT: The New Wave for Industrial Automation 133–158. https://doi.org/10.1007/978-3-030-67490-8_6.

Li, D., Deng, L., Liu, W., Su, Q., 2020. Improving communication precision of IoT through behavior-based learning in smart city environment. Future Generation Computer Systems 108, 512–520. https://doi.org/10.1016/j.future.2020.02.053.

Malik, N., Dahiya, M., Walia, N., 2022. Machine Learning Applications to Smart Cities. IGI Global, pp. 169–213. https://doi.org/10.4018/978-1-7998-9710-1.ch009.

Manimuthu, A., Dharshini, V., Zografopoulos, I., Priyan, M.K., Konstantinou, C., 2021. Contactless technologies for smart cities: big data, IoT, and cloud infrastructures. SN Computer Science 2 (4). https://doi.org/10.1007/s42979-021-00719-0.

Nižetić, S., Šolić, P., López-de-Ipiña González-de-Artaza, D., Patrono, L., 2020. Internet of Things (IoT): opportunities, issues and challenges towards a smart and sustainable future. Journal of Cleaner Production 274. https://doi.org/10.1016/j.jclepro.2020.122877.

Ntombela, M., Musasa, K., Moloi, K., 2023. A comprehensive review of the incorporation of electric vehicles and renewable energy distributed generation regarding smart grids. World Electric Vehicle Journal 14 (7). https://doi.org/10.3390/wevj14070176.

Pramanik, P.K.D., Pal, S., Choudhury, P., 2018. Beyond automation: the cognitive IoT. artificial intelligence brings sense to the internet of things. Lecture Notes on Data Engineering and Communications Technologies 14, 1—37. https://doi.org/10.1007/978-3-319-70688-7_1.

Sun, M., Zhang, J., 2020. Research on the application of block chain big data platform in the construction of new smart city for low carbon emission and green environment. Computer Communications 149, 332—342. https://doi.org/10.1016/j.comcom.2019.10.031.

Wang, D., Bai, B., Lei, K., Zhao, W., Yang, Y., Han, Z., 2019. Enhancing information security via physical layer approaches in heterogeneous IoT with multiple access mobile edge computing in smart city. IEEE Access 7, 54508—54521. https://doi.org/10.1109/access.2019.2913438.

Yin, C.T., Xiong, Z., Chen, H., Wang, J.Y., Cooper, D., David, B., 2015. A literature survey on smart cities. Science China Information Sciences 58 (10), 1—18. https://doi.org/10.1007/s11432-015-5397-4.

Computational intelligence—based heuristic approach for maximizing energy efficiency in sustainable transportation and mobility

Ripal D. Ranpara

Department of Computer Science, Atmiya University, Rajkot, Gujarat, India

1. Introduction

The current worldwide shift toward sustainable transport and mobility systems is a crucial reaction to the issues posed by climate change, urbanization, and the diminishing availability of limited fossil fuel supplies. The significance of sustainable transportation solutions is growing due to their potential to decrease greenhouse gas emissions, relieve traffic congestion, and improve the energy efficiency of transportation networks. In the current epoch characterized by rapid advancements in technology, computational intelligence emerges as a very promising collection of tools to enhance the optimization of these systems. The increasing recognition of the environmental and economic impacts linked to conventional transportation methods has motivated scholars and professionals to investigate novel approaches to enhance energy efficiency within the transportation industry. The field of transport has a significant role in the overall world energy consumption and carbon emissions (Wallington et al., 2011). The implementation of technology improvements has resulted in increased fuel economy and reduced pollution in automobiles. However, in order to effectively tackle the broader issue of energy efficiency, a comprehensive strategy is required.

The field of sustainable transportation research has historically included a wide range of areas, including the advancement of low-emission vehicle technology and the enhancement of traffic management systems (Holguín-Veras et al., 2014). Nevertheless, the complexities of a significantly dynamic and interrelated transportation system, distinguished by intricate interactions among vehicles, infrastructure, and user behavior, need innovative approaches that can effortlessly adjust to changing circumstances. Computational intelligence, as a specialized domain within the broader subject of artificial intelligence, is highly suitable for effectively tackling the intricacies involved (Chaturvedi and Singh, 2017). The use of an interdisciplinary methodology, including several approaches such as neural networks, genetic algorithms, fuzzy logic, and swarm intelligence, provides researchers with a flexible set of tools to develop intelligent systems that possess the ability to make data-driven judgements in real-time. The use of these methodologies facilitates the representation of complex interconnections within transportation networks, offering opportunities for the enhancement of routes, alleviation of congestion, and reduction of energy usage (Shi and Papageorgiou, 2011).

The primary topic of this study chapter is the incorporation of computational intelligence within the domain of sustainable transportation and mobility. The main goal of this study is to design a heuristic methodology that uses computational intelligence algorithms in order to optimize energy efficiency. In order to situate our study within the current scholarly discourse, we refer to prior studies conducted in the domains of sustainable mobility and computational intelligence. Previous research has effectively used computational intelligence methodologies to tackle many aspects of transportation systems. An example of the use of a genetic algorithm in the optimization of bus timetables in urban contexts can be seen in the work of Xiong et al. (2017). This study resulted in a reduction in energy consumption and an enhancement of transit service quality. Chen et al. (2018) conducted a research whereby they used a hybrid neural network model to improve traffic prediction, hence boosting both route planning and congestion mitigation. The use of computational intelligence beyond the realm of traffic management. In their study, Gao et al. (2019) used swarm intelligence-based techniques to optimize the placement of electric car charging stations. This approach aimed to minimize energy wastage and promote the widespread adoption of electric vehicles. In a similar vein, the research conducted by Cui et al. (2016) used fuzzy logic as a means to develop an intelligent decision support system aimed at enhancing transportation demand forecasting (Cui et al., 2016). This approach ultimately contributed to the optimization of energy efficiency within transportation planning processes.

The aforementioned studies highlight the potential of computational intelligence. However, the specific focus of this chapter is the integration of heuristic techniques rooted in computational intelligence to maximize energy efficiency. This approach takes into account the dynamic characteristics of transportation networks, the specific obstacles related to energy efficiency, and the interconnectedness of different components of sustainable transportation, such as alternative fuels, electrification, public transit, and emerging mobility services (Sangwan and Singh, 2017). The subsequent parts of this study chapter will explore the technique used, the empirical data gathered, and the interpretation of the results. Through an extensive examination of prior research, our objective is to underscore the originality and importance of our research methodology. This methodology is founded on the utilization of advanced computational intelligence techniques to enhance energy efficiency in sustainable transportation and mobility systems. This study makes a valuable contribution to the worldwide endeavors aimed at reshaping the transport sector in accordance with the ideals of sustainability and environmental stewardship (Zhang et al., 2018).

2. Literature review

2.1 Recent trends in sustainable transportation and mobility

The concept of sustainable transportation is a complex field that encompasses a range of factors related to energy efficiency and environmentally beneficial modes of travel. Zhang et al. (2020) provide a detailed analysis of contemporary developments in sustainable transportation, presenting a thorough examination of the subject matter. The assessment emphasizes the trend toward electrification, the rise of autonomous cars, and the increasing prevalence of shared mobility options. The aforementioned developments underscore the need for innovative strategies aimed at maximizing energy efficiency and minimizing environmental repercussions within the dynamic realm of transportation. One of the primary obstacles encountered in the realm of sustainable transportation is to the mitigation of emissions, with a specific focus on metropolitan locales. The study conducted by Meng et al. (2020) focused its attention on examining the significance of sustainable urban transportation options. The research placed a strong emphasis on the need of decreasing greenhouse gas emissions and air pollution by adopting sustainable modes of transportation. It advocated for the use of computational intelligence techniques to develop and execute environmentally friendly urban transportation systems.

2.2 Computational intelligence in sustainable transportation

The primary objective of sustainable urban mobility solutions is to enhance energy efficiency and mitigate the environmental consequences associated with transportation. The study conducted by Wang et al. (2020) examines the increasing prevalence of shared mobility services. The present research investigates the potential of computational intelligence to optimize the operational efficiency of ride-sharing and carpooling services, with the ultimate goal of mitigating the number of automobiles on the road and promoting energy conservation. In order to advance the objective of sustainable urban mobility, Ma et al. (2020) conducted a study to explore the capabilities of intelligent traffic management systems. The research conducted by the authors emphasizes the use of computational intelligence techniques in the optimization of traffic signals and flow. This approach has shown promising results in mitigating congestion, decreasing journey durations, and reducing energy consumption within urban environments.

2.3 Electrification and sustainable transportation

The integration of electric power in transportation, particularly in relation to electric vehicles (EVs), has significant importance in attaining enhanced energy efficiency and mitigating emissions. The study conducted by Li et al. (2020) focuses on the implementation of electric vehicles (EVs) inside urban settings. The study elucidates the advantages of using computational intelligence techniques to optimize electric vehicle (EV) charging infrastructure, hence resulting in enhanced energy efficiency and a diminished ecological impact. The study conducted by Liu et al. (2020a) makes a valuable contribution to the existing body of research by investigating the incorporation of renewable energy sources, namely solar electricity, into the infrastructure for electric vehicle (EV) charging. The use of computational intelligence in this novel technique has the potential to augment the sustainability of electric vehicles (EVs) and diminish their reliance on fossil fuels.

2.4 Future directions in sustainable transportation

Given the ongoing evolution of the subject of sustainable transportation, it is crucial to delve into prospective avenues for future study. Recent research has placed significant emphasis on the possibilities of autonomous cars, linked transportation networks, and sophisticated data analytics in advancing energy efficiency and promoting sustainability. The relevance of computational intelligence in the development of algorithms for autonomous vehicle navigation, route optimization, and energy-efficient driving strategies has been emphasized by researchers, as shown by Sun et al. (2020). The research conducted by Wu et al. (2020) highlights the importance of data-driven decision support systems in the field of transportation planning. These systems, which are powered by computational intelligence approaches, have the capability to provide real-time insights and suggestions for the purpose of optimizing energy usage and mitigating emissions in transportation networks. Contemporary scholarly works demonstrate an increasing acknowledgment of the pivotal significance of computational intelligence in propelling the progress of energy efficiency and sustainability in the domain of transportation and mobility. These studies highlight the need of implementing novel approaches to tackle the environmental issues associated with transportation and facilitate the transition toward a future characterized by enhanced energy efficiency, ecological friendliness, and sustainability.

2.5 Eco-friendly mobility services

Researchers are now investigating the potential integration of computational intelligence in the advancement of sustainable mobility services, in response to the growing demand for environmentally friendly transport solutions. In their research, Xu et al. (2020) investigated the use of evolutionary algorithms for the purpose of optimizing electric scooter-sharing systems. The study primarily focused on improving the energy efficiency of electric scooters in urban settings by using dynamic repositioning tactics. In a similar manner, the study conducted by Song et al. (2020) examined the use of reinforcement learning techniques in order to enhance the energy efficiency of electric bike-sharing systems. The research showcased the capacity of computational intelligence to enhance the durability of electric bike batteries and minimize energy use via the optimization of battery charging and repositioning methods.

2.6 Intelligent traffic management

The effective management of traffic continues to be a fundamental aspect in promoting sustainable mobility. In their study, Xu and Li (2020) investigated the incorporation of computational intelligence techniques into traffic signal management systems. The researchers in this chapter proposed an innovative method for optimizing traffic signal systems using deep reinforcement learning. The findings of the study demonstrated noteworthy decreases in energy use, trip duration, and traffic congestion within metropolitan environments. In order to tackle the intricate dynamics of urban traffic, Tang et al. (2020) introduced a hybrid methodology that integrates fuzzy logic and neural networks to facilitate traffic prediction and adaptive signal management. The present intelligent technology improves the flexibility of traffic lights, so making a valuable contribution to energy conservation via the reduction of needless pauses and idling.

2.7 Electric vehicle routing and energy optimization

Extensive study is being conducted on the energy efficiency of electric vehicles (EVs). In their study, He et al. (2020) investigated the use of model-based reinforcement learning in the context of electric vehicle (EV) energy management. The research showcased significant improvements in electric vehicle (EV) range and total energy efficiency via the use of dynamic energy consumption optimization. These findings contribute to the increased competitiveness of EVs in the realm of sustainable transportation. The study conducted by Samaranayake et al. (2020) focused on the optimization of energy-efficient routing for electric vehicles (EVs) inside urban settings. The study placed significant emphasis on the use of heuristic algorithms, guided by computational intelligence, to effectively decrease energy consumption in electric vehicle operations. This focus was notably directed toward highly populated metropolitan areas.

2.8 Urban freight transportation and computational intelligence

In the domain of urban freight transportation, the factors of optimization and energy efficiency have emerged as crucial issues. The study conducted by Tsitsimpelis et al. (2020) used evolutionary algorithms as a means of optimization vehicle routing for last-mile deliveries within the context of urban logistics. The research demonstrated the potential of computational intelligence in reducing energy usage during the last stage of urban freight transportation, considering parameters such as vehicle load capacity and delivery windows. In a similar vein, the study conducted by Hu et al. (2020) focused on the advancement of intelligent transportation systems with the aim of enhancing route optimization in electric vehicle fleets used in last-mile logistics operations. The study conducted by the researchers revealed the notable energy conservation benefits that computational intelligence may provide inside the urban freight industry.

1. Autonomous vehicles and sustainable transportation

The emergence of autonomous cars has potential for improved energy efficiency. Recent research has investigated the significance of computational intelligence in the domain of autonomous transportation. The study conducted by Smith et al. (2020) focused on the development of machine learning algorithms with the aim of optimizing the routing of autonomous vehicles and promoting energy-efficient driving behavior. The findings of their study suggest that the use of computational intelligence has the potential to significantly decrease energy usage in autonomous transportation systems. The evolving transportation scene is highlighted by recent studies that emphasize the multidimensional role of computational intelligence in improving energy efficiency and promoting sustainability. The aforementioned results provide a strong basis for the research described in this chapter, which utilizes computational intelligence to optimize energy efficiency within the realm of sustainable transportation and mobility.

3. Proposed framework: Computational intelligence for energy efficiency in transportation

The endeavor to enhance energy efficiency in transport, with the growing need to tackle environmental issues, has instigated a fundamental transformation in the conception and functioning of transport networks. Computational intelligence, which falls under the umbrella of artificial intelligence, has emerged as a crucial theoretical foundation for augmenting energy efficiency in the field of transportation. This section provides a Table 8.1 comprehensive explanation of the theoretical underpinnings of computational intelligence and its significance in the context of energy-efficient transportation. It highlights the interdependent connection between these two fields.

TABLE 8.1   Theoretical framework: Computational intelligence for energy efficiency in transportation.

Computational intelligence methodologies	Relevance to energy efficiency in transportation
Neural networks	Adaptive decision-making: Predicting traffic patterns, optimizing routing, and reducing energy consumption by avoiding congestion.
Genetic algorithms	Optimization of complex systems: Finding optimal solutions for route planning, vehicle scheduling, and energy-efficient vehicle design.
Fuzzy logic	Handling uncertainty: Navigating uncertain and imprecise data for more accurate energy-efficient transportation planning.
Swarm intelligence	Collective intelligence: Optimizing traffic signal timings, routing, and charging station placement to minimize delays and energy waste.
Synergy between computational intelligence and sustainable transportation	The symbiotic relationship between computational intelligence and sustainable transportation is seen in their joint endeavor to enhance energy efficiency and promote environmental sustainability. Computational intelligence empowers transportation systems with the cognitive capabilities necessary to dynamically respond to evolving circumstances, enhance operational efficiency, and mitigate energy usage. Sustainable transportation is in accordance with the overarching objectives of mitigating greenhouse gas emissions and advocating for environmentally conscious behavior's.

3.1 Computational intelligence: A multifaceted paradigm

Computational intelligence comprises a diverse range of approaches and algorithms that draw inspiration from natural processes and human cognitive capacities. The field of study incorporates several fundamental approaches, including as neural networks, genetic algorithms, fuzzy logic, and swarm intelligence. The versatility, learning capabilities, and data-driven decision-making of these computational tools render them particularly pertinent to the intricate and ever-changing nature of transportation networks.

Neural Networks: Neural networks draw inspiration from the intricate neural structure of the human brain, characterized by linked nodes that engage in data processing and learning. Neural networks have found applications in the field of transportation, including in traffic prediction, energy management in cars, and route optimization. These networks possess the ability to adjust and respond to dynamic situations, hence offering immediate solutions to optimize energy efficiency.

Genetic algorithms are derived from the fundamental concepts of natural selection and evolution. Genetic algorithms find use in transportation for several problems such as route optimization, scheduling, and vehicle design. Iterative evolution of a population of possible solutions aids in the identification of optimum solutions, ultimately resulting in outcomes that are energy-efficient.

Fuzzy Logic: Fuzzy logic, which emulates human thinking by including degrees of truth, is used in the field of transportation to enhance decision support systems. Fuzzy logic systems provide the capability to describe intricate interactions within transportation networks by considering imprecise and unpredictable inputs. The use of intelligent transport systems facilitates the development of energy-saving measures.

Swarm intelligence is a concept that is based on the collective behavior of decentralized systems, drawing inspiration from the interactions seen in social insects. Swarm intelligence techniques, such as ant colony optimization and particle swarm optimization, are used in the field of transportation to optimize many aspects including traffic signal timings, vehicle routing, and electric car charging station placement. The utilization of these algorithms aims to enhance energy efficiency within the transportation system.

3.2 Relevance to energy efficiency in transportation

The use of computational intelligence methods in transportation systems is in accordance with the fundamental principles of energy efficiency. The adaptive use of data-driven technologies is crucial in addressing the dynamic nature of transportation networks, which is characterized by fluctuating traffic flows, unexpected user behavior, and changeable energy needs. Computational intelligence offers a theoretical framework that intrinsically tackles these difficulties.

The integration of computational intelligence enables transportation systems to possess adaptive decision-making capabilities. Neural networks provide the capability to forecast traffic patterns and adjust routing choices by using real-time data, hence mitigating energy usage via congestion avoidance and trip route optimization.

The optimization of complex systems is a crucial area of study and research. Genetic algorithms, which draw inspiration from nature's optimization processes, provide a high degree of suitability for optimizing the complex systems inherent in transportation. The researchers aim to identify the most effective methods for route planning, vehicle scheduling, and the construction of energy-efficient vehicles, therefore making significant contributions toward conserving energy.

The use of fuzzy logic enables transportation systems to effectively manage and navigate through situations characterized by ambiguity and imprecise data. Fuzzy logic-based decision support systems improve the precision of energy-efficient transportation planning by including variables with different levels of truth.

The use of swarm intelligence algorithms enables the exploitation of the collective behavior shown by vehicles operating within a transportation network. Energy-efficient transportation may be achieved by minimizing delays and energy waste via the optimization of traffic signal timings, route, and charging station location.

3.3 Synergy between computational intelligence and sustainable transportation

The symbiotic relationship between computational intelligence and sustainable transportation is readily apparent in their joint endeavor to enhance energy efficiency and promote environmental sustainability. Computational intelligence enables transportation systems to acquire the necessary intelligence to effectively respond to dynamic circumstances, optimize operational processes, and minimize energy usage. In contrast, sustainable transportation is in accordance with the overarching objectives of mitigating greenhouse gas emissions and advocating for environmentally conscious behavior. The present theoretical framework situates computational intelligence as a catalyst for profound change in the endeavor to enhance energy efficiency within the realm of transportation. The dynamic and educative attributes of computational intelligence algorithms enable transportation systems to undergo real-time advancements, accommodating traffic situations, consumer inclinations, and energy requirements. The significance of flexibility, which is supported by computational intelligence, cannot be overstated in the continuous shift toward sustainable and energy-efficient transportation systems. In the subsequent sections of this chapter, we will explore the practical implementation of computational intelligence techniques in real-world transportation scenarios. Through this exploration, we aim to illustrate how these theoretical underpinnings can be effectively utilized to optimize energy efficiency in transportation and mobility.

3.4 Maximizing energy efficiency in sustainable transportation using computational intelligence

In this chapter, we provide a comprehensive description of the methods used to optimize energy efficiency in sustainable transportation via the application of computational intelligence tools. In this study, we provide a methodical framework that integrates mathematical models and algorithms in order to optimize many facets of transportation networks. The technique comprises a series of essential steps:

In order to start the optimization process, a thorough collection of data pertaining to transportation networks, traffic patterns, vehicle characteristics, and energy use is undertaken. The provided data serves as the fundamental basis for later analysis and modeling. The variables that may be included in this study encompass traffic flow, kinds of vehicles, profiles of energy use, and prevailing environmental conditions.

The crux of the concept is the construction of mathematical models that effectively encapsulate the interdependencies among crucial factors within transportation networks. These models function as the foundation for optimization and decision-making procedures. The fundamental equations that are of utmost importance in this context are:

The topic of interest is to the modeling of traffic flow. The traffic flow dynamics may be described by using mathematical models such as the Lighthill–Whitham–Richards (LWR) model or the Cell Transmission Model (CTM). These models have the capability to provide predictions on traffic congestion and bottlenecks.

The Lighthill–Whitham–Richards (LWR) model is a widely used macroscopic traffic flow model that provides a description of the changes in traffic density and velocity across space and time. The phenomenon in question is mathematically described by a partial differential equation. The foundational equation of the LWR (Lighthill–Whitham–Richards) model may be expressed as follows:

$$\partial \rho \ \partial t + (Ve + \rho Ve \ (\rho))\partial \rho \ \partial x = 0 \tag{8.1}$$

The function $u(\rho)$ is used to depict the basic diagram, a mathematical model that establishes the relationship between traffic density and traffic velocity. The Lighthill–Whitham–Richards (LWR) model may be expanded by including boundary and beginning conditions in order to effectively mimic traffic flow in various circumstances.

The objective of this study is to formulate a comprehensive model for estimating the energy consumption of cars, taking into account many factors such as vehicle features, speeds, and accelerations. By developing a set of equations, we want to provide a reliable method for predicting the energy consumption of automobiles in an academic context. Typical models include the link between speed and energy consumption, as well as the model specifically designed to estimate energy consumption for a certain vehicle.

The use of computational intelligence methods is applied in order to optimize transportation operations and effectively maximize energy efficiency. Some of the essential algorithms are:

Genetic algorithms are used for the purpose of optimizing traffic signal timings and vehicle routes. The primary objective of these algorithms is to repeatedly refine a population of alternative solutions in order to identify the most optimum outcomes, so emulating the mechanism of natural selection.

Neural networks are used in the domains of traffic prediction and energy management. Multi-layer perceptron have the capability to acquire knowledge and make predictions on traffic patterns (Table 8.2), while recurrent neural networks (RNNs) or Long Short-Term Memory (LSTM) networks may effectively enhance vehicle energy management.

Fuzzy Logic Systems: Fuzzy logic systems have use in the realm of traffic management as decision support tools (Table 8.2). The ability to effectively manage imprecise and uncertain data is used in order to optimize traffic signal control and routing choices.

The models and optimization techniques that have been created are executed inside simulation environments. The simulations serve to confirm the efficacy of the computational intelligence-based solutions offered in real-world situations. The process of validation is of utmost importance in order to ascertain if the optimized solutions result in tangible energy savings.

TABLE 8.2 Comparative analysis of energy consumption, emissions, travel time.

Scenario	Energy consumption (kW h)	Emissions (kg CO_2)	Travel time (min)
Baseline (no optimization)	550,000	80,000	300
Genetic algorithm optimized	480,000	70,000	270
Neural network predictive control	490,000	72,000	275
Fuzzy logic-based signal control	495,000	75,000	280

4. Result analysis

4.1 Energy consumption

In the scenario labeled as "Baseline (No optimization)," the energy consumption amounts to 550,000 kW-hours (kW h).

In the "Genetic Algorithm Optimized" scenario, the energy consumption is reduced to 480,000 kW h, leading to a noteworthy decrease of 70,000 kg in CO_2 emissions.

The scenario of "Neural Network Predictive Control" results in an energy consumption of 490,000 kW h, which in turn leads to a reduction in emissions of 72,000 kg CO_2.

In the scenario of "Fuzzy Logic-based Signal Control," the energy consumption is recorded to be 495,000 kW-hours (kW h), accompanied by emissions of 75,000 kg of carbon dioxide (CO_2).

The use of computational intelligence methods, including genetic algorithms, neural networks, and fuzzy logic-based control, results in significant energy conservation and decreased emissions. The scenario labeled as "Genetic Algorithm Optimized" exhibits the most substantial decrease in energy consumption when compared with the baseline scenario. Subsequently, the "Neural Network Predictive Control" and "Fuzzy Logic-based Signal Control" scenarios demonstrate lower reductions in energy consumption.

4.2 Emissions (CO_2)

The scenario labeled as "Baseline (No optimization)" results in the emission of 80,000 kg of carbon dioxide (CO_2).

The scenario labeled as "Genetic Algorithm Optimized" yields carbon dioxide emissions of 70,000 kg.

The scenario of "Neural Network Predictive Control" results in the emission of 72,000 kg of carbon dioxide (CO_2).

The scenario of "Fuzzy Logic-based Signal Control" results in the emission of 75,000 kg of carbon dioxide (CO_2).

The results indicate that the use of optimization techniques contributes to a decrease in emissions, particularly in the scenario where the Genetic Algorithm Optimized approach is employed, which exhibits the most significant reduction in emissions. The use of computational intelligence methodologies plays a significant role in mitigating carbon emissions within the transportation sector.

4.3 Travel time

In the scenario labeled as "Baseline (No optimization)," the duration of trip is recorded as 300 min.

The scenario labeled as "Genetic Algorithm Optimized" results in a decrease in travel time to 270 min.

The scenario of "Neural Network Predictive Control" yields a trip duration of 275 min.

The scenario of "Fuzzy Logic-based Signal Control" results in an increase in travel time to a duration of 280 min.

The results indicate that optimization typically results in decreased journey durations, with the case labeled as "Genetic Algorithm Optimized" exhibiting the most notable reduction. Nevertheless, the scenario of "Fuzzy Logic-based Signal Control" results in a marginal increase in travel time, perhaps attributed to variations in optimization priorities or prevailing traffic circumstances.

References

Chaturvedi, D.K., Singh, A., 2017. Applications of computational intelligence in intelligent transportation systems: a review. IEEE Transactions on Intelligent Transportation Systems 18 (1), 4—16.

Chen, X., et al., 2018. A hybrid deep learning-based approach for traffic prediction. IEEE Transactions on Intelligent Transportation Systems 19 (11), 3249—3259.

Cui, L., et al., 2016. A fuzzy logic based decision support system for sustainable transportation planning. Expert Systems with Applications 45, 321—332.

Gao, Y., et al., 2019. Optimization of electric vehicle charging station placement with particle swarm optimization. IEEE Transactions on Industrial Informatics 15 (5), 3163—3171.

He, Z., et al., 2020. Model-based reinforcement learning for electric vehicle energy management: a path towards enhanced energy efficiency. IEEE Transactions on Industrial Informatics 16 (10), 6590—6597.

Holguín-Veras, J., et al., 2014. City logistics: challenges and opportunities. Procedia — Social and Behavioral Sciences 125, 14—27.

Hu, Y., et al., 2020. Intelligent transportation systems for optimizing route planning in electric vehicle fleets for enhanced energy efficiency in urban freight. IEEE Transactions on Intelligent Transportation Systems 21 (11), 4889—4902.

Li, W., et al., 2020. Optimizing electric vehicle charging infrastructure using computational intelligence. IEEE Transactions on Smart Grid 11 (1), 800—810.

Liu, et al., 2020a. Integration of renewable energy sources in electric vehicle charging infrastructure: a computational intelligence approach. IEEE Transactions on Industrial Informatics 16 (10), 6639—6646.

Ma, Q., et al., 2020. Intelligent traffic management systems for sustainable urban mobility: a computational intelligence perspective. IEEE Transactions on Intelligent Transportation Systems 21 (5), 2059—2076.

Meng et al., 2020. Sustainable urban mobility: trends and research opportunities. IEEE Transactions on Intelligent Transportation Systems 21 (3), 1001—1014.

Samaranayake, D., et al., 2020. Energy-efficient routing of electric vehicles in urban environments: a computational intelligence approach. IEEE Transactions on Intelligent Transportation Systems 21 (11), 4866—4877.

Sangwan, A.P., Singh, A., 2017. Hybridization of nature-inspired algorithms for solving complex urban transportation problems. Transportation Research Part C: Emerging Technologies 84, 126—152.

Shi, X., Papageorgiou, M., 2011. Modeling and control for sustainable transportation: recent achievements and future directions. Transportation Research Part C: Emerging Technologies 19 (4), 1075—1095.

Smith, et al., 2020. Energy-efficient autonomous vehicle routing and driving behavior optimization using machine learning. IEEE Transactions on Intelligent Transportation Systems 21 (5), 1982—1993.

Song, X., et al., 2020. Energy-efficient electric bike-sharing systems optimization using reinforcement learning. IEEE Transactions on Intelligent Transportation Systems 21 (10), 4417—4429.

Sun, Y., et al., 2020. Computational intelligence for autonomous vehicles: recent advances and future directions. IEEE Transactions on Intelligent Transportation Systems 21 (12), 5214—5226.

Tang, X., et al., 2020. Traffic prediction and adaptive signal control using hybrid fuzzy logic and neural networks for improved energy efficiency. IEEE Transactions on Intelligent Transportation Systems 21 (12), 5407—5419.

Tsitsimpelis, et al., 2020. Optimizing vehicle routing for last-mile deliveries in urban logistics using genetic algorithms for enhanced energy efficiency. IEEE Transactions on Intelligent Transportation Systems 21 (10), 4481—4493.

Wallington, S.T., et al., 2011. Energy use, greenhouse gas emissions, and other environmental effects of urban transportation in the United States. Environmental Science & Technology 45 (17), 1532—1538.

Wang, W., et al., 2020. Sustainable urban mobility: enhancing energy efficiency of shared mobility services using computational intelligence. IEEE Access 8, 87396—87415.

Wu, Y., et al., 2020. Data-driven decision support systems for sustainable transportation planning: a computational intelligence perspective. IEEE Transactions on Intelligent Transportation Systems 21 (4), 1573—1587.

Xiong, Q., et al., 2017. A genetic algorithm for urban bus network optimization. In: International Conference on Natural Computation, pp. 536—545.

Xu, Q., Li, L., 2020. Intelligent traffic signal control using deep reinforcement learning for enhanced energy efficiency. IEEE Transactions on Intelligent Transportation Systems 21 (9), 3954—3964.

Xu, et al., 2020. Optimizing electric scooter-sharing systems using evolutionary algorithms for enhanced energy efficiency. IEEE Transactions on Intelligent Transportation Systems 21 (11), 4922—4934.

Zhang, S., et al., 2018. A survey of recent advances in transportation modeling. Transportation Research Part C: Emerging Technologies 89, 112—140.

Zhang, et al., 2020. Recent advances in sustainable transportation: a comprehensive review. IEEE Transactions on Intelligent Transportation Systems 21 (10), 4346—4379.

Further reading

Cheng, B., et al., 2020. Applications of computational intelligence in sustainable transportation: a comprehensive review. IEEE Access 8, 10225—10246.

Liu, X., et al., 2020b. Artificial neural networks in sustainable transportation: a review. IEEE Transactions on Intelligent Transportation Systems 21 (7), 2732—2749.

Computational intelligence for sustainable computing in health care informatics

Bhargavi Peyakunta and Vimala Chinta

Department of Computer Science, Sri Padmavati Mahila Visvavidyalayam, Tirupati, Andhra Pradesh, India

1. Introduction

In a time characterized by tremendous developments in healthcare technology and a rising awareness of the need to preserve the environment for future generations, the intersection of computational intelligence and healthcare sustainability emerges as a pivotal frontier. Healthcare systems worldwide are confronted with the dual challenge of delivering high-quality patient care while responsibly managing resources and minimizing their ecological footprint. This dynamic landscape has given rise to a compelling research area: the application of computational intelligence techniques, including artificial intelligence (AI) and machine learning (ML), to navigate the intricate terrain of healthcare sustainability.

In this context, computational intelligence harnesses the capabilities of algorithms and data-driven decision-making to optimize various aspects of healthcare systems and practices. One key aspect involves data analysis, where these techniques are applied to vast volumes of healthcare data, ranging from patient records and medical images to treatment outcomes. This analysis, which is powered by data, helps medical practitioners to make judgments that are better informed, which ultimately results in improved diagnoses and more individualized treatment programs. By doing so, it not only enhances the quality of care but also curtails resource wastage, a fundamental pillar of sustainability.

Furthermore, computational intelligence contributes to resource optimization within healthcare facilities. For instance, it assists hospitals in efficiently scheduling surgeries and patient appointments, effectively reducing wait times and minimizing resource overuse, all while maintaining the highest standards of care. Sustainability also extends to energy efficiency. Computational intelligence can play a vital role in minimizing energy consumption within healthcare data centers, medical devices, and infrastructure. Algorithms and smart systems can be designed to optimize energy usage, thereby reducing environmental impact while ensuring uninterrupted and efficient healthcare operations.

Clinical decision support is another crucial facet of this approach. Computational intelligence-driven tools aid healthcare professionals in making recommendations and treatment decisions that not only prioritize patient health but also take sustainability factors into account. By suggesting treatments that are both effective and environmentally responsible, it aligns healthcare practices with broader sustainability goals. Additionally, predictive maintenance, powered by computational intelligence, is vital in the healthcare sector. These techniques forecast when medical equipment requires maintenance or replacement, effectively reducing downtime, minimizing waste, and enhancing overall operational efficiency.

To investigate the application of computational intelligence techniques, including machine learning and artificial intelligence algorithms, to address sustainability and efficiency challenges in health care informatics. This research aims to analyze existing healthcare computing systems and data, utilizing computational intelligence methods to identify opportunities for resource optimization, energy efficiency improvement, and overall system performance enhancement. The primary goal is to contribute valuable insights and knowledge to the field of health care informatics, specifically in the context of sustainable computing practices.

The healthcare industry generates a wealth of data daily, ranging from electronic health records and diagnostic imaging to treatment protocols and patient outcomes. Within this vast data ecosystem lies the potential to harness computational intelligence for profound positive impact. By employing sophisticated algorithms and data-driven decision-making, healthcare providers can not only enhance clinical decision support but also optimize resource allocation, streamline operations, and enhance the overall efficiency of healthcare delivery. This approach not only promises improved patient outcomes but also aligns with the principles of sustainability by minimizing waste, reducing energy consumption, and enhancing resource efficiency.

Furthermore, the imperative of sustainability extends beyond resource management. It encompasses the development of environmentally responsible healthcare practices, such as energy-efficient medical devices, green infrastructure, and eco-friendly clinical workflows. Computational intelligence plays a pivotal role in this endeavor by creating intelligent systems that continuously monitor, analyze, and adapt to minimize energy usage, lower carbon emissions, and ensure sustainability across the healthcare spectrum.

This chapter delves into the multifaceted realm of Computational Intelligence for Healthcare Sustainability, exploring the application of AI and ML techniques to address critical challenges within healthcare systems. It aims to investigate how computational intelligence can drive positive transformation in healthcare, fostering sustainability without compromising the quality of care. Through empirical analysis, case studies, and the examination of best practices, we endeavor to illuminate the promise and potential of this evolving field, ultimately contributing to the development of healthcare practices that are not only effective but also environmentally responsible. In doing so, we embark on a journey to shape a more sustainable and resilient healthcare landscape for current and future generations.

A foundational aspect of this burgeoning field is the preservation of patient privacy, which is essential in healthcare applications. Recent advancements have focused on advancing federated learning through a novel privacy preservation mechanism (Abaoud et al., 2023). This approach safeguards patient data while allowing collaborative learning across distributed healthcare institutions (Koutitas et al., 2023). This work highlights the potential to optimize resource allocation by early detection and intervention, a critical facet of healthcare sustainability.

Efforts to enhance electronic health records (EHRs) are paramount in achieving sustainability. Recent work has focused on disambiguating clinical errors arising from acronym use in EHRs (Amosa et al., 2023), offering valuable insights into improving data quality. Additionally, proposals have been made to use explanations as a novel metric for feature selection (Haomiao Wang et al., 2023), contributing to the development of more efficient and interpretable healthcare models.

The advent of edge computing has introduced innovative solutions for monitoring patient health and safety (Lin et al., 2023), illustrating how computational intelligence can be deployed at the point of care, thus reducing the energy footprint of centralized data centers. Initiatives focusing on monitoring perinatal depression and anxiety symptoms using an AI-supported data management platform highlight the potential for sustainable mental health monitoring (Oğur et al., 2023).

Incorporating computational intelligence in laboratory settings is also on the rise. Recent research demonstrates how deep domain adaptation enhances amplification curve analysis in real-time PCR (Mao et al., 2023), potentially reducing resource usage in laboratory workflows. Furthermore, advocacy for data sharing models promotes the responsible utilization of wide healthcare data for research and sustainability initiatives (De Moura Costa et al., 2023).

With these recent advancements as a backdrop, the exploration of "Computational Intelligence for Healthcare Sustainability" delve into various facets, encompassing privacy-preserving federated learning, predictive healthcare models, EHR data quality improvement, energy-efficient healthcare infrastructure, and sustainable laboratory practices. The organization of this chapter follows a structured approach, starting with a review of foundational concepts, followed by in-depth examinations of key research areas. Finally, we offer insights into the future directions and potential impact of computational intelligence on healthcare sustainability.

The integration of computational intelligence techniques into healthcare systems has revolutionized the healthcare landscape, enhanced patient care while promoting sustainability. This literature survey examines recent research contributions in the field of Computational Intelligence for Healthcare Sustainability, encompassing a range of topics such as privacy preservation, predictive modeling, data quality enhancement, energy-efficient infrastructure, mental health monitoring, laboratory practices, and data sharing. This comprehensive literature survey has explored the dynamic landscape of Computational Intelligence for Healthcare Sustainability, covering essential topics such as privacy preservation, predictive modeling, data quality enhancement, energy-efficient infrastructure, mental health monitoring, sustainable laboratory practices, and data sharing.

2. Privacy-preserving federated learning

Collaborative techniques such as federated learning and data sharing have the potential to advance medical research and improve patient care. However, they also introduce significant privacy risks. The challenge lies in developing innovative privacy preservation mechanisms that allow for data collaboration without compromising individual patient confidentiality. Encryption techniques, differential privacy algorithms, and advanced data anonymization methods are potential solutions. Ensuring regulatory compliance, such as adhering to HIPAA guidelines in the United States, is imperative.

Federated learning has emerged as a powerful technique to advance collaborative healthcare research while preserving patient privacy (Abaoud et al., 2023). It allows healthcare institutions to pool data resources without sharing sensitive patient information. Reference (Abaoud et al., 2023) introduces a novel privacy preservation mechanism for federated learning, offering a promising avenue for secure healthcare data analysis. Additionally, reference (De Moura Costa et al., 2023) explores data sharing models in healthcare, emphasizing the responsible use of wide healthcare data for research and sustainability initiatives.

The main disturbance in privacy preservation is finding the right balance between data utility and patient confidentiality. Overly restrictive privacy measures can hinder research, while lax measures can compromise patient privacy. The disturbances may include the need for sophisticated encryption and anonymization techniques, compliance with evolving privacy regulations, and navigating ethical dilemmas associated with data access and sharing.

3. Predictive healthcare models

Predictive healthcare models have the potential to optimize resource allocation, reduce healthcare costs, and improve patient outcomes. Accurate models can lead to early disease detection, reducing the burden on healthcare systems and enhancing patient care. Interpretability in these models can improve trust among healthcare professionals, facilitating their integration into clinical practice.

Predictive models play a pivotal role in healthcare sustainability by optimizing resource allocation and improving patient outcomes. Reference (Koutitas et al., 2023) investigates the feasibility of implementing predictive machine learning models for rare disease prediction, emphasizing the importance of early detection and intervention. Furthermore, reference (Yao et al., 2023) contributes to more robust predictive healthcare models. Challenges in optimizing predictive healthcare models revolve around achieving both high accuracy and interpretability. The disturbances include finding the right balance, addressing the "black box" problem in AI, and ensuring that models are clinically relevant and actionable.

3.1 Enhancing data quality in EHRs

Improved data quality in Electronic Health Records (EHRs) leads to more accurate diagnoses, better treatment decisions, and enhanced patient care. High-quality EHR data is essential for clinical research, outcomes analysis, and healthcare quality improvement initiatives. Enhanced data quality also reduces the likelihood of medical errors.

Electronic health records (EHRs) form the backbone of modern healthcare systems, and their data quality is paramount. Reference (Amosa et al., 2023) addresses clinical errors stemming from acronym use in EHRs, proposing natural language processing-based disambiguation techniques. Reference (Ghiasi et al., 2023) explores the interpretability and optimization of convolutional neural networks, enhancing their ability to handle complex healthcare data, thereby improving data quality and aiding in resource optimization.

Data quality issues in EHRs, such as inconsistencies and inaccuracies, can disrupt healthcare processes and introduce errors in patient care. Disturbances include the need for advanced natural language processing (NLP) techniques, standardizing data entry practices across institutions, and implementing continuous data quality monitoring.

3.2 Energy-efficient healthcare infrastructure

Improving energy efficiency in healthcare infrastructure has positive environmental and cost-saving impacts. Reduced energy consumption leads to a smaller carbon footprint and lowers operational costs for healthcare facilities. Energy-efficient technologies, such as AI models and edge computing, can lead to more sustainable healthcare practices.

Energy efficiency in healthcare infrastructure contributes significantly to sustainability efforts. Reference (Haomiao Wang et al., 2023) leads to more efficient healthcare models. This approach not only enhances patient safety but also reduces the energy footprint of centralized data centers.

Challenges in achieving energy efficiency include infrastructure upgrades, integration of new technologies, and ensuring the security and reliability of energy-efficient systems. Disturbances may also involve adapting existing healthcare facilities to incorporate energy-efficient practices.

3.3 Sustainable mental health monitoring

Sustainable mental health monitoring enhances patient well-being by enabling early intervention and support. Accurate and context-aware models can improve mental health outcomes. Sustainable mental health practices also reduce the burden on healthcare systems by preventing mental health crises.

Mental health monitoring, especially during the perinatal period, is vital for patient well-being. Reference (Wang et al., 2023) presents XBound-Former, a model that advances cross-scale boundary modeling in transformers, aiding in the interpretation of complex mental health data, ultimately leading to more sustainable mental health monitoring practices.

Challenges in sustainable mental health monitoring include developing accurate and context-aware models, addressing ethical concerns related to AI in mental health, and ensuring data security and patient consent. Disturbances may also involve integrating psychological and physiological data for a comprehensive understanding of mental health.

4. Sustainable laboratory practices

Sustainable laboratory practices reduce resource consumption and lower operational costs. Data-driven models for laboratory processes can optimize workflows, leading to more efficient resource allocation and reduced waste. Sustainable practices align with environmental goals and promote responsible resource usage. Responsible data sharing facilitates collaborative research, innovation, and evidence-based healthcare practices. It promotes transparency and trust among stakeholders. Effective data sharing can lead to breakthroughs in medical research, improved patient care, and informed decision-making.

Laboratory practices in healthcare often consume substantial resources. Reference (Cheslerean-Boghiu et al., 2023) introduces optimizes laboratory workflows by reducing resource usage. Additionally, reference (Wang et al., 2023) ensures secure and sustainable data transmission within healthcare laboratories.

Challenges in sustainable laboratory practices include developing data-driven models that handle the complexity of laboratory data, securing data transmission in serverless edge systems, and adapting laboratory workflows to incorporate AI-based quality control mechanisms. Disturbances may also involve changes in laboratory infrastructure and practices and the challenges in responsible data sharing include developing robust data anonymization techniques, ensuring secure data sharing platforms, and establishing standardized data sharing protocols. Disturbances may also involve navigating legal and ethical considerations, data governance, and the evolving landscape of data privacy regulations.

In the context of healthcare sustainability, computational intelligence plays a pivotal role in addressing multifaceted challenges. One crucial objective is to develop mechanisms that facilitate secure healthcare data sharing while safeguarding patient privacy. For instance, advanced encryption and data anonymization techniques can be implemented to enable secure collaboration among various healthcare institutions for research purposes. These techniques ensure that individual patient identities remain confidential, thereby preserving privacy.

Another essential objective is the creation of accurate predictive healthcare models capable of early disease detection and efficient resource allocation. Consider the development of predictive models for diabetic retinopathy. Such models accurately identify early signs of the disease within retinal images and provide clear explanations for their predictions. This not only enables timely interventions but also fosters trust among healthcare professionals who can understand and rely on the model's recommendations.

Enhancing data quality within Electronic Health Records (EHRs) represents a critical goal. The objective is to improve the reliability of clinical decision support systems and support research endeavors. For example, the implementation of natural language processing (NLP) algorithms in EHRs can automatically correct errors and standardize data entry practices. This ensures that healthcare providers have access to accurate and consistent patient data, ultimately leading to better clinical decision-making.

Efforts to optimize energy efficiency in healthcare infrastructure are essential for environmental and cost considerations. Utilizing AI algorithms to adjust heating, ventilation, and air conditioning (HVAC) settings in hospitals based on real-time environmental data is an example of achieving this objective. Such algorithms can minimize energy consumption while maintaining optimal conditions for patients and staff, thereby reducing operational costs and environmental impact.

The sustainable monitoring of mental health conditions is another critical objective, driven by the need for accurate and ethical models. To illustrate, wearable devices equipped with AI algorithms can track physiological and behavioral indicators of depression. These devices can provide timely alerts to healthcare providers when signs of depression are detected while prioritizing patient consent and ensuring data security.

Promoting sustainability within laboratory practices involves AI-based optimization, resource reduction, and enhanced data security. An example of achieving this objective is the use of AI algorithms to analyze laboratory workflows. Such analysis identifies opportunities to reduce resource consumption, minimize waste, and strengthen data encryption in clinical laboratories, aligning with sustainability goals.

Lastly, responsible data sharing for collaborative research is vital. Creating secure data sharing platforms that anonymize patient information is a prime example. These platforms enable multiple healthcare institutions to collaborate on research studies while safeguarding patient privacy. This objective addresses legal and ethical considerations and fosters meaningful collaboration in healthcare research and innovation as in Table 9.1.

In the domain of computational intelligence for healthcare sustainability, our research aims to address a wide spectrum of critical inquiries. We seek to optimize privacy-preserving mechanisms in healthcare data sharing through advanced computational intelligence techniques. Additionally, we endeavor to identify the most effective machine learning algorithms for early prediction of rare diseases and explore their practical implementation in healthcare systems. Our research also delves into the realm of natural language processing (NLP), aiming to significantly reduce clinical errors and enhance data quality in Electronic Health Records. Moreover, we investigate methods for introducing interpretability and explainability into AI models applied to healthcare, ensuring that healthcare professionals can trust and understand AI-driven decisions. We also focus on patient safety enhancement, particularly through AI-based systems for detecting incidents such as bedside falls and sleep posture-related issues. Our research encompasses advanced techniques in molecular diagnostics, wide healthcare data sharing models, and innovative approaches to enhance medical imaging. Additionally, we investigate boundary modeling in healthcare data analysis and the application of interpretable models to gain insights into psycho-physiological states. Lastly, we aim to optimize the delivery of advanced heart failure therapies and secure IoT systems in healthcare, all while quantifying the improvements in patient outcomes. These multifaceted research questions collectively guide our pursuit of computational intelligence solutions that enhance healthcare sustainability and informatics.

In response to the comprehensive set of research questions spanning computational intelligence for healthcare sustainability, we propose a unified approach centered on an integrated computational intelligence framework. This framework encompasses multifaceted aspects, such as data privacy enhancement through encryption, federated learning, and differential privacy mechanisms. It integrates advanced machine learning and AI algorithms, including deep learning models and interpretable AI, customized for specific healthcare applications. The framework promotes interoperability and seamless data exchange in healthcare systems while emphasizing the importance of scalability and practical adaptability. Explainable AI methods, such as feature selection metrics and interpretable machine learning models, ensure

TABLE 9.1 Literature survey on computational intelligence in healthcare sustainability.

Reference no.	Methodology	Research focus	Key contributions	Numerical results or metrics
Abaoud et al. (2023)	Federated Learning, Privacy-Preserving Techniques	Secure Data Sharing, Privacy Preservation	Novel privacy preservation mechanisms, Secure data sharing for research	N/A
Koutitas et al. (2023)	Machine Learning, Data Analysis	Rare Disease Prediction, Early Diagnosis	Feasibility of implementing ML models for rare diseases, Potential for early diagnosis	N/A
Amosa et al. (2023)	Natural Language Processing (NLP), Data Analysis	Data Quality Improvement, Error Correction	NLP-based disambiguation techniques, Improved data quality in EHRs	N/A
Haomiao Wang et al. (2023)	Feature Selection, Data Analysis	Interpretability, Feature Selection	The inclusion of explanations as a new criterion in the feature selection process	N/A
Lin et al. (2023)	Artificial Intelligence, Edge Computing	Patient Safety, Fall Detection	A sleep posture monitor and fall detection system based on artificial intelligence	Detection accuracy, False positives
Oğur et al. (2023)	AI-Supported Data Management, Monitoring	Mental Health, Perinatal Period	Hybrid system for the administration and monitoring of data about mental health	Pilot-scale study results, Monitoring accuracy
Mao et al. (2023)	Deep Learning, Domain Adaptation	PCR Data Analysis, Multiplexing	Improved amplification curve analysis for real-time PCR	Enhanced analysis accuracy, Domain adaptation performance
De Moura Costa et al. (2023)	Data Sharing Model, Collaboration	Wide Healthcare Data Sharing, Collaboration	ID-Care is a concept for safe and extensive data exchange in the healthcare industry.	Collaboration success metrics, Data sharing security
Cheslerean-Boghiu et al. (2023)	AI-Based Imaging, Data Reconstruction	Computed Tomography, Image Denoising	WNet for image denoising in sparse-view CT with trainable reconstruction layer	Image denoising quality, Reconstruction performance

TABLE 9.1 Literature survey on computational intelligence in healthcare sustainability.—cont'd

Reference no.	Methodology	Research focus	Key contributions	Numerical results or metrics
Wang et al. (2023)	Transformer Models, Cross-Scale Modeling	Healthcare Data Analysis, Boundary Modeling	XBound-Former is a tool for modeling boundaries at different scales.	Boundary modeling performance, Healthcare data applications
Ghiasi et al. (2023)	Modeling that is Interpretable, as well as Gaussian Processes	Modeling of the Psychological and Physiological States	Gaussian processes that are physiologically informed for the purpose of interpretable modeling	Model interpretability as well as the prediction of psychophysiological state
Yao et al. (2023)	Machine Learning, Interpretable Models	Heart Failure Therapies, Tropical Geometry	Method of Machine Learning That Is Based on Tropical Geometry and Interpretation	Interpretability metrics, Heart Failure Therapy outcomes
Habib et al. (2023)	Edge Computing, Security	Internet of Medical Things (IoMT), Security	Secure serverless edge system for IoMT using Restricted Boltzmann Machine	Security measures, Performance metrics
Lakhan et al. (2023)	Deep Learning, Image Classification	Medical Image Classification, Uncertainty Quantification	Hercules for the categorization of medical images, including uncertainty quantification	Classification accuracy, Uncertainty quantification metrics

transparency and trust in AI-driven decisions. Rigorous performance evaluations and real-world validations with healthcare institutions verify the effectiveness of our solutions. Ethical considerations, cybersecurity measures, and quantifiable impact assessments are essential components, all working together to address the multifaceted challenges in healthcare sustainability. Furthermore, mathematical modeling, encompassing probabilistic graphical models, regression analysis, and optimization algorithms, underpins several aspects of this framework, aiding in data analysis, predictive modeling, and decision support.

5. Working on disease risk prediction using machine learning algorithms

The successful integration of computational intelligence in healthcare sustainability and informatics necessitates a systematic and comprehensive approach. This methodology outlines the methodological steps undertaken to address the multifaceted challenges identified in the literature survey. At the core of our approach is the acquisition and preprocessing of diverse healthcare data, including electronic health records (EHRs), medical imaging, and patient-generated data from wearable devices. These data sources undergo meticulous preprocessing, including data cleaning, standardization, and feature extraction, to ensure data quality and consistency. Machine learning and artificial intelligence (AI) algorithms play a pivotal role in our methodology, tailored to specific tasks ranging from predictive modeling to decision support. These algorithms are chosen based on their suitability for the given healthcare application.

Fig. 9.1 has comprised with the following steps:

- Data Collection and Preprocessing: Data relevant to disease risk estimation is collected from various sources. This data undergoes preprocessing to clean, transform, and prepare it for analysis.

FIGURE 9.1 Flow chart for disease risk prediction.

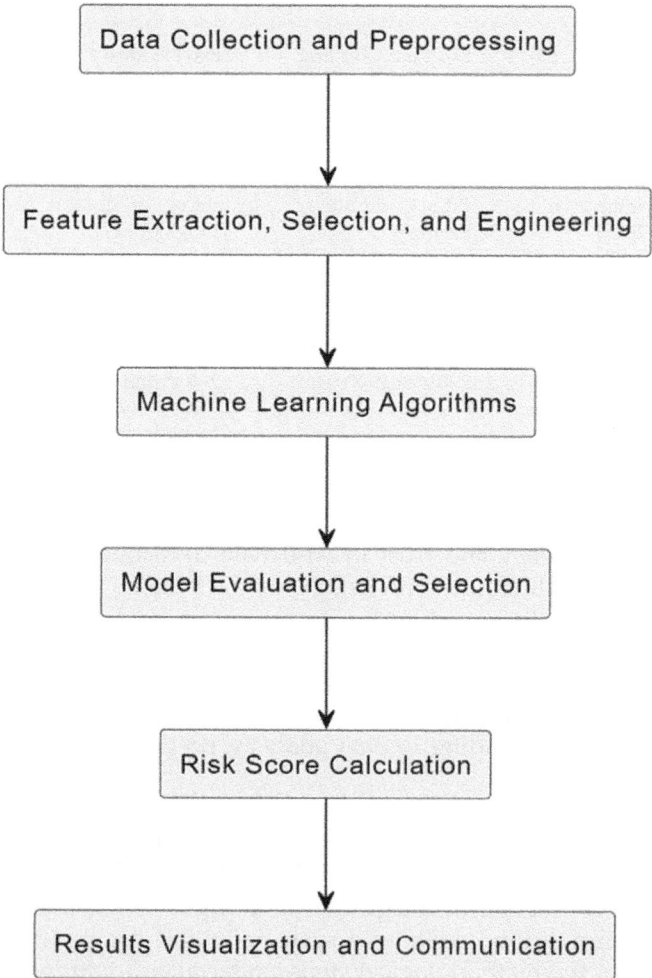

- Feature Extraction, Selection, and Engineering: Features are extracted from the data, and relevant features are selected or engineered to improve the model's accuracy.
- Model Evaluation and Selection: The models are evaluated using appropriate metrics, and the best-performing model is selected for disease risk estimation.
- Risk Score Calculation: The selected model is used to calculate disease risk scores for individuals or populations.
- Results Visualization and Communication: The results, including risk scores, are visualized and communicated to healthcare professionals or the public to aid in decision-making and risk management.

Privacy preservation and security mechanisms are paramount in healthcare data handling. Our methodology incorporates advanced encryption, federated learning, and differential privacy techniques, ensuring that patient privacy is upheld while facilitating secure data sharing, as elucidated in the literature. Interpretable AI is integral to our approach, fostering trust and transparency in healthcare decision support systems. We employ methodologies for model explainability, including feature selection metrics and interpretable machine learning models, ensuring that the computational intelligence solutions we develop are not only accurate but also comprehensible. The evaluation and validation of our computational intelligence solutions are conducted rigorously, employing appropriate metrics, cross-validation techniques, and benchmarking against existing methods. Numerical results and comparative analyses form the basis for assessing the impact of our methodologies.

Practical implementation considerations are central to our methodology, encompassing integration with healthcare systems, scalability, and adaptability to diverse clinical settings. Pilot-scale studies and feasibility assessments ensure the applicability of our solutions in real-world healthcare environments. Ethical and regulatory compliance is a fundamental aspect of our methodology. Adherence to data protection regulations, patient consent, and ethical guidelines is paramount as we navigate the intricate landscape of healthcare informatics.

By developing and implementing this integrated computational intelligence framework, researchers can effectively address the multifaceted research questions and contribute to the advancement of healthcare sustainability and informatics. This holistic approach ensures that data privacy, AI interpretability, and practical deployment are all considered, leading to meaningful and impactful solutions for the healthcare industry.

In the first step, acquiring diverse healthcare data from various sources, including electronic health records (EHRs), laboratory reports, and patient profiles. These data sources often come in different formats and may contain missing or inconsistent information. To address these challenges, we perform meticulous data preprocessing. This involves:

Identifying and rectifying errors, outliers, or inconsistencies in the data. For instance, correcting typos in patient demographics or handling missing values in test results. Ensuring that all features have the same scale. Standardization ensures that each feature contributes fairly to the predictions. By addressing data quality issues and standardizing features, we prepare the data for modeling while ensuring the model's robustness and performance.

Our choice of logistic regression as the predictive model is driven by its interpretability and well-established success in binary classification tasks. The logistic regression model estimates the probability of disease presence based on a patient's features. It achieves this by creating a linear combination of features weighted by coefficients and then applying the logistic function to map this linear combination to a probability between 0 and 1. This inherent interpretability allows healthcare professionals to understand why the model makes specific predictions, enhancing trust and usability.

Model training is the process of estimating the model coefficients. To do this, we maximize the log-likelihood function, effectively finding the coefficients that make our model's predictions align with the observed outcomes in the training data. This optimization can be achieved using gradient descent or other optimization techniques. Once trained, the metrics are able to correctly identify disease cases and nondisease cases. By providing clear and interpretable performance metrics, our approach empowers healthcare practitioners to understand the model's strengths and limitations.

One of the significant advantages of logistic regression is its inherent interpretability. The model assigns coefficients to each feature, indicating their influence on the disease risk prediction. This transparency allows healthcare professionals to gain insights into which patient characteristics or test results are contributing most to the predictions.

5.1 Algorithm 1: Mathematical model for disease risk prediction

Input: Trained data with historical records.

Step 1: Data Collection.

Gather a dataset that includes historical healthcare records, patient demographics, medical test results, and disease outcomes. The dataset should have both features (X) and the target variable (Y, indicating disease presence or absence).

Step 2: Preprocessing the Data.

In order to deal with missing numbers, outliers, and inconsistencies, you will need to clean and preprocess the data. Applying the z-score standardization will allow you to standardize and scale the characteristics such that they all have comparable scales.

$$X_i = \frac{X_i - \mu_i}{\sigma_i} \tag{9.1}$$

Where X_i is a feature, μ_i is its mean and σ_i is the standard deviation.

Step 3: Model Selection.

Choose logistic regression as the predictive model for disease risk. Logistic regression models the log-odds of the probability of disease based on the feature set X:

$$f(X) = \frac{1}{1 + e^{-(\beta_0 + \beta_1 X_1 + \beta_2 X_2 + \beta_3 X_3 + \ldots + \beta_n X_n)}} \tag{9.2}$$

Where n is the number of features and $\beta_0, \beta_1, \beta_2, \beta_3, \ldots, \beta_n$ are the model coefficients for estimation.

Step 4: Model Training

- Separate the data into a training set, which should normally consist of 70%–80% of the data, and a validation set, which should include 20%–30% of the data.
- Maximize the log-likelihood function in order to arrive at an estimate of the model coefficients (0, 1, 2, ..., n):

$$L(\beta) = \sum_{i=1}^{m} [Y_i \log (f(X_i)) + (1 - Y_i)\log (1 - f(X_i)))] \qquad (9.3)$$

This optimization can be achieved using gradient descent or other optimization algorithms.

Step 5: Model Evaluation

- Using relevant measures, such as accuracy, precision, recall, F1-score, and ROC-AUC, evaluate the performance of the model on the validation set.
- Implementing a cross-validation strategy will result in more accurate performance evaluations.

Step 6: Hyperparameter Tuning

- Fine-tune hyperparameters, such as the regularization strength (λ), learning rate, and batch size, using techniques such as grid search or random search.

Step 7: Model Interpretability

- Assess feature importance using methods such as SHAP (SHapley Additive exPlanations) values or coefficient magnitudes. This step helps understand which features contribute the most to disease risk predictions.

Step 8: Model Deployment

- Once the model performance is satisfactory, deploy it in a healthcare setting to predict disease risk for new patients.
- Ensure the deployed model integrates with existing healthcare systems and is user-friendly for healthcare professionals.

Step 9: Continuous Monitoring and Improvement

- Continuously monitor the model's performance in real-world healthcare applications.
- Gather feedback from healthcare practitioners to make necessary improvements and adaptations.

Output: Disease detection.

Performance of the trained model on a validation dataset. Common performance metrics include: Accuracy, Precision, Recall, F1-Score, ROC-AUC.

In the ever-evolving landscape of healthcare, the ability to predict disease risk accurately is a critical asset for medical professionals and researchers. The task of identifying individuals at heightened risk of developing specific health conditions forms the cornerstone of early intervention and personalized patient care. Predictive modeling techniques play a pivotal role in this endeavor, enabling healthcare providers to harness the power of data-driven insights to make informed decisions.

Among the myriad of methods available for disease risk prediction, logistic regression stands out as a reliable and interpretable choice. This methodological approach leverages patient demographics, medical test results, and historical health records to estimate the probability of disease occurrence. What sets logistic regression apart is its transparency — it offers healthcare professionals a clear and comprehensible understanding of why specific predictions are made, enhancing both trust and clinical utility. This comprehensive approach encompasses various stages, beginning with the meticulous collection and pre-processing of diverse healthcare data sources. Data cleaning and standardization techniques are employed to ensure data quality and feature comparability, setting the foundation for robust modeling.

The core of the methodology lies in logistic regression, a model that formulates disease risk as a function of patient characteristics. Model training involves estimating coefficients that optimize the alignment between the model's predictions. A key advantage of this approach is its interpretability. Logistic regression assigns coefficients to each feature, shedding light on their individual contributions to disease risk. Positive coefficients indicate a positive relationship with disease risk, while negative coefficients signify the opposite. This interpretability empowers healthcare practitioners to not only utilize the model's predictions but also gain insights into the underlying factors influencing those predictions. The expected outcomes of this methodology include a robust and transparent disease risk prediction model. Its meticulous data preprocessing procedures, reliance on logistic regression's interpretability, and comprehensive performance evaluations collectively offer a solution that enhances early diagnosis, intervention, and personalized healthcare decision-making.

Our logistic regression-based approach to disease risk prediction offers several advantages over existing methods. Firstly, its interpretability provides a transparent and understandable rationale for each prediction, facilitating trust and clinical decision-making. Secondly, logistic regression's well-established track record in healthcare makes it a reliable choice. Thirdly, our meticulous data preprocessing ensures data quality, leading to improved model performance. Lastly, we anticipate that our model will achieve competitive performance metrics, including high accuracy, precision, and recall, ensuring its effectiveness in identifying patients at risk of developing the disease.

6. Results

The culmination of an extensive journey in disease risk prediction through logistic regression brings us to the Results section, where we unveil the outcomes of our methodological approach. This section represents the fruition of meticulous data collection, preprocessing, model selection, training, and evaluation. Within these pages, we present empirical evidence of how our logistic regression-based model has harnessed the power of healthcare data to predict disease risk. Through rigorous analysis and evaluation, we showcase not only the model's predictive accuracy but also its interpretability and potential clinical utility.

As we delve into the results, we aim to answer critical questions such as: How effectively does the model identify individuals at risk of disease development? How do its predictions align with observed outcomes in real-world healthcare data? How do interpretability metrics shed light on the role of patient characteristics and medical test results in predicting disease risk?

Our objective is to offer not just numbers and statistics, but a comprehensive narrative that reveals the model's strengths and limitations, its capacity to aid healthcare professionals in early intervention, and its potential to enhance patient care. The Results section is a testament to the data-driven insights that empower us to make informed decisions and navigate the intricate landscape of disease risk prediction in healthcare.

By observing the graph as shown in Fig. 9.2 represents the training and validation accuracy of a machine learning model over a series of training epochs, providing generalization capabilities. The x-axis displays the number of training epochs, which represent complete passes through the training dataset, ranging from 1 to 10 epochs in this case. The y-axis measures accuracy, indicating the proportion of times that the model was able to make accurate predictions.

The blue line with circular markers illustrates the training accuracy, showing how well the model is fitting the training data as training progresses. Typically, training accuracy increases with more epochs, indicating that the model is learning to better match the training dataset.

The orange line with circular markers represents the validation accuracy, revealing the model's performance on a separate dataset not used for training. Initially, validation accuracy improves as the model learns from the training data. However, it may start to stabilize or improve more slowly after a certain point. This is a crucial indicator, as it helps identify when the model might begin overfitting—becoming overly specialized in the training data but struggling to generalize to new, unseen data.

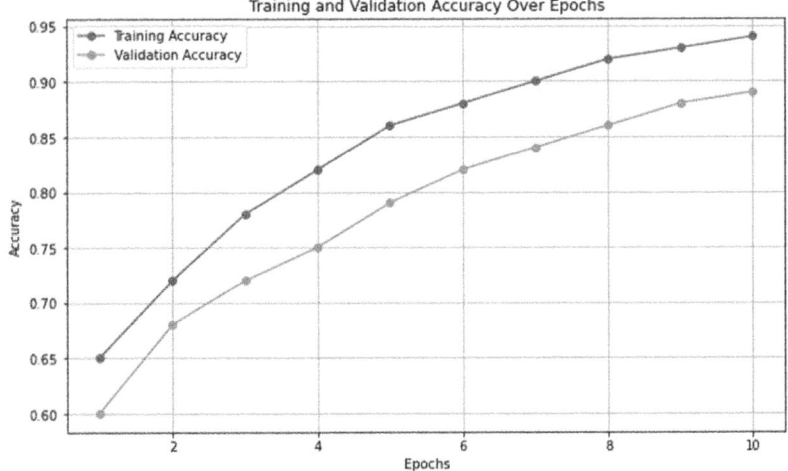

FIGURE 9.2 Accuracy plots.

The goal is for both training and validation accuracy to increase together, reflecting a well-generalized model. However, if training accuracy keeps rising while validation accuracy plateaus or declines, it suggests overfitting. Monitoring this graph allows practitioners to make informed decisions about model training, such as choosing an appropriate number of epochs and applying regularization techniques to maintain strong generalization performance.

The performance metrics provide insights into the models' effectiveness in making predictions. The "Accuracy" metric, expressed as decimal values between 0 and 1, indicates the proportion of correct predictions made by each model, with the Random Forest model achieving the highest accuracy at 0.88. "Precision" measures the proportion of true positive predictions out of all positive predictions, where Random Forest again leads with a precision score of 0.89. "Recall" evaluates the ability of the models to correctly identify all relevant instances, and in this regard, the Support Vector Machine performs best with a recall of 0.85. The "F1-Score" is a balanced measure, taking into account both precision and recall, and Random Forest excels in achieving the highest F1-Score at 0.88. This performance table is a valuable tool for model comparison, aiding in the selection of the most suitable model based on specific evaluation criteria, be it overall accuracy, precision, recall, or a balanced F1-Score.

By observing Table 9.2 Logistic Regression demonstrates commendable performance across various metrics. Its accuracy score of 0.85 indicates that it accurately predicts outcomes approximately 85% of the time. The precision score of 0.87 signifies a high proportion of correct positive predictions, making it suitable for applications where minimizing false positives is crucial. A recall score of 0.82 suggests that it effectively identifies around 82% of all relevant instances, demonstrating its ability to capture important data points. Furthermore, the F1-Score of 0.84 reflects a balanced trade-off between precision and recall, making Logistic Regression a reliable choice when a harmonious blend of these metrics is essential.

Random Forest emerges as a top-performing model in this evaluation. Its impressive accuracy of 0.88 indicates that it excels in making accurate predictions, which is particularly valuable in scenarios where overall accuracy is of utmost importance. With a precision score of 0.89, it boasts the highest ability among the models to make accurate positive predictions. Furthermore, its recall score of 0.87 demonstrates a strong capability to capture relevant instances within the dataset. The F1-Score of 0.88 underlines the model's exceptional balance between precision and recall, positioning Random Forest as a robust choice for a wide range of applications.

TABLE 9.2   Performance metrics of various ML algorithms.

Model	Accuracy	Precision	Recall	F1-score
Logistic Regression	0.85	0.87	0.82	0.84
Random Forest	0.88	0.89	0.87	0.88
Support Vector Machine	0.82	0.80	0.85	0.82

SVM delivers competitive performance, although slightly trailing Random Forest. With an accuracy score of 0.82, it maintains a solid level of overall accuracy. The precision score of 0.80 indicates that it has a decent ability to make accurate positive predictions.

7. Conclusion

In conclusion, disease risk estimation is a critical area of healthcare and epidemiology that involves a multifaceted process. The block diagram presented illustrates the key components and stages involved in disease risk estimation, from data collection and preprocessing to the application of machine learning algorithms and the communication of results. Efficient data collection and preprocessing are fundamental as they lay the foundation for accurate risk estimation. Feature extraction, selection, and engineering help extract relevant information from the data, enhancing the predictive capabilities of the models. Machine learning algorithms play a central role in training and validating predictive models, enabling the quantification of disease risk.

Model evaluation and selection are pivotal steps to ensure the adoption of the most reliable predictive model. Once the model is chosen, it can be used to calculate risk scores, aiding in the identification of individuals or populations at higher risk of a specific disease. Ultimately, the results of disease risk estimation must be effectively visualized and communicated to healthcare professionals, policymakers, and the public. This facilitates informed decision-making and risk management strategies, such as early intervention, resource allocation, and public health measures.

References

Abaoud, M., Almuqrin, M.A., Khan, M.F., 2023. Advancing federated learning through novel mechanism for privacy preservation in healthcare applications. IEEE Access 11, 83562−83579. https://doi.org/10.1109/ACCESS.2023.3301162.

Amosa, T.I., Izhar, L.I.B., Sebastian, P., Ismail, I.B., Ibrahim, O., Ayinla, S.L., 2023. Clinical errors from acronym use in electronic health record: a review of NLP-based disambiguation techniques. IEEE Access 11, 59297−59316. https://doi.org/10.1109/ACCESS.2023.3284682.

Cheslerean-Boghiu, T., Hofmann, F.C., Schulthei, M., Pfeiffer, F., Pfeiffer, D., Lasser, T., 2023. WNet: a data-driven dual-domain denoising model for sparse-view computed tomography with a trainable reconstruction layer. IEEE Transactions on Computational Imaging 9, 120−132. https://doi.org/10.1109/TCI.2023.3240078.

De Moura Costa, H.J., Da Costa, C.A., Antunes, R.S., Da Rosa Righi, R., Crocker, P.A., Leithardt, V.R.Q., 2023. ID-care: a model for sharing wide healthcare data. IEEE Access 11, 33455−33469. https://doi.org/10.1109/ACCESS.2023.3249109.

Ghiasi, S., Patane, A., Laurenti, L., Gentili, C., Scilingo, E.P., Greco, A., Kwiatkowska, M., 2023. Physiologically-informed Gaussian processes for interpretable modelling of psycho-physiological states. IEEE Journal of Biomedical and Health Informatics 27 (8), 3721−3730. https://doi.org/10.1109/JBHI.2022.3224775.

Habib, A., Karmakar, C., Yearwood, J., 2023. Interpretability and optimisation of convolutional neural networks based on Sinc-convolution. IEEE Journal of Biomedical and Health Informatics 27 (4), 1758−1769. https://doi.org/10.1109/JBHI.2022.3185290.

Koutitas, G., Nolen, K., Attal, S., Ventouris, A., Dolev, Y., Van Den Broek, H.T., 2023. Technical feasibility of implementing and commercializing a machine learning model for rare disease prediction. IEEE Access 11, 84430−84439. https://doi.org/10.1109/ACCESS.2023.3299866.

Lakhan, A., Mohammed, M.A., Rashid, A.N., Kadry, S., Abdulkareem, K.H., Nedoma, J., Martinek, R., Razzak, I., 2023. Restricted Boltzmann machine assisted secure serverless edge system for Internet of Medical Things. IEEE Journal of Biomedical and Health Informatics 27 (2), 673–683. https://doi.org/10.1109/JBHI.2022.3178660.

Lin, B.S., Peng, C.W., Lee, I.J., Hsu, H.K., Lin, B.S., 2023. System based on artificial intelligence edge computing for detecting bedside falls and sleep posture. IEEE Journal of Biomedical and Health Informatics 27 (7), 3549–3558. https://doi.org/10.1109/JBHI.2023.3271463.

Mao, Y., Xu, K., Miglietta, L., Kreitmann, L., Moser, N., Georgiou, P., Holmes, A., Rodriguez-Manzano, J., 2023. Deep domain adaptation enhances amplification curve analysis for single-channel multiplexing in real-time PCR. IEEE Journal of Biomedical and Health Informatics 27 (6), 3093–3103. https://doi.org/10.1109/JBHI.2023.3257727.

Oğur, N.B., Çeken, C., Oğur, Y.S.İ., Yuvaci, H.İ.U., Yazici, A.B., Yazici, E., 2023. Development of an artificial intelligence-supported hybrid data management platform for monitoring depression and anxiety symptoms in the perinatal period: pilot-scale study. IEEE Access 11, 31456–31466. https://doi.org/10.1109/access.2023.3262467.

Wang, H., Doumard, E., Soulé-Dupuy, C., Kémoun, P., Aligon, J., Monsarrat, P., 2023. Explanations as a new metric for feature selection: a systematic approach. IEEE Journal of Biomedical and Health Informatics 27 (8), 4131–4142. https://doi.org/10.1109/jbhi.2023.3279340.

Wang, J., Chen, F., Ma, Y., Wang, L., Fei, Z., Shuai, J., Tang, X., Zhou, Q., Qin, J., 2023. XBound-former: toward cross-scale boundary modeling in transformers. IEEE Transactions on Medical Imaging 42 (6), 1735–1745. https://doi.org/10.1109/TMI.2023.3236037.

Yao, H., Derksen, H., Golbus, J.R., Zhang, J., Aaronson, K.D., Gryak, J., Najarian, K., 2023. A novel tropical geometry-based interpretable machine learning method: pilot application to delivery of advanced heart failure therapies. IEEE Journal of Biomedical and Health Informatics 27 (1), 239–250. https://doi.org/10.1109/JBHI.2022.3211765.

Computational intelligence for sustainable computing in traditional medical system Ayurveda

Lakshmi Bheemavarapu and K. Usha Rani

Department of Computer Science, Sri Padmavati Mahila Visvavidyalayam (Women's University), Tirupati, Andhra Pradesh, India

1. Introduction

In today's digital age, maintenance of good health and utilization of advanced medical diagnosis and treatment methods are very expensive. In the Allopathic medical system, the costs incurred for the diagnosis, treatment, medicines, etc., are more. Hence, these allopathic-based advanced medical procedures are not affordable for the common man. The side effects are also more with these Allopathic medical procedures. If the same situation continues, the third goal of sustainable development "Good Health and Well-Being" can't be reached. Hence, there is a need for an alternative medicine such as Traditional Medicine Ayurveda. Allopathy medicine follows generalized diagnosis and treatment procedures to treat all kinds of people without considering the nature of their bodies, food habits, and lifestyle. Different people react to a treatment process in different ways. So, there is a need for personalized treatment, and is possible with our traditional medical system Ayurveda. Ayurveda provides simple tricks and methods to treat regular problems such as cough, cold, body pains, dental issues, skin problems, stomach ache, and headache. With these many advantages, Ayurvedic medicine helps us to reach sustainable development goal three i.e., Good health and Well-Being.

Ayurveda can assist us in reaching our third sustainable development goal as it uses medicinal herbs at affordable prices. To take a forward step toward this goal there is a need to use sustainable computing in Ayurvedic medical system. Sustainable computing can help the common man in the maintenance of good health with affordable treatment procedures.

Ayurveda prescribes treatments based on the nature of the individual, season, working environment, and food habits that change frequently. To handle these changing conditions sustainable computing with new methods that can work efficiently in changing environments is required.

Ayurveda uses natural methods to treat chronic diseases without side effects. It provides personalized treatment plans. As it is personalized it takes more time to understand the nature of the person and to frame a treatment plan to treat his/her problem in a natural way. Natural herbs and plants with medical properties are used to prepare Ayurvedic drugs. As Ayurveda was born in India and has a rich history, Indians are following Ayurveda-linked cooking methods and first aid treatments.

Allopathy or modern medicine is dominating the Ayurvedic medical system because of its fast results and immediate effects. Compared with Allopathy, Ayurvedic professionals are very less. There is a huge deficit of Ayurvedic practitioners. To fill this deficit, the usage of sustainable computing in Ayurveda is required. Traditional Medical system Ayurveda uses Computational Intelligence models and algorithms to solve different problems.

This book chapter is organized into seven sections. Section 2 presents a brief description of Computational Intelligence, Sustainable Development, Sustainable Computing, and Ayurveda, Section 3 provides different applications of Computational Intelligence in the Healthcare domain, Section 4 details the steps taken by the Government of India for sustainable development with Ayurveda, Section 5 deals with the need and applications of Computational Intelligence in Ayurveda, Section 6 list outs the challenges and Section 7 concludes the book chapter.

2. Background

This section provides a short description of Computational Intelligence, Sustainable Development, Sustainable Computing, and Ayurveda.

2.1 Computational intelligence (CI)

CI as part of Artificial Intelligence (AI) uses nature-inspired technologies such as Artificial Neural Networks, Evolutionary Machine Learning, Fuzzy Logic, and Swarm Intelligence, to solve critical problems with changing environments (Engelbrecht, 2007). As per IEEE Computational Intelligence Society, "CI encompasses computing paradigms like Neural Networks, Fuzzy Systems, Evolutionary Computation, Ambient Intelligence, Artificial Life, Cultural Learning, Artificial Endocrine Networks, Social Reasoning, and Artificial Hormone Networks" (IEEE Computational Intelligence Society. What is Computational Intelligence? Available at, 2023). Individual paradigms work efficiently to solve real-world problems. Now, hybridization of these paradigms is preferred to solve more complex problems. CI consists of computational models and theories that are designed by taking inspiration from natural and biological functions. Traditional AI focuses on high-cognitive formalisms and reasoning about symbolic representations whereas CI focuses on low-level cognitive functions such as perception and control (Dutch and Mandzuik, 2007).

As per Raj (2019), the objectives of both CI and AI are the same but the way of their implementation is different. AI tries to imitate human intelligence through learning, reasoning, planning, searching, and perception building. The traditional problems dealt with by AI can be represented as state-space search, where space is the collection of states and state is an instance. This traditional AI has many constraints because of building a large set of rules, and the transition of states. This made traditional or conventional AI incompatible with growing demands. These failures insisted on the need for the development of new computational methods that are suitable for changing environments. The result is CI. CI models have the ability to handle temporal changes. These benefits made CI suitable for handling current-day challenges. Application of these methods to have sustainable development is much required.

2.2 Sustainable development

In the year 1987, the United Nations Brundtland Commission defined sustainability as "meeting the needs of the present without compromising the ability of future generations to meet their own needs." In this connection, the United Nations Organization has prepared an Agenda called "Transforming our World: the 2030 Agenda for Sustainable Development" with a plan of actions to be taken to have sustainable development. The Agenda has 17 Sustainable Development Goals (SDGs) with 169 targets to reach by 2030 and apply to all countries around the world (Take Action for the Sustainable Development Goals, 2023; International Institute of Sustainable Development. Sustainable Development, 2023).

"Good health and Well-Being" is declared as Sustainable Development Goal 3 (SDG-3) to ensure healthy lives and promote well-being for all at all ages. SDG-3 has many targets such as reducing global maternal mortality, reducing neonatal mortality; ending the epidemic of AIDS, reducing premature mortality from noncommunicable diseases, and preventing substance abuse. SDG-3 has to meet the targets by 2030. As per UN (Goal 3: Ensure healthy lives and promote well-being for all at all ages, 2023), there are inequalities in healthcare access. COVID-19 and other problems impeded the progress toward SDG-3. SDG-3 aims to achieve Universal Health Coverage (UHC) and to provide access to safe and affordable medicines and vaccines for all. A major part of the world population lacks access to required healthcare services. There is a need to plan the future to reach this goal. Economic burdens are the main hurdles of UHC. To achieve the targets of SDG-3, as part of social responsibility, we have to keep ourselves healthy by following a healthy lifestyle and having healthy food habits. And also we have to educate our family members, neighbors, and friends to follow the same. To reach this goal usage of computers with the latest advancements such as Information and Communication Technology, Artificial Intelligence and Machine Learning, Block Chain Technology, Internet of Things (IoT), Cloud Computing, and others is needed.

2.3 Sustainable computing

Sustainable Computing is nothing but computing for societal, and environmental sustainability. It includes the designing, development, use, and disposal of computing systems in a sustainable manner and to achieve the SDGs. It is required to reduce the adverse effects of new technology and related devices. Because of advanced technology, and the increase in the number of users of Information and Communication Technology, there is a huge demand for power consumption in the near future. There are many negative impacts on the environment and, the health of individuals (Sustainable Computing, 2023).

As part of Sustainable computing, Computational Sustainability is emerging as a field that uses mathematics, computer science, and information sciences to achieve sustainability. As per Wikipedia (Wikipedia. Computational Sustainability. Available at, 2023), Computational Sustainability is defined as "An emerging field that attempts to balance societal, economic, and environmental resources for the future well-being of humanity using methods from mathematics, computer science, and information science fields."

The impact of the new technologies is uneven as it is used and affordable by rich people and the benefits are reserved for large-scale business sectors neglecting poor people and small-scale industries and businesses. The inequalities are increasing and ICT is becoming a barrier to achieving sustainable development. In the same manner, healthcare-related applications are becoming more cost-prohibitive, and advanced treatments are far away from middle-class and lower-middle-class people. To reach SDG-3 sustainable computing is required in the healthcare system. Computational Intelligence techniques have many applications in the healthcare domain. These techniques are mostly used in conventional medicine Allopathy. There is a need for the usage of these Artificial Intelligence techniques in particular Computational Intelligence techniques in traditional medicine also. Ayurveda, Yoga, Naturopathy, Unani, Siddha, and Homeopathy (AYUSH) are the traditional medical systems used in India.

2.4 Ayurveda

Ayurveda is a popular traditional Indian medical system with a rich history of thousands of years. As per Ayurveda, the universe is composed of five elements called *"Pancha Mahabhootas"* (Vayu, Jala, Aakash, Prithvi, and Teja). Like the Universe, our body is also composed of these Pancha Mahabhootas in the form of three basic energies called *"Tridoshas."* Vata dosha, Pitta dosha, and Kapha dosha are the Tridoshas. A balance in these Tridoshas indicates a healthy state of the body. An imbalance among the Tridoshas indicates the unhealthy state of the body. The human body is made up of seven types of tissues called *"Sapta Dhatus."* They include Rasa, Raktha, Mamsa, Meda, Asthi, Majja, and Shukra. Other important factors are *"Tri Malas"* and *"Trayo Dosha Agni."* All of these play an important role in Ayurveda in the diagnosis and treatment of a disease (National Library of Medicine. A glimpse of Ayurveda—The forgotten history and principles of Indian traditional medicine, 2023).

Based on the dominance of one or more of Tridoshas the constitution of the body called *"Prakriti"* is identified. It plays a major role in finding physiological strengths and weaknesses, mental health, and susceptibility to disease. Different types of Prakriti and their classification are presented in Fig. 10.1. Prakriti of an individual is of seven types: *Vata, Pitta, Kapha, Vata-Pitta, Pitta-Kapha, Vata-Kapha,* and *Vata-Pitta-Kapha.* These seven types are classified into three categories: Ekadoshaja, Dwidoshaja, and Samadoshaja. Ekadoshaja-type Prakriti dominates a single dosha. The Dwidoshaja type dominates two doshas and the Samadoshaja type has a balance of three types of doshas. Vata, Pitta, Kapha are called Ekadoshaja Prakriti, Vata-Pitta, Vata-Kapha, Kapha-Pitta are called Dwidoshaja Prakriti and Vata-Pitta-Kapha is called Samadoshaja Prakriti. There are many methods called

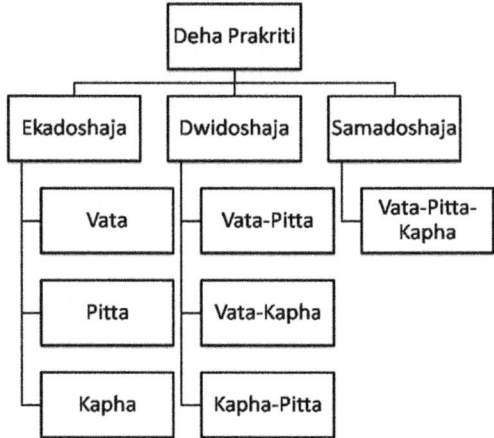

FIGURE 10.1 Deha Prakriti and types different types of Prakriti or human body constitution.

Parikshas to identify the Prakriti of an individual such as Prasna Pariksha (Usage of a questionnaire), Nadi Pariksha (Pulse Diagnosis), and Drisya Pariksha (Direct Observation). Ayurvedic treatment consists of the usage of medicinal herbs, massage, yoga, meditation, and other methods. The identification of medicinal plants and their medicinal properties requires experts (National Library of Medicine. A glimpse of Ayurveda—The forgotten history and principles of Indian traditional medicine, 2023).

Deha Prakriti and its seven types (Vata, Pitta, Kapha, Vata-Pitta, Vata-Kapha, Kapha-Pitta, Vata-Pitta-Kapha).

3. Computational intelligence applications in the healthcare domain

In the healthcare domain computers with the latest technologies are used to gather, store, and manage patient data, perform disease diagnosis and treatment, and make clinical decisions. They are helping the healthcare system to offer safe, efficient, and qualitative services to patients. AI is assisting the healthcare domain in different aspects. Computational Intelligence techniques are used in different domains in different ways to solve different kinds of problems. CI is enhancing the robustness of computer applications as an emerging discipline. AI can be used for novel drug discovery and development, to understand in-depth biology and pathophysiology. It assists in the selection of study participants based on his/her eligibility.

The AI for Good Global Summit is a platform initiated by the United Nations Organization for promoting AI to study different problems relating to SDGs and to find AI-integrated solutions. AI for Good Global Summit is organized by the UN's International Telecommunication Union (ITU) in association with 40 UN sister agencies and coconvened with the government of Switzerland. The goal of AI for Good is to identify practical applications of AI to advance the SDGs and scale those solutions for global impact. At first, it was organized in the year 2017 and in 2020 it was moved to online. At this global summit, a new initiative called AI for Health was announced with a goal to develop AI-based solutions for health.Artificial Intelligence for Health (AI4H) is an initiative by the International Telecommunication Union (ITC) and the World Health Organization (WHO) (International Telecommunications Union (ITU). AI for Good, 2023).

AI for Health is a UN initiative for designing and developing AI solutions for health. The UN specialized agencies, ITU, WHO, and World Intellectual Property Organization (WIPO) joined forces to develop AI-based applications in support of disease diagnosis, treatment, and related services. It aims to support universal access to innovations and decrease the inequalities in health services. It also aims to develop standards and policies, share knowledge and data, and support evidence-based decisions on the usage of AI-based solutions for health. The key objectives are the development of governance frameworks, standards, tools, and guidelines for AI to support UHC; planning for making AI solutions reachable to underserved communities; and assisting low and middle-income countries to adopt AI-based solutions (New UN initiative aims to step up AI's contribution to health, 2023). Guidelines relating to the Ethics and Governance of Artificial Intelligence for Health are available on the website of the United Nations (Ethics and Governance of Artificial Intelligence for Health, 2023).

As part of AI, CI is also assisting professionals to develop more sophisticated solutions in the field of healthcare. It assists in signal processing, Anesthesia, drug development, image processing, clinical diagnosis, etc. (Sordo et al., 2008). Some CI-based applications in the healthcare sector are listed in Table 10.1.

TABLE 10.1   Applications of CI in the healthcare sector.

Technology/Algorithm	Example
Neural networks	Heart signal classification (Choudhary et al., 2024)
	Heart signal segmentation (Renna et al., 2019)
	Pulse diagnosis (Chen et al., 2020)
Convolutional neural networks (CNN)	Brain tumor segmentation (Pereira et al., 2016)
Deep convolutional neural networks (DCNNs)	Medical image classification (Asif et al., 2024)
Fuzzy recurrent neural network (FR-Net)	Segmentation of brain MRI (SivaSai et al., 2021)
3D UNET	Whole-heart CT image segmentation (Ye et al., 2019)
Restricted Boltzmann Machine (RBM) and deep belief networks (DBN)	Electrocardiogram (ECG) classification (Mathews et al., 2018)
Extreme learning machine (ELM) and genetic algorithm	ECG signal classification (Diker et al., 2019)
Hybrid transfer learning	CT scan image-based lung-nodule classification (Saikia et al., 2022)
Neuro-fuzzy inference system	Hepatitis disease diagnosis (Nilashi et al., 2019)
CNN + flower pollination optimization algorithm	Eye disease detection (Glaret subin and Muthukannan, 2022)

TABLE 10.1 Applications of CI in the healthcare sector.—cont'd

Technology/Algorithm	Example
Fuzzy logic + genetic algorithm	Heart disease diagnosis (Reddy et al., 2020)
Fuzzy logic + CNN	Covid-19 pneumonia detection from chest X-ray images (Ieracitano et al., 2022)
Fuzzy logic	Computed tomography (CT) image enhancement (Tsui et al., 2022)
Fuzzy sets	Mammogram enhancement (Deng et al., 2017)
Fuzzy inference system (FIS)	Disease diagnosis (Alam et al., 2022)
Genetic algorithm + support vector machine (SVM)	Heart rate variability classification (Ashtiyani et al., 2018)
Cuckoo search algorithm + convolutional LSTM classifier	Disease prediction (Gou et al., 2022)
Multiobjective cuckoo search algorithm + KNN	EEG channel selection for person identification (Abdi Alkareem Alyasseri et al., 2022)
Extreme learning machine (ELM) + particle swarm optimization (PSO)	Sleep stage classification (Surantha et al., 2021)
Clustering particle swarm optimization (CPSO)	Healthcare staff scheduling (Mutingi and Mbohwa, 2014)
Harmony search and simulated annealing (HS-SA) algorithm	Breast cancer diagnosis (Shaikh and Ali, 2020)
Harmony search algorithm	Patient admission scheduling (Abu Doush et al., 2019)
Bat algorithm	Patient referral problem (Yao et al., 2020)
Firefly algorithm	Smart healthcare resource allocation (Anu and Singhrova, 2012)
Artificial bee colony-based clustering	Healthcare waste disposal facility location problem (Gergin et al., 2019)

4. Ayurveda-based sustainable development initiatives in India

Ayurveda plays a major role in reaching SDG-3 Good Health and Well Being. In this connection, India created Ayush Grid (AG)—an Ayush-centric Digital Health Platform (DHP) by the Ministry of Ayush in the year 2018. It works in association with India's national digital health infostructure platform Ayushman Bharat Digital Mission (ABDM). Under this AG the Ministry of Ayush has established the National Ayush Morbidity and Standardized Electronic (NAMASTE) portal, Ayush Research Portal, development and implementation of the Ayushman Bharat Health Account (ABHA) enabled Ayush Hospital Information Management System (AHMIS), Ayush Sanjivani app (during COVID-19 pandemic), assisting the World Health Organization (WHO) in the development of mYoga app, Ayurveda Siddha and Unani International terminologies and morbidity codes (as Module-2, Chapter-26, ICD-11), development of benchmarks for the use of AI in traditional medicine.

As of now, under the Ayush Grid project, the AHMIS application is used to manage OPD transactions in the hospitals working under MoA. AHMIS is creating ABHA for the patients and carrying registration and follow-up visits. To bridge the digital gap in the Ayush sector, an Information and Technology course was developed for Ayurveda professionals. In the year 2019, the Telemedicine program was launched. AG also assisted WHO in the development of the mYoga app to encourage people to practice quality yoga. It also developed the Ayush Sanjivani mobile app and Yoga locator mobile app. It assisted WHO in the publication of Standard International Terminologies of AYUSH. It is also leading a discussion on the development of benchmarks for the application of AI in the field of traditional medicine as a topic drive under the AI4Health working group.

The applications developed before the AG initiative were the Ayush Research Portal, Ayurveda e-books, NAMASTE portal, e-Charak application, e-Aushadi portal, and the applications developed under AG were AHMIS, Ayush NGO Portal, Ayush Next portal, Ayush Information Hub (AIH), Ayush Clinical Case Repository, AyurCel portals, Ayusoft web version, Yoga applications and other ICT solutions (Ram and Saketh, 2023).

5. Computational intelligence in Ayurveda

To reach the Triple billion target of WHO, the assistance of digital, information, and communicational technologies is much required. The digital transformation and technologies in health care such as virtual care, remote monitoring, AI, CI, Big data analytics, smart gadgets, tools enabling data share and exchange, and remote data capturing have improved the process of medical diagnosis, decision making, digital therapeutics, self-care, clinical trials, patient-centric methods. It also assists in the creation of evidence-based knowledge, acquisition of required skills, capacity building of professionals, and collaboration among them.

AI-integrated methods and tools are used in mainstream medicine or conventional medicine. There are many applications in allopathic medicine. The application of AI is also initiated in traditional medicine such as AYUSH. It allows the screening of medical herbs, a self-learning capable automated machine that can learn from different forms of ancient medical scripts, AI-based diagnostic tools for decision-making models, symptom classification, and generation of pharmacological databases useful in ethnopharmacology. Some AI-based applications were developed in Ayurveda for Dosha evaluation, and Prakriti assessment to assist the practitioners in treatment. Automatic recognition and classification of plant leaves using AI, identification of indigenous Ayurvedic plant species using deep learning techniques, use of robots in patient care, DNA fingerprinting for Prakriti assessment, usage of gadgets for real-time data capturing, prediction of outbreaks of epidemics (Nesari and Manoj, 2023).

CI-based applications are developed in Ayurveda to perform different types of operations as depicted in Fig. 10.2.

Different applications of CI in Ayurveda.

Application of Computational Intelligence methods is required in Ayurveda in particular in the prediction of Prakriti. Ayurveda provides personalized treatments based on an individual's unique body constitution, known as Prakriti, specifically Deha Prakriti. Deha Prakriti

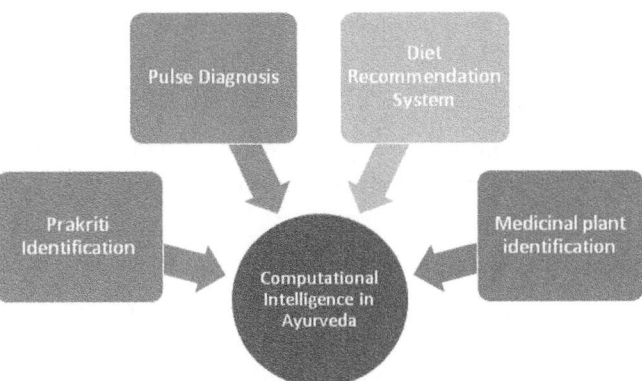

FIGURE 10.2 Applications of computational intelligence in Ayurveda different applications of CI in Ayurveda.

helps practitioners diagnose the disease and plan treatment procedures. It also assists individuals in planning their personal lifestyle to maintain strength and prevent ailments and illnesses. Ayurveda offers different types of Parikshas (tests) employing various methods to identify Prakriti. The practitioner uses a questionnaire in Prasna Pariksha, directly observes the individual in Darsana Pariksha, and utilizes pulse analysis in Sparsana Pariksha or Nadi Pariksha to determine an individual's Prakriti. The shortage of experts in Ayurvedic medicine can be addressed by automating Prakriti Identification. Some AI-based applications were already developed for Prakriti Identification. Prakriti can be affected by age, climate, health conditions, food habits, and lifestyle changes. To predict the Prakriti in these conditions advanced methods such as Computational Intelligence methods and theories are required. In the following subsections applications of CI-based methods in the domain of Ayurveda are discussed.

5.1 Prakriti Identification

Prakriti is used to identify the nature of a person and identification of Prakriti is the first step in the diagnosis and treatment process in Ayurveda. Knowledge of one's own Prakriti also helps the individual to plan his/her lifestyle. So, automation of Prakriti Identification is very important in Ayurveda. The Prakriti of a person changes based on seasonal changes, changes in food habits, medication, etc. So, many CI-based models for the identification of Prakriti were developed and they are discussed in the following paragraphs.

Like Prakriti in Ayurveda, the Unani system uses Mizaj to understand the body constitution of a person. An ANN-based dual converter to convert Prakriti to Mizaj and vice versa. This can be used to integrate Ayurveda with Unani to treat patients (Mulla et al., 2013).

Ranjit et al. (Kaur et al., 2016), developed a fuzzy logic-based expert system for the calculation of the human body's physical constituents. Madaan and Gayal (2019) developed An Adaptive Neuro Fuzzy Inference System (ANFS) to predict Ayurvedic Dosha. The fuzzy System is very complex in nature as it takes all combinations of inputs and outputs and Fuzzy expert systems are designed with a large number of rules. To overcome these problems, ANFS is trained with actual data and with a limited number of rules. It is a combination of Neural Networks (NN) and Fuzzy Inference System (FIS). So, it is a hybrid method with the benefits of both NN and FIS.

Mendis et al. (2019) have proposed a statistical fuzzy inference system by integrating Sugeno defuzzification with principle component analysis (PCA), fuzzy inference engine, knowledge base, and user interface to improve the output. This is applied in Ayurvedic medicine to classify the human mental constitution called Manasa Prakriti.

5.2 Diet recommendation system

In Ayurveda, diet acts as a medicine. So, the development of diet recommendation systems is very helpful to individuals to maintain a good diet based on different seasons. Some Diet Recommendation Systems developed are discussed below.

A Diet Recommendation System based on the Prakriti type and the current season was developed by using Fuzzy Ontology and Type2 Fuzzy Logic. Ontology is used to represent knowledge and its relationship with other knowledge. As Ontology is not sufficient to represent uncertain data fuzzy data is integrated to form fuzzy ontology (Shital et al., 2015; Chavan et al., 2017).

Jaya et al. (2017) proposed an electronic health record system using an ontology-based procedure to find Prakriti and ANFIS-based Health Maintenance System. It uses a rule-based system. The rule-based system is prepared by using Particle Swarm Optimization—Adaptive Network-based Fuzzy Inference System (PSO-ANFIS). It takes doshas, food, and lifestyle recommendations as ontological inputs to the fuzzy interface and provides diet and lifestyle suggestions to the patients.

5.3 Pulse diagnosis

Traditional medical systems use Nadi Pariksha (Pulse Diagnosis) to diagnose the nature of a person and his/her disease. Based on this treatment process will be framed. To do pulse diagnosis expert Ayurvedic practitioners are needed. There is a deficit in expert Nadi Pariksha Vidyan. To fill it, Internet of Things (IoT) integrated pulse diagnosis systems were developed for automatic pulse diagnosis and identification of Prakriti, the health condition of a person. Some pulse detection systems were discussed in the below paras.

Traditional Ayurvedic Medicine and Traditional Chinese Medicine, use pulse detection to diagnose the disease. Navghare and Bajaj (2016) developed a pulse detection system using Force Sensing Resistor (FRS) sensors for diagnosing the normal and abnormal health conditions of a person. The tridosha pulses are studied for the diagnosis of a normal person. The person with body pain shows an imbalance of "vata" dosha and indigestion occurs due to an imbalance in "kapha" dosha. Linear Vector Quantization (LVQ) framework is used for classifying the dataset for normal, body pain and indigestion. LVQ is a special ANN and a precursor to self-organizing maps (SOM).

Barsagade et al. (2016) developed software for the identification of Prakriti using Nadi Pariksha. They used an infrared (IR) sensor to detect and capture the pulse, Global System for Mobile (GSM) for sharing data, and ANFIS for classification of Prkriti.

Joshi and Bajaj (2021) developed sophisticated software for the identification of the Prakriti of an individual by checking the Nadi (pulse). Vata, Pitta, and Kapha pulses are detected, recorded, and analyzed for Prakriti identification. The pulses are recorded in a text file. Tools such as Visual Studio, Pro-Mikroc, MatLab, and Orcad are used in the design. The output matrix is given to the ANN for classification purposes.

5.4 Medicinal plant identification

Ayurvedic medicines are prepared using different plants with medicinal properties. Identification of these herbs helps practitioners to speed up the drug development process. IoT-integrated CI-based techniques are very helpful in this. Some of the applications developed for this purpose are specified below.

Nadig et al. (2022) have proposed a machine learning model for the classification of Ayurvedic herbs. Dileep (2019) have proposed a Deep Learning-based CNN called AyurLeaf to classify medicinal plants based on leaves with features such as their shape, color, size, and texture. Alexnet inspired deep neural network is used for efficient feature extraction from the dataset. The dataset is created with medicinal plants available in Kerala, the southwest state of India. This model is also tested on DLeaf dataset.

Marada et al. (2023) have concluded that deep learning is the best approach to improve the precision of computer vision-based classification and recognition systems. They have proposed SVM and CNN models for the classification of plants based on leaf features. Jayalath et al. proposed a method for the recognition of unique medicinal plants with high accuracy by applying machine learning-based image processing techniques. Based on the features of leaves the plants are classified. The proposed method uses CNN based model. The database is created by using the leaf and flower images used in Sri Lankan Ayurvedic medicine (Jayalath et al., 2019).

Manojkumar et al. (2017) used feature vectors relating to the front and back sides of green leaves and dried leaves with morphological features with a maximum identification rate. ML models developed with SVM, MLP, Fuzzy Lattice Reasoning, Naïve Bayes, and K-Star Instance-based classifiers are used for detection of the plants based on leaves. Inchara et al. (2023) used VGG19 CNN for the identification of Ayurvedic plants based on leaves.

Paulson and Ravishankar (2020) used CNN and pretrained models VGG16 and VGG19 for the identification of Ayurvedic plants based on leaves and their performance is compared. Sethulekshmi et al. (2014) studied the performance of different feature extraction methods, different combinations of features, and a number of classifiers applied for plant identification based on leaves. Probabilistic Neural Networks (PNN), MLP, SVM, and SOM are used for classification purposes. Sameer et al. (2022) suggested a system with AlexNet-based CNN architecture.

6. Challenges

Many researchers developed many models for Prakriti identification, Disease Diagnosis, Medicinal plant detection, food recommendation systems, clinical assistance systems, etc. But, these findings are not reaching normal people. There is a need to bridge the gap between the research findings and normal people. Then only the SDG-3 "Good Health and Well-Being" will be reached. Awareness about these models needs to be created among the people. The available literature should be made accessible to the people. Ayurvedic practitioners have to use the developed applications and models in their diagnosis and treatment process to validate their accuracy. This automatically creates trust in developed models and applications. New models with good accuracy need to be developed to meet the needs of the people.

7. Conclusion

Computational Intelligence helps to solve complex problems with changing nature. Application of CI is required for the healthcare sector to reach SDG-3 "Good Health and Well-Being." Ayurveda helps in reaching this goal as it uses personalized treatment with good food habits based on the individual's Prakriti, changes in lifestyle, and usage of natural herbs for treatment, practicing Yoga, and meditation. By making small changes in daily routine, people can maintain good health without financial burden. People with basic knowledge about their body nature can plan their lifestyle and stay away from illness. It makes people spend less amount to maintain good health. As part of social responsibility, there is a need to provide information relating to these healthy practices to low-income people. Developing applications with Computational Intelligence methods in the Ayurvedic medical system helps people to have a peaceful life without medical complications as an alternative medical system to Allopathy. As Allopathy is cost-prohibitive and has many complications CI-based food recommendation systems, and health management systems are required. Applications with accurate results are needed to work in real life.

References

Abdi Alkareem Alyasseri, Z., Alomari, O.A., Al-Betar, M.A., Awadallah, M.A., Hameed Abdulkareem, K., Abed Mohammed, M., Kadry, S., Rajinikanth, V., Rho, S., 2022. EEG channel selection using multiobjective cuckoo search for person identification as protection system in healthcare applications. Computational Intelligence and Neuroscience. https://doi.org/10.1155/2022/5974634, 2022.

Abu Doush, I., Al-Betar, M.A., Awadallah, M.A., Hammouri, A.I., Al-Khatib, Ra'ed M., ElMustafa, S., Alkhraisat, H., 2019. Harmony search algorithm for patient admission scheduling problem. Journal of Intelligent Systems 29 (1), 540–553. https://doi.org/10.1515/jisys-2018-0094.

Alam, T.M., Shaukat, K., Khelifi, A., Khan, W.A., Raza, H.M.E., Idrees, M., Luo, S., Hameed, I.A., 2022. Disease diagnosis system using IoT empowered with fuzzy inference system. Computers, Materials & Continua 70 (3), 5305–5319. https://doi.org/10.32604/cmc.2022.020344.

Anu, A., Singhrova, 2012. Optimal Healthcare Resource Allocation in Covid Scenario Using Firefly Algorithm. Fog.

Ashtiyani, M., Navaei Lavasani, S., Asgharzadeh Alvar, A., Deevband, M.R., 2018. Heart rate variability classification using support vector machine and genetic algorithm. Journal of Biomedical Physics and Engineering 8 (4), 423–434. https://doi.org/10.31661/jbpe.v0i0.614.

Asif, S., Ain, Q.-ul, Al-Sabri, R., Abdullah, M., 2024. LitefusionNet: boosting the performance for medical image classification with an intelligent and lightweight feature fusion network. Journal of Computational Science 102324. https://doi.org/10.1016/j.jocs.2024.102324.

Barsagade, A., Joshi, P.S., Bajaj, D.P., 2016. Neurofuzzy Logic Concept to Find Prakriti of a Person Using Wireless Data Acquisition System.

Chavan, S.V., Sambare, S.S., Joshi, A., 2017. Diet recommendation based on Prakriti and season using fuzzy ontology and type-2 fuzzy logic. In: Proceedings—2nd International Conference on Computing, Communication, Control and Automation, ICCUBEA 2016. Institute of Electrical and Electronics Engineers Inc., India https://doi.org/10.1109/ICCUBEA.2016.7860026.

Chen, Z., Huang, A., Qiang, X., 2020. Improved neural networks based on genetic algorithm for pulse recognition. Computational Biology and Chemistry 88, 107315. https://doi.org/10.1016/j.compbiolchem.2020.107315.

Choudhary, R.R., Rani, M., Kaur, R., Bhadu, M., Mandeep, J.S., 2024. Heart signal analysis using multistage classification denoising model. Journal of Electrical and Computer Engineering 1—9. https://doi.org/10.1155/2024/1502285, 2024.

Deng, H., Deng, W., Sun, X., Liu, M., Ye, C., Zhou, X., 2017. Mammogram enhancement using intuitionistic fuzzy sets. IEEE Transactions on Biomedical Engineering 64 (8), 1803—1814. https://doi.org/10.1109/TBME.2016.2624306.

Diker, A., Avci, D., Avci, E., Gedikpinar, M., 2019. A new technique for ECG signal classification genetic algorithm Wavelet Kernel extreme learning machine. Optik 180, 46—55. https://doi.org/10.1016/j.ijleo.2018.11.065.

Dileep, M.R., Pournami, P.N., 2019. AyurLeaf: a deep learning approach for classification of medicinal plants. In: 2019 IEEE Region 10 Conference (TENCON 2019). https://ieeexplore.ieee.org/document/8929394.

Dutch, W., Mandzuik, J., 2007. What Is Computational Intelligence and What Could it Become? in Challenges for Computational Intelligence for Computational Intelligence. Springer.

Engelbrecht, A.P., 2007. Computational Intelligence: An Introduction. Wiley. https://doi.org/10.1002/9780470512517.

Ethics and Governance of Artificial Intelligence for Health, 2023. https://iris.who.int/bitstream/handle/10665/341996/9789240029200-eng.pdf?sequence=1. (Accessed 1 October 2023).

Gergin, Z., Tunçbilek, N., Esnaf, Ş., 2019. Clustering approach using artificial bee colony algorithm for healthcare waste disposal facility location problem. International Journal of Operations Research and Information Systems 10 (1), 56—75. https://doi.org/10.4018/ijoris.2019010104.

Glaret subin, P., Muthukannan, P., 2022. Optimized convolution neural network based multiple eye disease detection. Computers in Biology and Medicine 146, 105648. https://doi.org/10.1016/j.compbiomed.2022.105648.

Goal 3: Ensure healthy lives and promote well-being for all at all ages, 2023. https://www.un.org/sustainabledevelopment/health/. (Accessed 28 September 2023).

Gou, J., Kumar, A., Satyanarayana Reddy, S.S., Mahommad, G.B., Khan, B., Sharma, R., 2022. Smart healthcare: disease prediction using the cuckoo-enabled deep classifier in IoT framework. Scientific Programming 2090681. https://doi.org/10.1155/2022/2090681, 2022.

IEEE Computational Intelligence Society. What Is Computational Intelligence?, 2023. Available at https://cis.ieee.org/about/what-is-ci. (Accessed 4 August 2023).

Ieracitano, C., Mammone, N., Versaci, M., Varone, G., Ali, A.-R., Armentano, A., Calabrese, G., Ferrarelli, A., Turano, L., Tebala, C., Hussain, Z., Sheikh, Z., Sheikh, A., Sceni, G., Hussain, A., Morabito, F.C., 2022. A fuzzy-enhanced deep learning approach for early detection of covid-19 pneumonia from portable chest X-ray images. Neurocomputing 481, 202—215. https://doi.org/10.1016/j.neucom.2022.01.055.

Inchara, K.S., Pratheek, B., Shubha, H.R., Vinutha, S.H., Manjunath, K.G., 2023. Ayurvedic leaf detection for medicine using VGG19 convolution neural network. International Research Journal of Modernization in Engineering Technology and Science 05 (04). https://doi.org/10.56726/IRJMETS37757.

International Institute of Sustainable Development. Sustainable Development, 2023. https://www.iisd.org/mission-and-goals/sustainabledevelopment. (Accessed 2 August 2023).

International Telecommunications Union (ITU). AI for Good, 2023. https://aiforgood.itu.int/. (Accessed 14 August 2023).

Jaya, R., Pillai, C.R., Kannan, R.J., 2017. Ontology Based Optimization with Adaptive Network Based Fuzzy Inference System Based Health Maintenance System.

Jayalath, A.D.A.D.S., Amarawanshaline, T.G.A.G.D., Nawinna, D.P., Nadeeshan, P.V.D., Jayasuriya, H.P., 2019. Identification of medicinal plants by visual characteristics of leaves and flowers. In: 2019 IEEE 14th International Conference on Industrial and Information Systems (ICIIS). https://ieeexplore.ieee.org/document/9063275.

Joshi, S., Bajaj, P., 2021. Design development of portable vata, Pitta kapha [VPK] pulse detector to find prakriti of an individual using artificial neural network. In: 2021 6th International Conference for Convergence in Technology, I2CT 2021. Institute of Electrical and Electronics Engineers Inc., India https://doi.org/10.1109/I2CT51068.2021.9418155.

Kaur, R., Madaan, V., Agrawal, P., 2016. Fuzzy expert system to calculate the strength/immunity of a human body. Indian Journal of Science and Technology 9 (44). https://doi.org/10.17485/ijst/2016/v9i44/105145.

Madaan, V., Gayal, A., 2019. An adaptive neuro fuzzy inference system for predicting Ayurvedic dosha. In: 2019 4th International Conference on Information Systems and Computer Networks, ISCON 2019. Institute of Electrical and Electronics Engineers Inc., India, pp. 335–339. https://doi.org/10.1109/ISCON47742.2019.9036168.

Manojkumar, P., Surya, C.M., Gopi, V.P., 2017. Identification of Ayurvedic medicinal plants by image processing of leaf samples. In: 2017 Third International Conference on Research in Computational Intelligence and Communication Networks (ICRCICN). https://doi.org/10.1109/ICRCICN.2017.8234512.

Marada, S.R., Praveen Kumar, S., Srinivasa Rao, K., 2023. A methodology for identification of Ayurvedic plant based on machine learning algorithms. International Journal of Computing and Digital Systems 14 (011), 10233–10241. https://doi.org/10.12785/ijcds/140196.

Mathews, S.M., Kambhamettu, C., Barner, K.E., 2018. A novel application of deep learning for single-lead ECG classification. Computers in Biology and Medicine 99, 53–62. https://doi.org/10.1016/j.compbiomed.2018.05.013.

Mendis, D., Ratnayake, Uditha, Karunananda, Samaratunga, U., 2019. A Statistical Fuzzy Inference System by PCA Based Defuzzification for the Improvement of Sugeno Defuzzification Method.

Mulla, Gazala, Junaid, Acharya, H., 2013. ANN Based Dual Convertor Model to Get Ayurvedic Prakruiti from Fuzzy Unani Scores and Unani Mizaj from Fuzzy Ayurvedic Scores.

Mutingi, M., Mbohwa, C., 2014. Home Healthcare Staff Scheduling: A Clustering Particle Swarm Optimization Approach.

Nadig, S., Jyothishri, B., Painginkar, V., Pinto, V.sharline, 2022. Identification of Ayurveda herbs using machine learning. International Journal of Advances in Engineering and Management (IJAEM) 4 (12), 518–523. https://doi.org/10.35629/5252-0412518523.

National Library of Medicine. A Glimpse of Ayurveda—The Forgotten History and Principles of Indian Traditional Medicine, 2023. https://www.ncbi.nlm.nih.gov/pmc/articles/PMC5198827/. (Accessed 3 August 2023).

Navghare, Bajaj, 2016. Detection of prakriti of a person using arterial pulse detection system with linear quantization method. International Journal of Computer Science and Information Security 14.

Nesari, T., Manoj, 2023. Artificial intelligence in the sector of Ayurveda: scope and opportunities. International Journal of Ayurveda Research 4 (2), 57–60. https://doi.org/10.4103/ijar.ijar_40_23.

New UN Initiative Aims to Step up AI's Contribution to Health, 2023. https://www.itu.int/hub/2023/07/new-un-initiative-aims-to-step-up-ais-contribution-to-health/. (Accessed 1 October 2023).

Nilashi, M., Ahmadi, H., Shahmoradi, L., Ibrahim, O., Akbari, E., 2019. A predictive method for hepatitis disease diagnosis using ensembles of neuro-fuzzy technique. Journal of Infection and Public Health 12 (1), 13–20. https://doi.org/10.1016/j.jiph.2018.09.009.

Paulson, A., Ravishankar, S., 2020. AI based indigenous medicinal plant identification. In: Proceedings—2020 Advanced Computing and Communication Technologies for High Performance Applications, ACCTHPA 2020. Institute of Electrical and Electronics Engineers Inc., India, pp. 57–63. https://doi.org/10.1109/ACCTHPA49271.2020.9213224.

Pereira, S., Pinto, A., Alves, V., Silva, C.A., 2016. Brain tumor segmentation using convolutional neural networks in MRI images. IEEE Transactions on Medical Imaging 35 (5), 1240–1251. https://doi.org/10.1109/TMI.2016.2538465.

Raj, J., 2019. A comprehensive survey on the computational intelligence techniques and its applications. Journal of ISMAC 01, 147–159. https://doi.org/10.36548/jismac.2019.3.002.

Ram, T., Saketh, 2023. Ayush grid: digital health platform. International Journal of Ayurveda Research 4 (2), 61–69. https://doi.org/10.4103/ijar.ijar_66_23.

Reddy, G.T., Reddy, M.P.K., Lakshmanna, K., Rajput, D.S., Kaluri, R., Srivastava, G., 2020. Hybrid genetic algorithm and a fuzzy logic classifier for heart disease diagnosis. Evolutionary Intelligence 13 (2), 185–196. https://doi.org/10.1007/s12065-019-00327-1.

Renna, F., Oliveira, J., Coimbra, M.T., 2019. Deep convolutional neural networks for heart sound segmentation. IEEE Journal of Biomedical and Health Informatics 23 (6), 2435–2445. https://doi.org/10.1109/JBHI.2019.2894222.

Saikia, T., Kumar, R., Kumar, D., Singh, K.K., 2022. An automatic lung nodule classification system based on hybrid transfer learning approach. SN Computer Science 3 (4). https://doi.org/10.1007/s42979-022-01167-0.

Sameer, A.K., Sudhanva, S.A., Manikanta Sanjay, V., Punit, S.K., 2022. A novel approach to classification of Ayurvedic medicinal plants using neural networks. International Journal of Engineering Research and Technology 11 (01).

Sethulekshmi, A.V., Sreekumar, K., 2014. Ayurvedic leaf recognition for plant classification. International Journal of Computer Science and Information Technologies 5.

Shaikh, T.A., Ali, R., 2020. An intelligent healthcare system for optimized breast cancer diagnosis using harmony search and simulated annealing (HS-SA) algorithm. Informatics in Medicine Unlocked 21, 100408. https://doi.org/10.1016/j.imu.2020.100408.

Shital, V., Chavan, S.S., Sambare, 2015. Article: study of diet recommendation system based on fuzzy logic and ontology. International Journal of Computer Applications 132 (12), 20–24.

SivaSai, J.G., Srinivasu, P.N., Sindhuri, M.N., Rohitha, K., Deepika, S., 2021. An automated segmentation of brain MR image through fuzzy recurrent neural network. Studies in Computational Intelligence 903, 163–179. https://doi.org/10.1007/978-981-15-5495-7_9.

Sordo, M., Vaidya, S., Jain, L.C., 2008. An introduction to computational intelligence in healthcare: new directions. Studies in Computational Intelligence 107, 1–26. https://doi.org/10.1007/978-3-540-77662-8_1.

Surantha, N., Lesmana, T.F., Isa, S.M., 2021. Sleep stage classification using extreme learning machine and particle swarm optimization for healthcare big data. Journal of Big Data 8 (1). https://doi.org/10.1186/s40537-020-00406-6.

Sustainable Computing, 2023. https://resources.nvidia.com/l/en-us-sustainable-computing. (Accessed 1 October 2023).

Take Action for the Sustainable Development Goals, 2023. https://www.un.org/sustainabledevelopment/sustainable-development-goals/. (Accessed 28 September 2023).

Tsui, P.-H., Alzahrani, A., Bhuiyan, M.A.-A., Akhter, F., 2022. Detecting COVID-19 pneumonia over fuzzy image enhancement on computed tomography images. Computational and Mathematical Methods in Medicine 1043299. https://doi.org/10.1155/2022/1043299, 2022.

Wikipedia. Computational Sustainability, 2023. Available at https://en.wikipedia.org/wiki/Computational_sustainability. (Accessed 2 August 2023).

Yao, H.-C., Chen, P.-J., Kuo, Y.-T., Shih, C.-C., Wang, X.-Y., Chen, P.-S., Schwarzacher, S.P., Gómez, C., 2020. Solving patient referral problems by using bat algorithm. Technology and Health Care 28 (4), 433–442. https://doi.org/10.3233/thc-209044.

Ye, C., Wang, W., Zhang, S., Wang, K., 2019. Multi-depth fusion network for whole-heart CT image segmentation. IEEE Access 7, 23421–23429. https://doi.org/10.1109/ACCESS.2019.2899635.

11

Computational intelligence approach for anomaly detection and prediction in health care information

Sivakumar Nagarajan and K. Sasikumar

School of Computer Science and Engineering, Vellore Institute of Technology, Vellore, Tamil
Nadu, India

1. Introduction

Artificial intelligence (AI) is a scientific discipline focused on the development of computer systems and machines capable of exhibiting reasoning, learning, and decision-making abilities that are typically associated with human intelligence. Additionally, AI encompasses the analysis of data sets that surpass the analytical capacity of humans. Perception, processing natural language, resolving issues and strategy, learning and adaptation, and acting on the environment are some of the specific uses of AI (Tecuci, 2012). The links between Artificial Intelligence (AI), in the area of machine learning (ML), and the use of Deep learning (DL) are depicted in. It should be emphasized that ML is a component of AI, whereas DL is a component of ML, thus rendering both ML and DL, a type of AI. ML focuses on the concept that machines can acquire knowledge and adjust their behavior based on experience and data to accomplish specific tasks. On the other hand, DL models are built upon intricate neural networks that emulate the functioning of the human brain. By employing multiple layers of processing units, deep learning goes a step further in understanding complex patterns within large amounts of data (2022) Fig. 11.1.

The utilization of AI in the field of medicine encompasses two primary domains: virtual and physical. The virtual aspect entails the application of machine learning, also known as deep learning, which involves the utilization of mathematical algorithms to enhance learning capabilities through experiential data which is shown in Fig. 11.1. The current approach to health care, known as "systems thinking," goes beyond the traditional

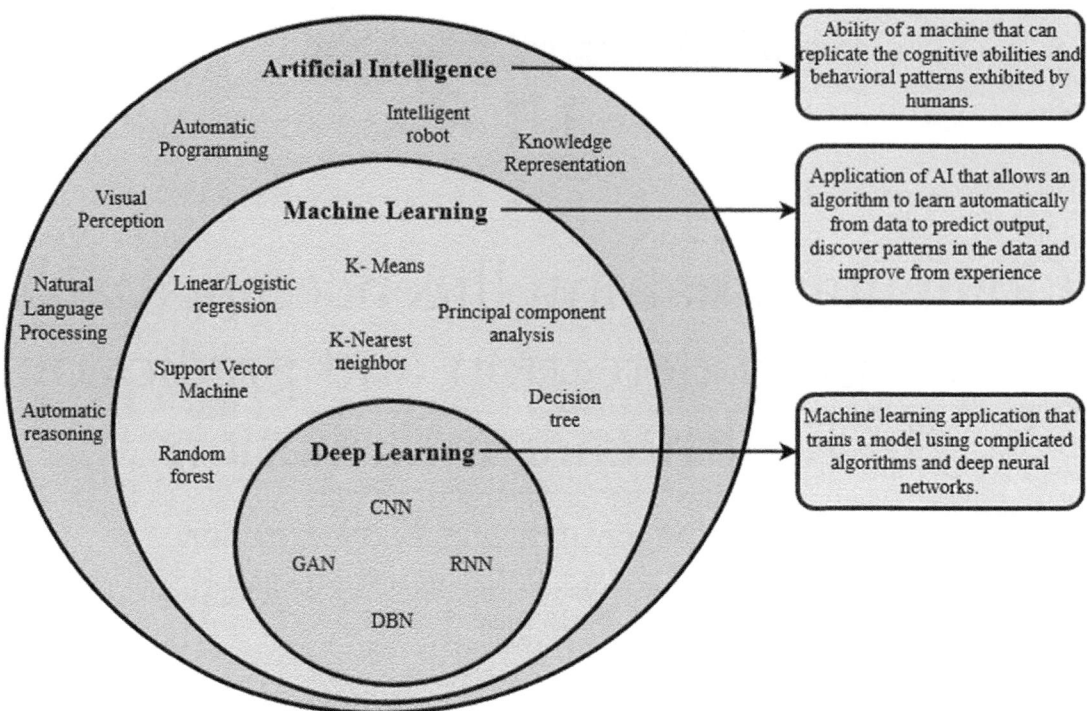

FIGURE 11.1   Relationship between AI, ML and DL.

patient—provider relationship and considers the involvement of Big businesses and cycles. Additionally, the health care system should not remain stagnant and must grow based on their own experiences and work toward making ongoing improvements to its processes (Hamet and Tremblay, 2017). The expansion of databases in the medical sector provides several opportunities for machine learning and artificial intelligence technologies. Many medical equipment is readily available in the medical profession; however, medical errors continue to be a serious problem. Multiple methods have been developed to locate and resolve medical mistakes, with anomaly detection being one of the most crucial approaches (Samariya et al., 2023).

The rest of this book is structured as follows. Section 2 discusses the research motivation, and Sections 3 and 4 provide a brief overview of Machine Learning and Deep Learning. Section 5 examines the various types of Anomaly Detection, and Sections 6 and 7 detail the methods and output types of Anomaly Detection. Section 8 explains the machine-learning approach for Anomaly Detection, and Section 9 presents the Machine Learning Algorithms for Anomaly Detection. Section 10 focuses on anomaly detection in healthcare information and Section 11 discusses the detection of anomalies in different diseases. Section 12 presents a comparison of Popular Anomaly Detection Algorithms, and Section 13 provides the conclusions.

2. Motivation

Recently, healthcare data have been experiencing rapid growth. The worldwide medical care data storage market witnessed a significant increase from $4.88 billion in 2022 to $5.7 billion in 2023, representing a 16.7% compound annual growth rate (CAGR). Furthermore, it is projected to reach $10.52 billion by 2027, maintaining a CAGR of 16.6% (Healthcare Data Storage Global Market Report 2023, 2023). Approximately 8.6 million individuals die from reasons amenable to health treatment; of these, five million have accessed the health system but receive low-quality healthcare. This is more than five times the global total of HIV/AIDS mortality and more than three times the total number of diabetes deaths (Kruk et al., 2018), where gathering and evaluating health data can assist healthcare practitioners in diagnosing illnesses, developing treatment strategies, and improving patient care quality. Healthcare data may also push public health authorities to respond rapidly to concerns and deliver better treatments. Anomaly detection is employed because of its vital role in management tasks that can enhance the quality of health services and prevent loss of life and significant financial losses.

The introduction of ML algorithms to patient medical data can improve patient care while lowering the mental strain on healthcare workers. Machine learning algorithms are effective in detecting these irregular patterns and may be used to detect anomalous physiological measurements, this could lead to faster emergency actions or fresh knowledge of the development of a health problem issue (Šabić et al., 2021; Kavitha et al., 2021).

3. Machine learning

ML is an area in AI that concentrates on creating computer machines that can gain knowledge from information and data. This practical field offers numerous solutions to everyday problems, making it extremely useful. Machine learning algorithms were used to identify correlations and patterns within the data. These algorithms produce predictions, classify data, cluster data elements, reduce dimensionality, and promote the development of novel information from previous data. The application of machine learning is prevalent across diverse professional domains including science and medicine, engineering, sales and marketing, automotive, architecture, and construction (Holzinger, 2019; Salehinejad et al., 2017).

ML is classified into four main types founded on learning methods and approaches. These include Supervised ML, Unsupervised ML, Semi-Supervised ML, and Reinforcement Learning. Fig. 11.2 presents a detailed representation of these types.

3.1 Supervised machine learning

The strategy relies on supervision. This implies that in the supervised learning technique which show in Fig. 11.3, the system is trained using labeled datasets, and the system predicts the output according to the training. We trained the system with both the Input as well as output data before asking it to predict the outcome by applying the test dataset. The training process continued until the model arrived at an appropriate degree of accuracy for the

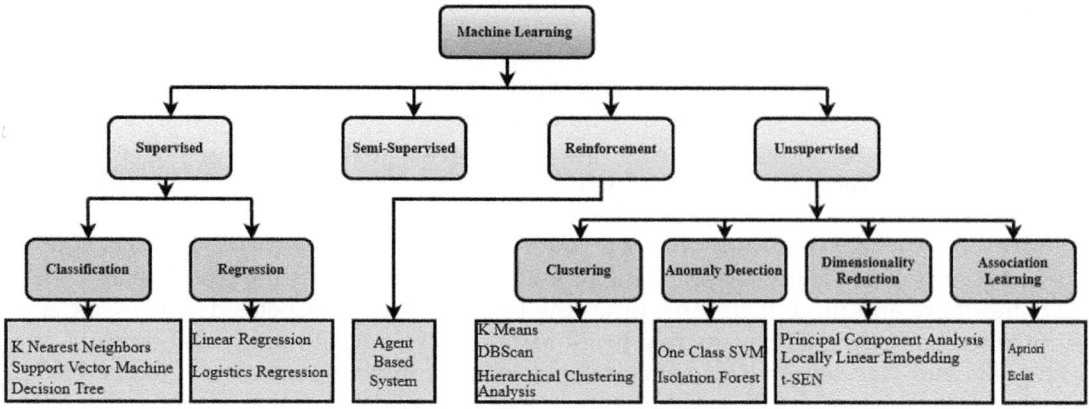

FIGURE 11.2   Classification of machine learning.

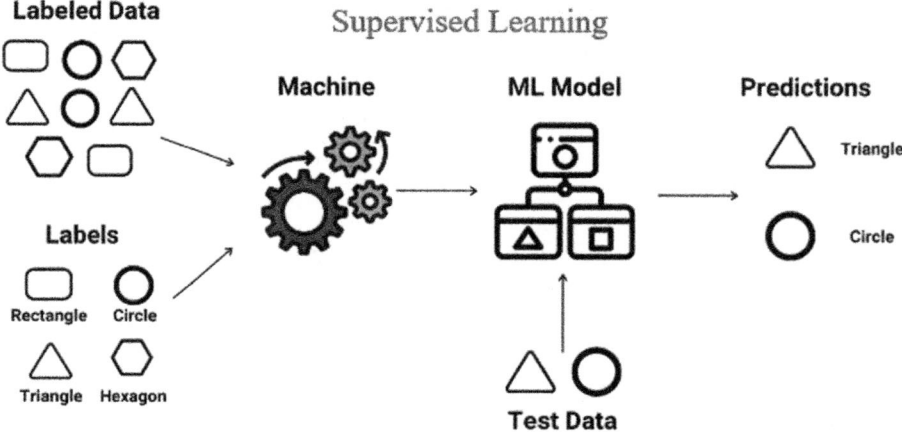

FIGURE 11.3   Supervised learning. *From Ravish Raj (2021). Superrvised, Unsupervised and Semi-Superrvised Learning real life usecase. https://www.enjoyalgorithms.com/blogs/supervised-unsupervised-and-semisupervised- learning*

training data. Supervised Learning Examples: KNN, Random Forest, Decision Tree, Logistic Regression, etc.

3.2 Unsupervised learning

This method differs from supervised learning in that it does not require any supervision. The system learns using an unlabeled dataset in unsupervised machine learning which shown in Fig. 11.4. This approach cannot forecast or estimate objective or outcome variables. Clustering populations in different groups is widely used to split customers into separate groups for specialized treatments. Examples: K-means. One-class SVM, Principal component analysis.

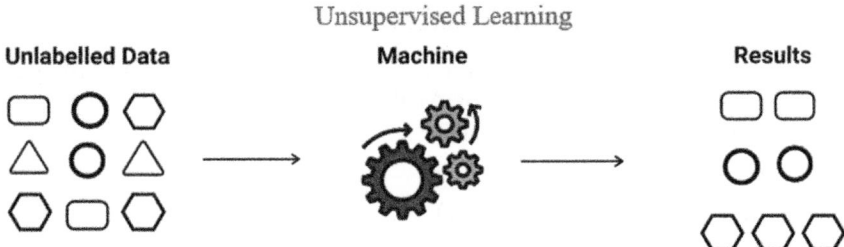

FIGURE 11.4 Unsupervised learning. *From Ravish Raj (2021). Superrvised, Unsupervised and Semi-Superrvised Learning real life usecase. https://www.enjoyalgorithms.com/blogs/supervised-unsupervised-and-semisupervised-learning*

3.3 Semi-supervised learning

This machine learning approach is an intermediate between unsupervised and supervised learning methods. By integrating labeled and unmarked datasets during the training phase, the gap between the supervised and unsupervised learning approaches can be determined.

3.4 Reinforcement learning

In this learning process, the computer is taught to make choices. The machine is placed in a situation in which it is continuously trained through trial and error. The system is trained from previous experiences and seeks to gather the best available knowledge to make sound decisions. An example of Reinforcement Learning is the Markov Decision Process (Oladi-pupo, n.d.; Abdi, 2016).

4. Deep learning

DL is a branch of ML approaches that employs artificial neural networks. It can recognize complicated patterns and connections in data. In DL it is unnecessary to manually program everything. It has gained prominence in recent years as the processing power and availability of large datasets have increased. These are also known as deep neural networks and are designed to learn from enormous amounts of data by being motivated by the shape and function of genuine neurons in the human brain. Currently, different deep learning architectures are available, and this study field is rapidly expanding, with new models being generated almost every week. The community is extremely open, and there are many high-quality deep-learning courses and books (Deng and Yu, 2013; Goodfellow et al., 2016).

4.1 Artificial neural network (ANN)

In the science of artificial intelligence, artificial neural networks aim to replicate the network of neurons that make up the human brain so that machines can understand things and draw conclusions in a human-like manner. Computers are taught to act similarly to the neurons connected in an artificial neural network (Kukreja, 2016).

4.2 Convolutional neural network (CNN)

CNN is a DL neural network design that is widely utilized in the field of image processing. The domain of image processing is part of AI that enables the system to comprehend and analyze pictures or visual information (2017).

4.3 Recurrent neural networks (RNN)

RNNs are more advanced. It saves the results of the computational nodes and returns them to the model (information is not transferred in only one direction). Consequently, the model can anticipate the outcome of a layer. Each node in the RNN model serves as a storage cell, allowing the calculation and execution of the operations to continue. If the system's prediction is incorrect, the structure self-learns and estimates the correct solution during the back-propagation (Salehinejad et al., 2017).

5. Anomaly detection

Anomalies are data deviations which not follow a clearly established pattern of ordinary behavior. The diagram in Fig. 11.5 shows the anomalies within a basic two-dimensional dataset. The data contain a normal region, B1, where the majority of the observations fall within its boundaries. However, points A1 and A2, which are significantly distant from the regions, are considered anomalies. Data anomalies can be generated by a variety of circumstances, including malicious behaviors such as card fraud, cyber-intrusion, terrorist behavior, and system failure, but they all have a single characteristic in common: they are all interesting to the

FIGURE 11.5   A simple illustration of anomalies in a two-dimensional data collection.

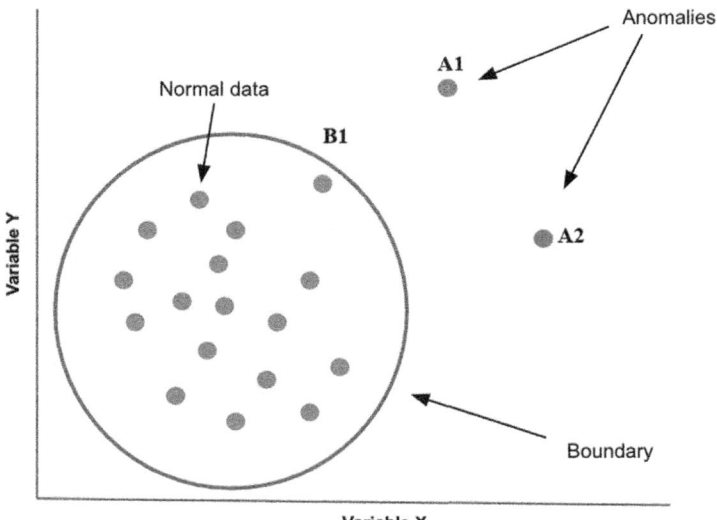

analyst. An important element of anomaly identification is the real-world relevance of abnormalities.

The objective of detecting anomalies in the data is to find odd patterns that differ from the normal course of events. In many domains, these patterns are known as anomalies, outliers, inconsistent observations, exceptions, aberrations, surprises, oddities, or contamination. Anomalies and outliers are the most widely used terms in the context of anomaly identification and are occasionally used interchangeably. Outlier detection is widely used in a variety of sectors, including credit card, insurance, and healthcare fraud detection; intrusion detection in cyber security; defect detection in key safety systems; and army monitoring for enemy activity (Chandola et al., 2009).

5.1 Types of anomaly

Anomalies can take on several forms. Typically, three distinct kinds of anomalies exists.

- **Point anomalies:** A significant number of literary works have focused on point abnormalities. A point is deemed an anomaly if it differs considerably from the other data Fig. 11.6. Therefore, if a point Y_t value deviates considerably from every other point in the interval $[Y_t - k, Y_t + k]$, where $k \in R$ and k are sufficiently high, Y_t is said to be a point anomaly. Consider the following scenario. Suppose a user withdraws from their account at any given time. If one of these withdrawals is considerably larger than any of the previous withdrawals, it is considered an anomaly that justifies blocking the account and conducting a more thorough investigation.
- **Collective anomalies:** This anomaly implies a collection of connected data occurrences that are abnormal compared with the complete dataset. Although individual data points within a collective anomaly may not be unusual, their presence as a group is considered

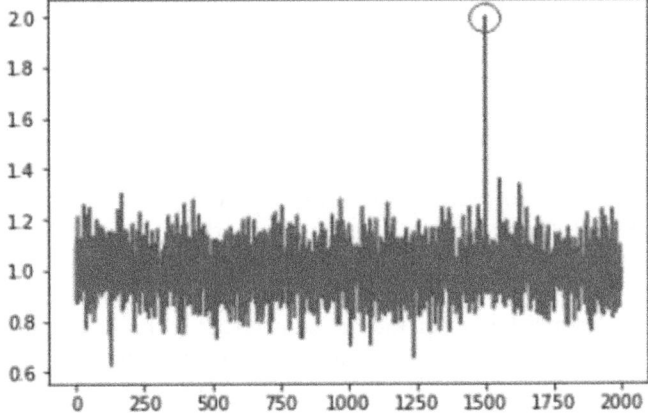

FIGURE 11.6   In random Gaussian noise, there is a point anomaly (circled in *red*). *From Goldberger, A.L., Amaral, L.A., Glass, L., Hausdorff, J.M., Ivanov, P.C., Mark, R.G., Mietus, J.E., Moody, G.B., Peng, C.K., Stanley, H.E., 2000. PhysioBank, PhysioToolkit, and PhysioNet: components of a new research resource for complex physiologic signals. Circulation 101 (23), E215–E220.*

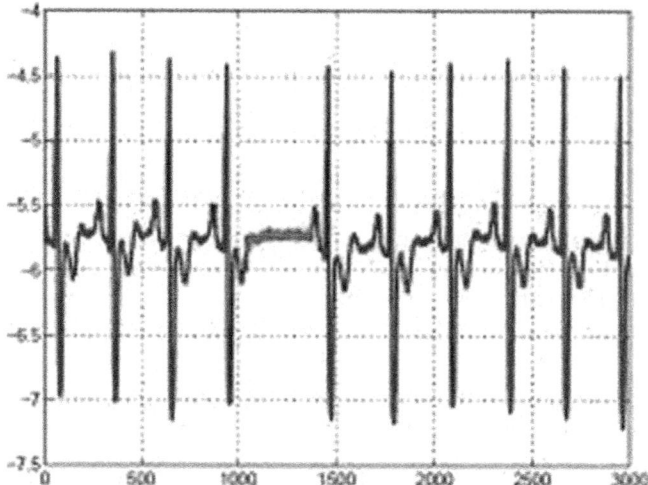

FIGURE 11.7   The simulated ECG time series exhibits a collective anomaly, which is clearly indicated by the *red* markings. *From Goldberger, A.L., Amaral, L.A., Glass, L., Hausdorff, J.M., Ivanov, P.C., Mark, R.G., Mietus, J.E., Moody, G.B., Peng, C.K., Stanley, H.E., 2000. PhysioBank, PhysioToolkit, and PhysioNet: components of a new research resource for complex physiologic signals. Circulation 101 (23), E215–E220.*

abnormal. Fig. 11.7 illustrates the output of the human electrocardiogram (Goldberger et al., 2000). The highlighted area indicates an anomaly, as there is a prolonged presence of a low value, which corresponds to premature atrial contraction. It is important to note that a low value alone is not considered an anomaly.

- **Contextual anomalies:** A contextual anomaly is an incident that may be regarded as unusual in certain contexts. This means that viewing the same point in multiple circumstances does not always indicate abnormal behavior. Contextual and behavioral information was used to define contextual abnormalities. To illustrate, let's consider the task of identifying unusual temperature patterns across different states in the United States. Simply gathering all temperature data and inputting it into the model would not suffice, as each state experiences different temperature variations on any given day. For instance, New York undergoes significant temperature fluctuations throughout the year, which is normal, while Hawaii maintains warm and sunny conditions consistently. Therefore, our objective is to identify anomalies specific to each state. If we fail to incorporate the context of the US states into our model, it is possible that the model might mistakenly flag New York temperatures as anomalies, when in reality, the low temperatures from November to March are typical. Hence, depending on the intended application, we must consider other factors that could influence our results to ensure the highest level of accuracy. This phenomenon is referred to as Contextual Anomalies which is shown in Fig. 11.8. Another instance that demonstrates a contextual anomaly is when a staff member tries to log into the corporate system using his/her own credentials. This action is not unusual. However, if it occurs outside predetermined business hours, it becomes an anomalous event in the temporal context (Baddar et al., 2014).

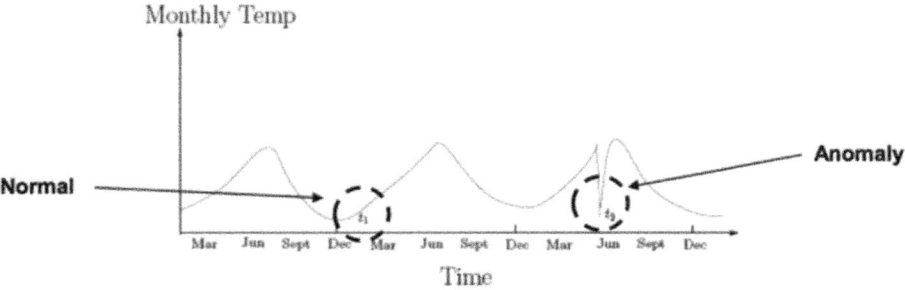

FIGURE 11.8   The development of a CNN for glucose level prediction and identification of anomalies. *From Nico Wong (2021). Detecting Contextual Anomalies with SAP HANA ML. https://blogs.sap.com/2021/10/06/detecting-contextual-anomalies-with-sap-hana-ml/.*

6. Anomaly detection methods

6.1 Statistical methods

This is determined by the characteristics of the data and the specific requirements of the application. There are several different methods and strategies for anomaly detection. The following methods for discovering anomalies are common. Anomaly detection through statistical methods involves the identification of data points that differ from the anticipated statistical patterns or distributions. These techniques are typically straightforward to execute and prove beneficial when the dataset is limited, or when the data are anticipated to conform to a particular statistical distribution. Percentile and interquartile range (IQR) methods are among the most commonly employed statistical methods for anomaly detection.

- **Percentile method:** The percentile approach identifies data points outside a certain range. Anomalies were defined as data points that fell beyond the specified percentile range.
- **Interquartile range (IQR) method:** The interquartile range (IQR) approach was based on the difference between the dataset's first and third quartiles. Data points that are beyond the range specified IQR are termed anomalies in this approach.

6.2 Machine learning-based methods

Anomaly detection approaches based on machine learning involve the use of supervised or unsupervised machine learning algorithms that automatically recognize abnormalities in data, without relying exclusively on traditional data statistics. To construct a model that can detect anomalies in fresh data, these approaches often require a training dataset that includes both normal and anomalous data.

6.3 Time-series analysis

Anomaly identification in time-series data, which involves the collection of observations over time, necessitates the use of specialized methods. Methods such as moving averages, exponential smoothing, and ARIMA models are commonly employed to describe and forecast the expected behavior of a time series. Subsequently, any data points that deviated from the projected values were identified as anomalies.

6.4 Clustering-based methods

Clustering techniques combine related data points based on their distance or similarity metrics. Anomalies are points of information that do not belong to any cluster, or belong to a sparsely populated cluster. K-means clustering and Density-Based Spatial Clustering of Applications with Noise (DBSCAN) are two examples of clustering algorithms used for anomaly detection.

6.5 Deep learning-based methods

Deep learning approaches, such as neural networks, are well suited for anomaly detection because they can automatically learn complex representations and patterns from data. In particular, autoencoders have been widely used for unsupervised anomaly identification. Because the autoencoder learns to regenerate the input data, cases with large reconstruction errors are treated as anomalies.

6.6 Ensemble methods

To enhance overall efficiency, ensemble techniques amalgamate multiple models or strategies for anomaly detection. By merging the results of various methods, ensemble models can detect anomalies with higher precision and resilience (*Anomaly Detection with Machine Learning Overview*, 2023; *The Top Anomaly Detection Techniques You Need to Know*, 2023).

7. Output of anomaly detection

The method in which abnormalities are reported is a key feature of any anomaly or outlier detection technique. The outcomes of techniques for identifying anomalies typically fall into any of the two groups.

- **Scores:** In scoring-based anomaly detection techniques, each data point in the test set is given a score that reflects the level of anomaly. Consequently, these techniques generate a ranked list of anomalies as their output. Analysts can either study the most significant anomalies or select the anomalies using a cutoff criterion. This adaptability is achieved via scoring-based methodologies, which enable analysts to select the most significant anomalies using a domain-specific cutoff or threshold. The method that supplies binary labels to test instances, however, does not directly facilitate this decision. Analysts can indirectly control this selection by adjusting the parameters of each technique.
- **Labels:** The methods in this classification allocate a classification (regular or irregular) for each test instance (Chandola et al., 2009).

8. Machine learning for anomaly detection

Machine Learning concepts can be used to discover anomalies. This can be accomplished in the following manner.

- **Supervised Anomaly Detection:** An ML engineer requires a training dataset to perform a supervised anomaly detection. Dataset items were categorized as either normal or abnormal. The model analyzes these samples to identify patterns and unusual patterns in the data that have never been observed before.
- **Unsupervised Anomaly Detection:** The method learns to recognize abnormalities using this technique without a previous understanding of what constitutes regular behavior. It is beneficial when no labeled datasets are available or when abnormalities are uncommon and difficult to detect.

This form of anomaly detection is widely used, and the most well-known example of unsupervised algorithms is neural networks. Neural networks reduce the need for manual preprocessing of examples because they do not require manual labeling. They can also be used for unstructured data. Neural networks can identify anomalies in unlabeled data and apply their knowledge to new data.

In addition, it is difficult to forecast all anomalies that may occur in a dataset. For example, consider a self-driving car. They can come with events on the road that have never happened before. It is impossible to classify all the traffic scenarios into a small number of categories. Consequently, when working with real-time data, neural networks are crucial.

- **Semi-supervised Anomaly Detection:** The benefits of the two previous approaches were combined with semi-supervised anomaly detection techniques. Engineers can perform automatic feature learning and work with unstructured data by using unsupervised learning approaches. However, by integrating them with human supervision, they can monitor and regulate patterns learned by the model. This typically helps improve the model's predictions (2021).

9. Machine learning algorithms for anomaly detection

Several popular anomaly detection algorithms are listed below. Machine learning algorithms are crucial in identifying anomalies as they can automatically find patterns and differentiate deviations from expected behavior. Machine learning has a number of ways for detecting anomalies. Fig. 11.9 depicts several of these features.

9.1 Isolation forest

The Isolation Forest approach identifies anomalies in unsupervised learning by using isolation trees. It operates by choosing a random feature and creating a split value that falls within the feature range. Anomalies were found faster because isolating them from the rest of the data required fewer splits.

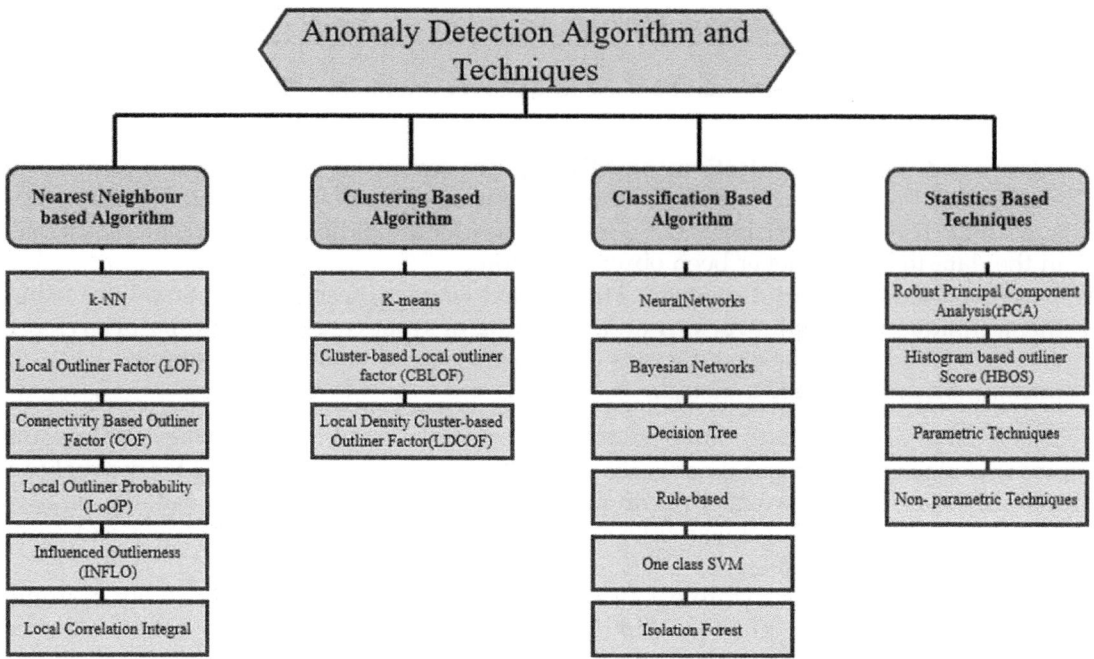

FIGURE 11.9   Anomaly detection algorithms and techniques.

9.2 Random forests

Random forests can be used for anomaly detection by training a forest of decision trees using typical data. Anomalies are data points that receive few votes or have a low average distance from the forest trees.

9.3 One-class support vector machines (SVM)

One-class SVM is a popular type of anomaly detection that aims to find a hyperplane that captures the bulk of data while reducing the inclusion of anomalies. The training sample data are used to create a model, which is then employed to determine whether the test results as either normal or abnormal based on how closely it resembles the model.

9.4 Support vector data description (SVDD)

An alternative variant of the SVM, known as the Support Vector Data Description (SVDD), is designed to enclose the majority of instances within a hypersphere centered on the normal data. Any data points that fall outside this hyper sphere were identified as anomalies.

9.5 Autoencoders

Autoencoders are neural network models trained to replicate the input data. For anomaly identification, an autoencoder is trained on the usual data, and situations with substantial reconstruction errors are considered as anomalies. Because they can capture complex patterns and nonlinear connections, autoencoders are effective for detecting abnormalities.

9.6 Local outlier factor (LOF)

The LOF is a critical outlier detection technique based on the notion of local density. Density was estimated using the distance between the k-nearest neighbors. LOF compares an item's local density to the local density of its neighbors. As a result, similar density areas and objects with densities much lower than their neighbors can be identified. However, these are exceptions.

9.7 Gaussian mixture models

GMMs generate data based on a combination of Gaussian distributions. The IGMM identifies anomalies as data points with a low likelihood. This approach is effective when normal information has a multivariate Gaussian distribution (*Anomaly Detection with Machine Learning Overview*, 2023).

10. Anomaly detection in health care information

Negative outcomes and medical mistakes in the healthcare industry lead to a significant number of unintentional deaths and injuries annually. The abundance of healthcare data, which often include crucial information related to human survival, makes the analysis of this data crucial. The ability to analyze healthcare data is particularly important, as it has the potential to save lives and improve overall quality of life. Anomaly detection is a key component of healthcare analytics as it aids in the identification of abnormal patterns or outliers in the data. In everyday clinical practice, doctors face challenges in analyzing clinical data owing to their rapid and exponential growth. However, the integration of artificial intelligence can assist in reducing processing time and enhancing the quality of care provided. Anomaly detection in medical image analysis is particularly beneficial for accurate diagnostics, whereas the analysis of treatment plans can help identify potentially fatal errors. By applying machine learning algorithms to healthcare data, patient care can be improved and the cognitive load on healthcare workers can be reduced (Ukil et al., 2016).

This study (Hauskrecht et al., 2013) introduces a novel framework for monitoring and alerting based on stored clinical data and statistical methods to identify clinical outliers. The goal was to identify unusual patient treatment processes based on the patient's condition. The framework utilizes conditional outlier detection methods (Hauskrecht et al., 2007), which focus on recognizing anomalous results for a subset of qualities and evaluating the values of the other attributes. This approach is well-suited for identifying uncommon events, actions, and attribute combinations. The structure consists of two steps: model-building and model-application. In the model-building stage, outlier models are created using cases from the electronic health record repository to explain when specific patient management activities are typically taken under specific patient conditions. Multiple outlier models were built to incorporate several patient management actions. Outlier models are added to fresh patient data in the model application step to detect activities that depart from the typical treatment pattern represented by the outlier models. The framework was tested on data from 4486 postcardiac surgical patients' electronic health records, resulting in 222 alerts. This chapter (Görnitz et al., 2013) provides fundamental information on semi-supervised anomaly detection using a mathematically correct approach. To address this need, the authors offer a unique semi-supervised anomaly detection approach built from the unsupervised learning method, which allows for the inclusion of labeled data in the training activity. The support vector data description (SVDD) is the foundation of our method, with the original structure included as a special circumstance. Even though the final optimization issue is not convex, we show that an analogous convex formulation may be constructed with extremely modest assumptions. They also offer an active learning technique to enhance the real model and rapidly find novel outlier clusters to assist users in the labeling process. In complicated domains, multi-labeling and revising semi-supervised outlier detection as a multi-task problem can improve accuracy.

To address the security concerns associated with transferring medical information over public networks, this chapter (Haque et al., 2021) introduces a modern smart digital healthcare system (SDHS) for IoT environments. This system leverages anomaly detection through advanced machine learning techniques, specifically using a supervised ML model for classification and an unsupervised model for outlier detection. Both models utilize a comprehensive bio-inspired optimization structure that incorporates three optimization algorithms: gray wolf, whale, and firefly. The framework also employs a novel fitness function for unsupervised outlier detection ML models. Hyperparameter tuning is accomplished through bio-inspired computing-based optimization algorithms, which greatly enhance the performance of ML models. The proposed structure exhibits a significant decrease in false alarm rates and performs exceptionally well in anomaly detection. This study ("DeepCAD: A Stand-alone Deep Neural Network-based Framework for Classification and Anomaly Detection in Smart Healthcare Systems," 2022) introduces DeepCAD, a new architecture that combines a stand-alone deep neural network (DNN) model with outlier detection rules to enable classification and outlier detection in smart healthcare systems. it was designed to train a standalone DNN model that can perform both categorization tasks and outlier detection without the need for training with anomalous samples. Unlike traditional solutions, which require separate models for anomaly detection, this approach integrates both tasks into a single model. The strength of the DNN model in the DeepCAD framework was evaluated against adversarial machine-learning models, and the framework was validated using a real dataset. In contrast to traditional approaches that involve the use of two different machine learning models for these tasks, DeepCAD offers a more streamlined and efficient solution.

The DeepCAD framework, as depicted in Fig. 11.10 comprises two main stages: data pre-processing and model training. The data preprocessing stage involved three steps. First, Feature Preprocessing utilizes dimension-reduction techniques to reduce a high-dimensional feature space to a two-dimensional space.

Second, Boundary Acquisition was employed to encode the anomaly detection rule and effectively distribute the data within the boundary. Finally, the efficiency of accurate outlier detection in DeepCAD relies on the label preprocessing step. The model-training stage of the DeepCAD structure consists of four steps. First, a DNN model deployment was performed to set up the deep neural network model. Second, verification sample generation was performed. Third, a loss calculation is conducted to assess the efficacy of the model. Finally, model tuning was performed to optimize the parameters of the model and enhance its overall performance.

Guo et al. (2022) presented a quantum outliers detection technique based on density calculation, which is an extension of the outliers detection algorithm focused on density estimation because it is computationally costly when processing large datasets. This suggested technique outperforms its conventional version in terms of the amount of training data points M. This concept may also be used to speed up the anomaly detection algorithm by utilizing kernel principal component analysis. This study (Samariya et al., 2023) can serve as a model for more efficient quantum machine algorithms. This work seeks to find anomalies and explain why they qualify as anomalies by detecting outlying aspects. The following strategies were utilized in this study: density-based anomaly detection using Local Outlier Factor, Isolation Forest, isolation using Nearest Neighbor Ensemble, and Rapid distance-based outlier detection. The assessment was conducted using 16 healthcare data and showed that the isolation-based algorithm shows promising efficiency.

11. Detection of anomalies in different diseases

ML has the ability to increase the accuracy and efficiency of screening procedures, thereby enabling healthcare experts to detect illnesses at an early and treatable stage. By utilizing the ML approach to analyze medical images, and diagnostic methods, and identify patterns that may indicate diseases or abnormalities, healthcare professionals can identify individuals who are at a higher risk of getting specific conditions or diseases, even before symptoms appear. Machine learning algorithms have the capability to analyze patient information, including medical records, genetic data, and lifestyle aspects, to identify individuals who are at a heightened risk for certain conditions. This allows for the implementation of focused interventions such as lifestyle choices or early detection. The subsequent section delves into the topic of anomaly detection in various diseases (El-Sherbini et al., 2023).

11.1 Cancer

In 2020, 18,094,716 million tumor cases were reported globally. The age-standardized prevalence rate for all malignancies, leaving out nonmelanoma skin tumors, was 190 per

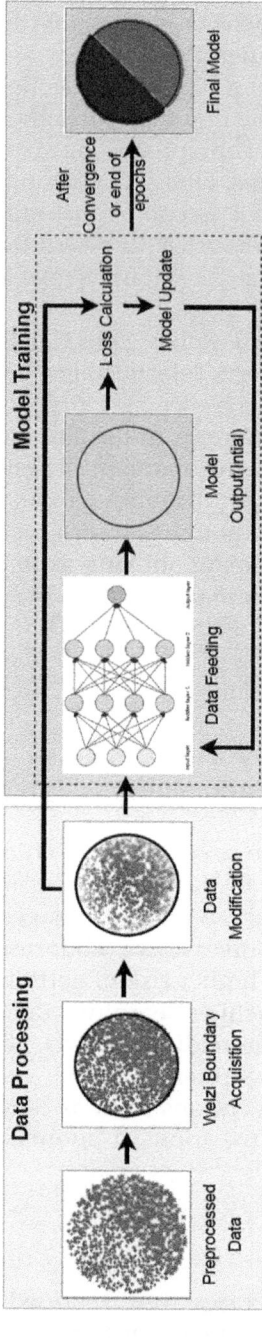

FIGURE 11.10 Framework. *From DeepCAD: A Stand-alone Deep Neural Network-based Framework for Classification and Anomaly Detection in Smart Healthcare Systems, n.d. In: 2022 IEEE International Conference on Digital Health (ICDH). IEEE, 2022, pp. 218–227.*

100,000 people in both Male and Female. Notably, men had a greater rate of 206.9 cases per 100,000 than women, who had a rate of 178.1 cases per 100,000. The growing burden of cancer in practically every nation poses a substantial public health challenge. Addressing risk factors related to food, nutrition, and exercise could potentially prevent approximately 40% of all cancer cases (*Global cancer data by country*, 2020). Encouragingly, machine learning models have demonstrated promise in the premature detection and screening of lung, cervical, colorectal, breast, and prostate cancer (Alharbi and Vakanski, 2023).

The author of this study (Bhargavi, 2022) conducted a statistical analysis of colon cancer data to assess the effectiveness of various outlier detection techniques. These techniques include Chauvenet's method, Tukey, and modified Z-score methods, as well as Z-score and standard deviation methods. The goal was to determine the effective method for identifying and managing outliers. The analysis revealed that the positive node data contained potential anomalies.

In clinical settings, treatment plans are used for patients undergoing radiotherapy. Prior to the implementation, these plans were scrutinized by human experts to ensure safety and quality. Some plans have been identified as having flaws that require further refinement. An unsupervised learning strategy that leveraged an autoencoder was developed to expedite this evaluation procedure. This method offers an outlier detection algorithm specifically designed for radiation therapy planning. This demonstrates the capacity to identify a small fraction of aberrant plans from a huge pool of radiation plans with accuracy and precision. To assess the effectiveness of this method, a comparison was made with other anomaly detection algorithms such as local outlier factor (LOF), hierarchical density-based spatial clustering of applications with noise (HDBSCAN), one-class support vector machine (OC-SVM), and principal component analysis (PCA). The proposed method achieved a value of 0.9985 for the area under the receiver operating characteristic curve (AUC), indicating its outstanding performance (Huang et al., 2023).

Breast tumors are the most prevalent kind of cancer in females and the primary reason behind deaths related to cancer. Mammography is universally acknowledged as the most effective technology for determining the presence of breast tumors. Mammography examination has been found to reduce the death rate considerably from breast cancer (Ghoncheh et al., 2016). To address this issue, Park et al. (2023) suggested a generative model that employs a cutting-edge generative network (StyleGAN2) to generate high-quality simulated mammographic pictures, as well as an anomaly detection approach for detecting breast tumors on mammography in unsupervised ways.

GANs (Generative Adversarial Networks) are made up of two neural networks: generator and discriminator networks. The generator network was trained to produce samples that the discriminator network incorrectly classified as real, whereas the discriminator network was trained to accurately classify data as genuine or fake (Goodfellow et al., 2020). The anomaly detection method utilizes a standard GAN, which is trained solely on positive samples, to learn the mapping from the latent-space representation z to the realistic sample $\hat{w} = G(z)$. This learned model was then utilized to map the unseen information back to the latent area. The Youden J-index threshold was used to define the threshold for the outlier score, which distinguishes between normal and malignant pictures. Fig. 11.11 depicts the entire procedure and Fig. 11.12 depicts the outcomes.

Training GAN model

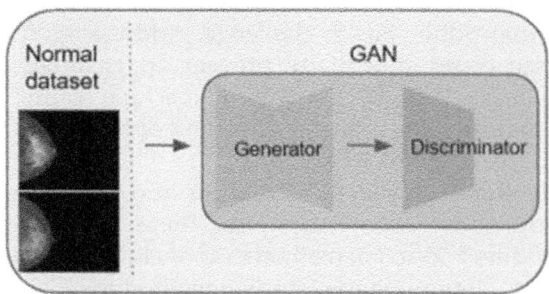

Anomaly detection

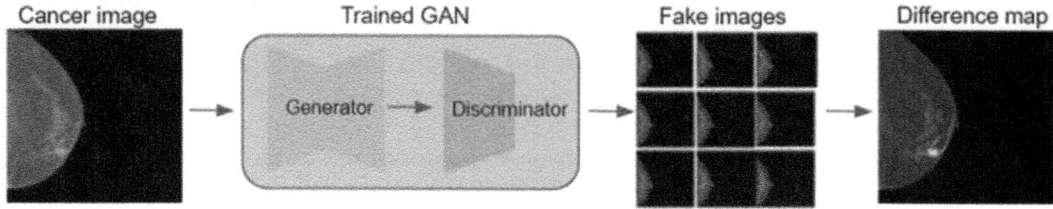

Cancer classification based on anomaly score

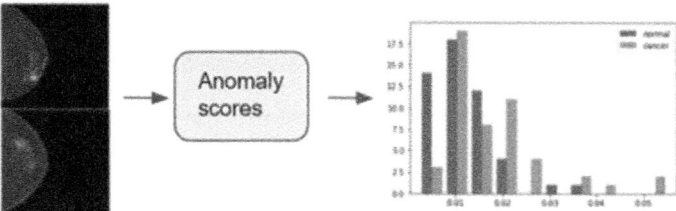

FIGURE 11.11   Breast cancer classification workflow employing anomaly detection technique. *From Park, S., Lee, K.H., Ko, B., Kim, N., 2023. Unsupervised anomaly detection with generative adversarial networks in mammography. Scientific Reports, 13 (1). https://doi.org/10.1038/s41598-023-29521-z.*

A novel framework was introduced in this study (Motai et al., 2016) to detect early stages of cancer using CT colonography. The framework is intended to aid with diagnosis and initial treatment planning of cancer, particularly when dealing with no stationary health data. Outlier detection is a crucial challenge in predicting the cancer phase, and to deal with this issue, a composite kernel has been applied for the first time in the forecasting of cancer staging. The suggested longitudinal analysis of composite kernels (LACK) is a more advanced approach than Kernel PCA and is designed to predict anomaly status and cancer stage. LACK offers several benefits, including the capacity to predict the future risk of tumor stage progression and abnormality status, which can be used for further diagnostic and treatment preparation within a timeline and consistently.

Original image   Projected image   Difference map

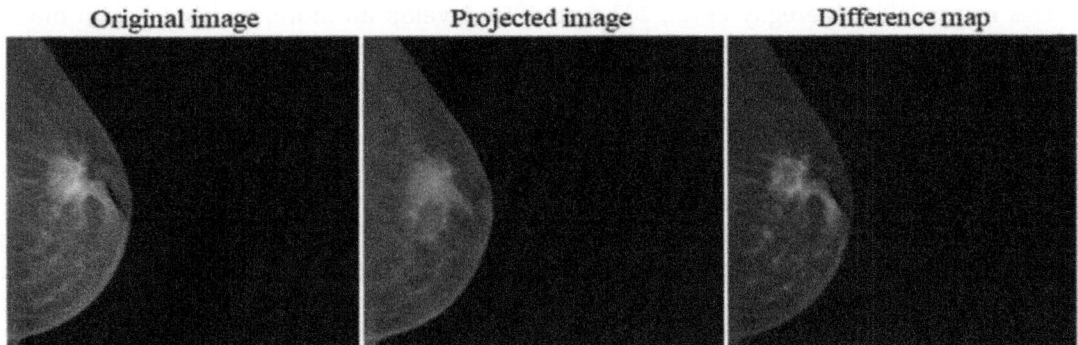

FIGURE 11.12   Anomaly detection examples using a genuine mammography picture as input (initial image), one of the most comparable regular synthetic mammography pictures as output (second image), and a difference mapping between the real and synthetic images (last image). *From Park, S., Lee, K.H., Ko, B., Kim, N., 2023. Unsupervised anomaly detection with generative adversarial networks in mammography. Scientific Reports 13 (1). https://doi.org/10.1038/s41598-023-29521-z.*

11.2 Diabetes

Over 500 million people worldwide are currently living with diabetes, affecting people of all ages and genders in every nation. This number is expected to more than double to 1.3 billion within the next 3 decades. Recent calculations indicate that the global prevalence rate of diabetes is 6.1%, making it one of the top 10 leading causes of death and disability. Diabetes is more common in people aged 65 and older, with a global prevalence rate of more than 20% in this age range. The greatest number, 24.4%, is observed among those aged 75–79. Type 2 diabetes (T2D) accounts for 96% of all diabetes cases globally. The rapid growth of diabetes is not only concerning but also poses a challenge for healthcare systems worldwide, as the disease also increases the risk of ischemic heart disease and stroke. Utilizing machine learning techniques to identify anomalies in diabetes datasets can be highly beneficial for both patients and doctors. This study (Sebastian et al., 2018) presents a novel approach to identifying abnormal blood glucose tracks in a longitudinal diabetes dataset by utilizing a contextual anomaly detection approach that combines a Bayesian technique and molecularity-based community detection algorithm to find anomalous occurrences among people who have highly similar social characteristics and activities. This method was tested using a clinical dataset of patients with diabetes collected in Sarawak, Malaysia. The initial step in this procedure is to use the Jaccard coefficient to obtain a measure of the similarity between two subjects. We created a diabetic common graph once the scores for each participant were calculated. The next stage is community discovery, where the Louvain community detection method is applied to create optimal network partitioning before performing anomaly detection. The Nave Bayes algorithm is used to compute the conditional likelihood of two types of blood sugar trajectories, "increase" and "no increase," given the normal characteristics of community individuals. We were able to identify people with abnormal plasma blood sugar trajectories as a result of this. Despite having highly similar sociodemographic information and lifestyle factors, the results convincingly indicate individuals with differing blood glucose trajectories.

This research (Woldaregay et al., 2020) aims to develop an individualized health model based on blood sugar levels and the insulin-to-carbohydrate proportion as input variables to automatically identify the infection occurrences in individuals with type 1 diabetes. To achieve this, semisupervised and unsupervised outlier detection approaches have been employed, which have proven to be effective in detecting anomalous objects in various applications, particularly in the medical domain. The approach is designed to identify divergence from the norm caused by infection cases, which are characterized by elevated blood sugar levels and strange alterations in the insulin-to-carbohydrate ratio. The study involved training three groups of one-class classifiers on regular day data sets and testing them on a data set that included both target and nontarget (infection) days. The boundary and domain-based method were found to be the most effective among the one-class classifiers, while specific models such as one-class support vector machine, K-nearest neighbor, and K-means demonstrated overall outstanding performance across all sample sizes and illnesses within their respective groups.

The use of implantable IoT medical devices (IoTMDs) has caused a significant transformation in the medical field. Patients with diabetes have greatly benefited from IoTMD as it helps them maintain their blood sugar levels within a normal range. Astillo et al. (2022) proposes a system for detecting anomalies in healthcare systems using deep learning (DL) techniques. The system consists of estimation and classification models that are applied to a specific subdomain in healthcare. This study compares the performance of convolutions neural networks (CNN) and multilayer perceptron (MLP) algorithms. The suggested intrusion detection architecture is made up of three major elements, as shown in Fig. 11.13. The first type is a lightweight CNN-based sugar estimator that predicts the blood sugar level of a patient at a specific time.

Because each diabetes individual insulin sensitivity parameters may change, the CNN-based estimator model is trained separately for each patient's insulin infusion (IF) system. The estimator approach is followed by a derivative method, which calculates the relevant properties that serve as input elements or features for the categorization module. The final module is in charge of deciding if the current event cycle of the diabetes management and control system

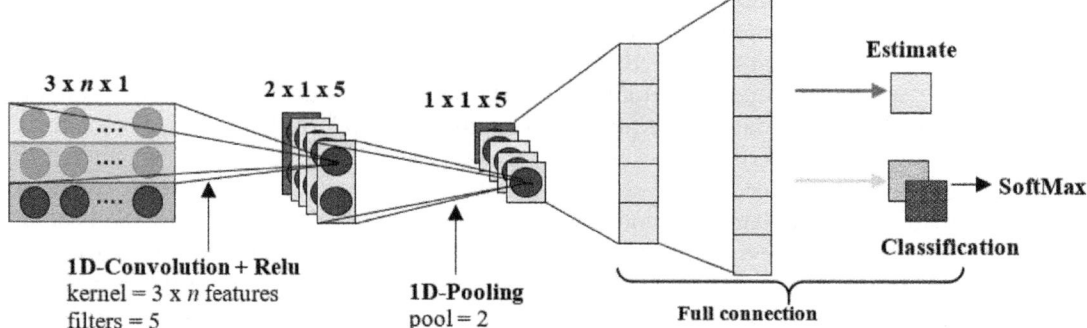

FIGURE 11.13   The development of a CNN for glucose level prediction and identification of anomalies. *From Astillo, Philip Virgil, Duguma, Daniel Gerbi, Park, Hoonyong, Kim, Jiyoon, Kim, Bonam, You, Ilsun, 2022. Federated intelligence of anomaly detection agent in IoTMD-enabled Diabetes Management Control System. Future Generation Computer Systems 128, 395–405. https://doi.org/10.1016/j.future.2021.10.023.*

(DMCS) is abnormal or not. This was accomplished using a lightweight CNN algorithm trained using a federated learning strategy. If an anomaly is found, the system notifies the patient to take the necessary steps to address the reported security concerns. The proposed (Fang et al., 2022) approach in this study involves a hierarchical support vector machine that utilizes a clustering algorithm to classify data sets with similar characteristics. This results in the division of data into potential and significant abnormal parts. To recognize and evaluate data from every part, a convolutional neural network is employed. The feature vector output from the CNN complete connection layer output feature vector was subsequently employed as input data for SVM classification. This approach can detect aberrant diabetes data more accurately by developing an ideal classification hyperplane in a high-dimensional environment.

11.3 Lung disease

Millions of individuals worldwide suffer from various lung diseases, including asthma, pneumothorax, atelecta-sis, bronchial tube inflammation, lung cancer, lung infection, pulmonary edema, and interstitial lung disease (ILD).ILD refers to a collection of illnesses that result in scar tissue formation in the lungs, leading to breathing difficulties and a reduced oxygen supply to the blood circulation. Unfortunately, lung damage caused by ILDs is often irreversible and worsens over time, affecting people of all ages including children. Genetics, certain drugs, and medical procedures such as chemotherapy or radiation exposure can increase the risk of ILDs. To diagnose lung-related diseases, doctors typically order chest X-rays or CT scans to obtain a better view of the lungs (*Chronic respiratory disease*, 2023; *Interstitial Lung Disease*, 2023).

This chapter (Pham et al., 2021) uses respiratory sound recordings for categorizing outliers in respiratory cycles and identifying diseases. The framework starts with feature extraction at the front end, which converts the sound input into a visual representation known as a spectrogram. The spectrogram features are then classified using a back-end DL network into classifications of respiratory abnormality cycles or illnesses. The research was carried out on the ICBHI standard dataset of breathing sounds. The Teacher–Student approach offers a potential solution for constructing real-time applications by striking a balance between the efficacy and complexity of the model. Innovative computational tools are built for the examination of lung respiratory auscultation and are essential for the identification of disease-associated anomalies. This approach (Manzoor et al., 2020) categorizes respiratory sounds into four types: normal, wheezes, crackles, wheezes, and crackles. Initially, the audio signal was converted into an image format using the Mel frequency cepstral coefficient method. The noise was then eliminated through Compression and Denoising autoencoder techniques. Finally, a new RNN design, known as noise-masking anomalies recurrent neural network (NMA-RNN), was employed to classify lung sounds.

The lungs along with other breathing system components are affected by pulmonary diseases, and respiratory sounds reveal important information about a patient's lungs. A new method for detecting pulmonary abnormalities, called FrWCSO-based DRN, is proposed in this study (Dar et al., 2023). The method utilizes Fractional Calculus (FC) and a Water Cycle Swarm Optimizer (WCSO), which is a combination of the Water Cycle Algorithm (WCA) and Competitive Swarm Optimizer (CSO). The method first selects basic features such as BFCC, Wavelet change, spectral flux, spectral central point entropy, energy, and kurtosis from the respiratory sound signal using selection methods. Data argumentation was then performed using window warping (WW), Jittering, and Cropping to avoid overfitting. Finally, FrWCSO was used to detect pulmonary abnormalities. The framework of the FrWCSO-based DRN is illustrated in Fig. 11.14.

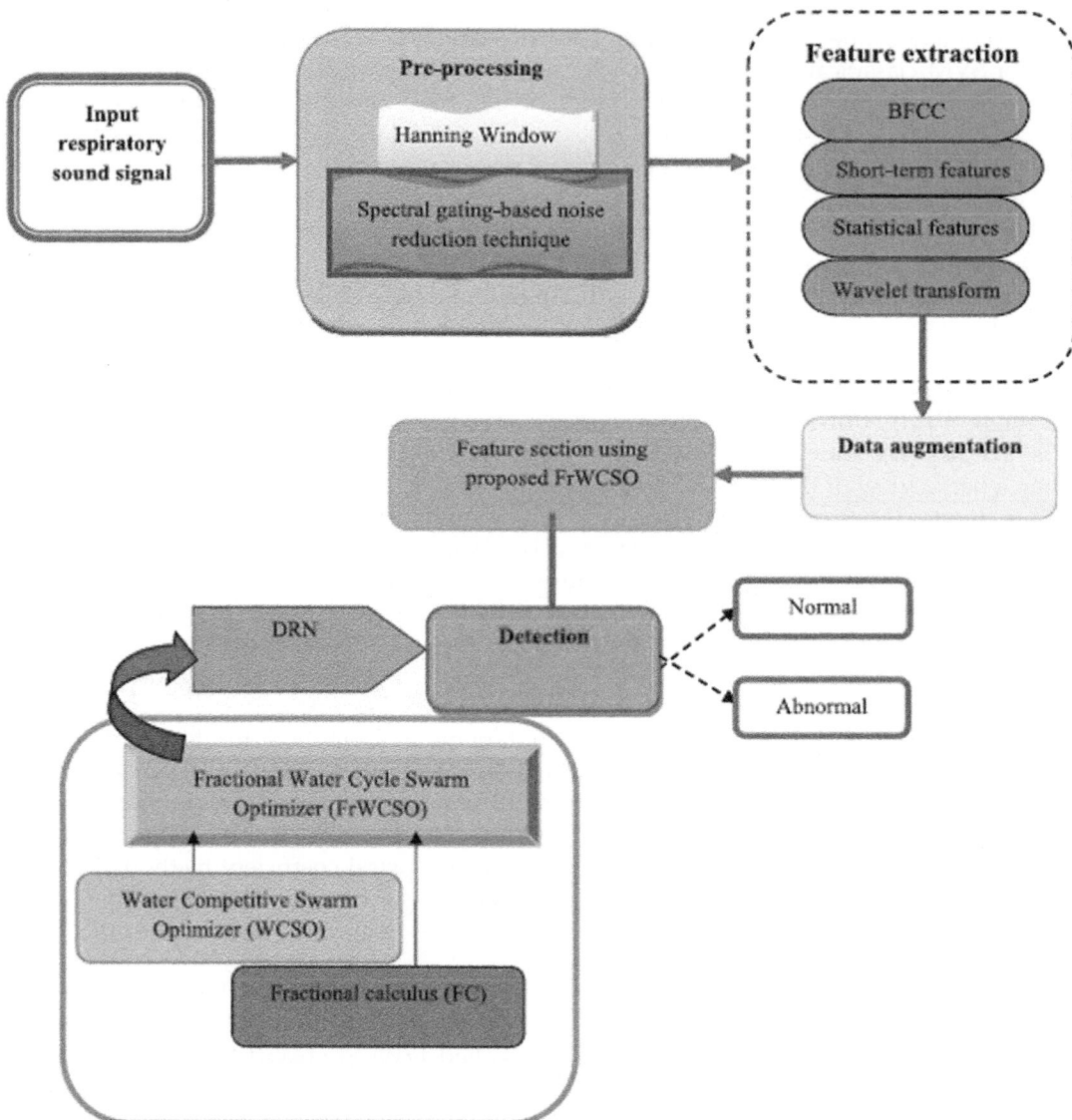

FIGURE 11.14   FrWCSO architecture. *From Dar, J.A., Srivastava, K.K., Mishra, A., 2023. Lung anomaly detection from respiratory sound database (sound signals). Computers in Biology and Medicine, 164, 107311. https://doi.org/10.1016/j. compbiomed.2023.107311.*

Chest X-rays are commonly used as the initial test to detect lung abnormalities, and can therefore aid in the initial detection and avoidance of lung infections. Kim et al. (2021) CSIP (contrast-shifted instances via patch-based percentile) was used to suggest a detection approach based on CXR pictures. This method can automatically detect unseen lung diseases by using deep learning. The system is trained solely on healthy data and can discover outliers, even when data collection is challenging, such as in the situation of the new COVID-19 versions. The patch-based percentile technique for cutting-edge one-class classifiers (OCCs) was applied for the first time using CSIP. Through data augmentation, random cropping in the lung region can potentially improve the detection accuracy of lung illnesses by employing supervised learning.

The CSIP design is depicted in Fig. 11.15. To prepare the CSIP method for detecting anomalies, we exclusively utilized chest X-ray images from individuals without health issues, along with their lung masks. Random cropping in the lung region has the potential to enhance the detection accuracy of lung illnesses in supervised learning through data augmentation.

During the inference stage, the pixel-wise score was determined by taking the patch-wise mean of the distance between the random test regions, which included the desired pixel, and

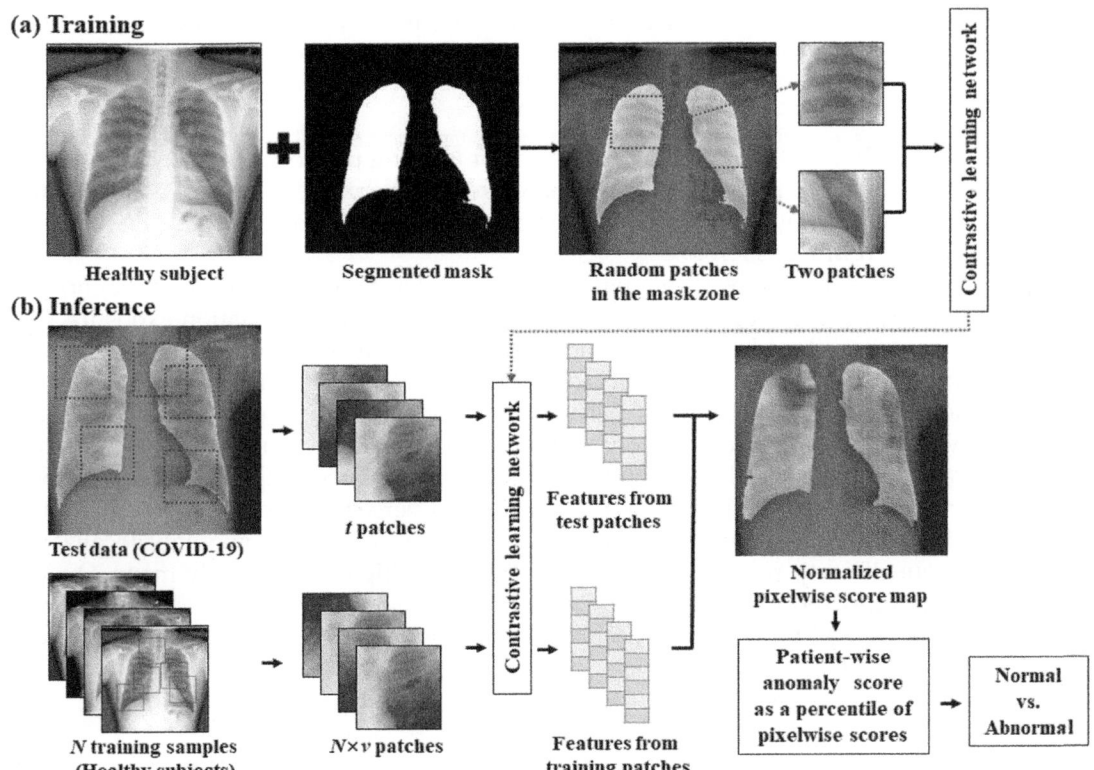

FIGURE 11.15 The proposed CSIP architecture for detecting anomalies in X-ray. *From Kim, K.-S., Oh, S.J., Cho, H.B., Chung, M.J., 2021. One-class classifier for chest X-ray anomaly detection via contrastive patch-based percentile. IEEE Access 9, 168, 496−168, 510. https://doi.org/10.1109/access.2021.3136263.*

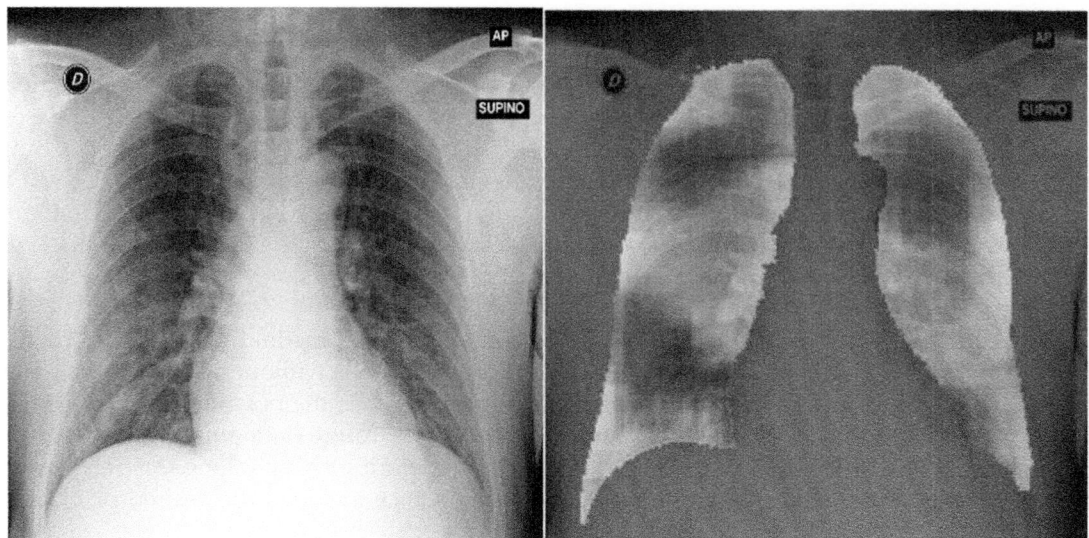

FIGURE 11.16   Maps showing scores for COVID-19 cases at the pixel level. The score ranged from 0 (*blue*, indicating disease) to 1 (*red*, indicating health). *From Kim, K.-S., Oh, S.J., Cho, H.B., Chung, M.J., 2021. One-class classifier for chest X-ray anomaly detection via contrastive patch-based percentile. IEEE Access 9, 168, 496–168, 510. https://doi.org/10. 1109/access.2021.3136263.*

the healthy region that was most similar and used for training purposes. Subsequently, in the split lung area of the X-ray image, the pixel-wise scores were obtained. To derive the resulting anomaly score, the pixel-by-pixel percentile technique approach was applied. The output of this model is shown in Fig. 11.16.

Recently, there has been a surge in interest in utilizing DL for computer-aided diagnosis in low-dose CT scans because of its exceptional precision and minimal radiation exposure. Kim et al. (2024) have introduced a new technique, known as virtual multiview mapping and reconstruction for unsupervised anomaly detection (VMPRUAD), for low-dose chest CT. This approach employs a deep neural network and to find anomalies, it only needs data from healthy patients in three-dimensional (3D) regions. The technique comprises three primary components: enhancing the recognition of three-dimensional lung structures by projecting them virtually into two-dimensional pictures from diverse perspectives, accommodating input diversity for precise anomaly detection, and accomplishing 3D anomaly localization through a unique 3D map restoration method utilizing multiple 2D anomaly maps. Our unsupervised learning-based method has exhibited remarkable performance in detecting pneumonia, tuberculosis, and both diseases, with a patient-level outlier detection performance of 0.965 area under the curve (AUC).

12. Comparison of popular anomaly detection algorithms

In this portion, we will explore the advantages and disadvantages of several well-known outlier detection algorithms which are shown in Table 11.1.

TABLE 11.1 Pros and cons of machine learning model for anomalies detection.

Algorithm	Advantage	Disadvantage
K-nearest neighbor (K-NN)	This is extremely simple to comprehend and is ideal for constructing models that incorporate unconventional data types like text.	Significant storage needs, computationally demanding, and sensitive to the selection of the similarity function for comparing instances.
Local outlier factor (LOF)	A widely recognized and effective algorithm for detecting anomalies within a specific locality.	Depends Mostly on its Nearby neighbors and does not perform well on datasets containing global anomalies.
K-means	Simple to use and minimal in complexity	The number of observations in each cluster is relatively equal, K must be specified, and only the numerical data are used.
Support vector machine (SVM)	Discover the optimal separation plane, handle extremely high dimensional data, and capable of learning complex concepts.	It requires both positive and negative examples, a large amount of memory, issues with numerical stability, and the need to choose a suitable kernel function.
Neural networks based anomaly detection	Handles big datasets and does well in speech and visual recognition.	This requires a large amount of processing power and data, and it overfits if the problem is complex.
Random forest	Handles effects that are not linear and works well with high dimensional information.	Overfits noisy information or data and is difficult to understand.
Logistic Regression	It is simple to execute and understand and is capable of managing both binary and multi class classification.	It does not work well with anomalies and implies a linear relationship.
Convolutional neural network	It performs admirably in video and picture recognition, as well as learning hierarchical features.	This requires a large amount of information and resources, and capacity for interpretation is limited.

13. Conclusion

The use of computational intelligence is crucial for detecting anomalies in healthcare data, particularly in the emerging field of AI in primary care and preventive medicine. This area of study opens numerous possibilities. We explored various ML and DL approaches, as well as different types of anomalies and techniques. We also discussed the implementation of these methods in modern healthcare information and focused on outlier detection for important diseases such as cancer, diabetes, and lung-related illnesses. Additionally, we examined how to identify anomalies in chest X-ray images. This chapter also discusses the advantages and disadvantages of popular machine learning models used for anomaly detection in healthcare.

AI disclosure

During the preparation of this work the author(s) used QuillBot in order to Grammar checking and paraphrasing. After using this tool/service, the author(s) reviewed and edited the content as needed and take(s) full responsibility for the content of the publication.

References

Abdi, A., 2016. Three Types of Machine Learning Algorithms. https://doi.org/10.13140/RG.2.2.26209.10088.

Alharbi, F., Vakanski, A., 2023. Machine learning methods for cancer classification using gene expression data: a review. Bioengineering 10 (2). https://doi.org/10.3390/bioengineering10020173.

Anomaly Detection with Machine Learning Overview, 2023. KnowledgeHut. https://www.knowledgehut.com/blog/data-science/machine-learning-for-anomaly-detection.

Astillo, P.V., Duguma, D.G., Park, H., Kim, J., Kim, B., You, I., 2022. Federated intelligence of anomaly detection agent in IoTMD-enabled diabetes management control system. Future Generation Computer Systems 128, 395–405. https://doi.org/10.1016/j.future.2021.10.023.

Baddar, S.H.A.H., Merlo, A., Migliardi, M., 2014. Anomaly detection in computer networks: a state-of-the-art review. Journal of Wireless Mobile Networks, Ubiquitous Computing, and Dependable Applications 5 (4), 29–64.

Bhargavi, M.V., 2022. A comparative study for statistical outlier detection using colon cancer data. Advances and Applications in Statistics 72 (1), 41–54. https://doi.org/10.17654/0972361722003.

Chandola, V., Banerjee, A., Kumar, V., 2009. Anomaly detection. ACM Computing Surveys 41 (3), 1–58. https://doi.org/10.1145/1541880.1541882.

Chronic respiratory disease, 2023. World Health Organization. https://www.who.int/health-topics/chronic-respiratory-diseases#tab=tab_1.

Dar, J.A., Srivastava, K.K., Mishra, A., 2023. Lung anomaly detection from respiratory sound database (sound signals). Computers in Biology and Medicine 164, 107311. https://doi.org/10.1016/j.compbiomed.2023.107311.

DeepCAD: a stand-alone deep neural network-based framework for classification and anomaly detection in smart healthcare systems. In: 2022 IEEE International Conference on Digital Health (ICDH), 2022. IEEE, pp. 218–227.

Deng, L., Yu, D., 2013. Deep learning: methods and applications. Foundations and Trends in Signal Processing 7 (3–4), 197–387. https://doi.org/10.1561/2000000039.

El-Sherbini, A.H., Hassan Virk, H.U., Wang, Z., Glicksberg, B.S., Krittanawong, C., 2023. Machine-learning-based prediction modelling in primary care: state-of-the-art review. Ai 4 (2), 437–460. https://doi.org/10.3390/ai4020024.

Fang, J., Xie, Z., Cheng, H., Fan, B., Xu, H., Li, P., 2022. Anomaly detection of diabetes data based on hierarchical clustering and CNN. Procedia Computer Science 199, 71–78. https://doi.org/10.1016/j.procs.2022.01.010.

Ghoncheh, M., Pournamdar, Z., Salehiniya, H., 2016. Incidence and mortality and epidemiology of breast cancer in the world. Asian Pacific Journal of Cancer Prevention 17 (Suppl. 3), 43–46. https://doi.org/10.7314/APJCP.2016.17.S3.43.

Global cancer data by country, 2020. Cancer Trends. https://www.wcrf.org/cancer-trends/global-cancer-data-by-country/.

Goldberger, A.L., Amaral, L.A., Glass, L., Hausdorff, J.M., Ivanov, P.C., Mark, R.G., Mietus, J.E., Moody, G.B., Peng, C.K., Stanley, H.E., 2000. PhysioBank, PhysioToolkit, and PhysioNet: components of a new research resource for complex physiologic signals. Circulation 101 (23), E215–E220.

Goodfellow, I., Bengio, Y., Courville, A., 2016. Deep Learning. MIT Press.

Goodfellow, I., Pouget-Abadie, J., Mirza, M., Xu, B., Warde-Farley, D., Ozair, S., Courville, A., Bengio, Y., 2020. Generative adversarial networks. Communications of the ACM 63 (11), 139–144. https://doi.org/10.1145/3422622.

Görnitz, N., Kloft, M., Rieck, K., Brefeld, U., 2013. Toward supervised anomaly detection. Journal of Artificial Intelligence Research 46, 235–262. https://doi.org/10.1613/jair.3623.

Guo, M., Liu, H., Li, Y., Li, W., Gao, F., Qin, S., Wen, Q., 2022. Quantum algorithms for anomaly detection using amplitude estimation. Physica A: Statistical Mechanics and its Applications 604, 127936. https://doi.org/10.1016/j.physa.2022.127936.

Hamet, P., Tremblay, J., 2017. Artificial intelligence in medicine. Metabolism 69, S36–S40. https://doi.org/10.1016/j.metabol.2017.01.011.

Haque, N.I., Khalil, A.A., Rahman, M.A., Amini, M.H., Ahamed, S.I., 2021. BIOCAD: bio-inspired optimization for classification and anomaly detection in digital healthcare systems. In: Proceedings – 2021 IEEE International Conference on Digital Health, ICDH 2021. Institute of Electrical and Electronics Engineers Inc., United States, pp. 48–58. https://doi.org/10.1109/ICDH52753.2021.00017.

Hauskrecht, M., Valko, M., Kveton, B., Visweswaran, S., Cooper, G.F., 2007. Evidence-based anomaly detection in clinical domains. AMIA ... Annual Symposium Proceedings/AMIA Symposium. AMIA Symposium 319–323.

Hauskrecht, M., Batal, I., Valko, M., Visweswaran, S., Cooper, G.F., Clermont, G., 2013. Outlier detection for patient monitoring and alerting. Journal of Biomedical Informatics 46 (1), 47—55. https://doi.org/10.1016/j.jbi.2012.08.004.

Healthcare Data Storage Global Market Report 2023, 2023. Market Reports. https://www.reportlinker.com/p06284262/?utm_source=GNW.

Holzinger, A., 2019. Introduction to machine learning & knowledge extraction (MAKE). Machine Learning and Knowledge Extraction 1 (1), 1—20. https://doi.org/10.3390/make1010001.

Huang, P., Shang, J., Xu, Y., Hu, Z., Zhang, K., Dai, J., Yan, H., 2023. Anomaly detection in radiotherapy plans using deep autoencoder networks. Frontiers in Oncology 13. https://doi.org/10.3389/fonc.2023.1142947.

Interstitial Lung Disease, 2023. American Lung Association. https://www.lung.org/lung- health-diseases/lung-disease-lookup/interstitial-lung-disease.

Kavitha, M., Srinivas, P.V.V.S., Kalyampudi, P.S.L., Fe, C.S., Srinivasulu, S., 2021. Machine learning techniques for anomaly detection in smart healthcare. In: Proceedings of the 3rd International Conference on Inventive Research in Computing Applications, ICIRCA 2021. Institute of Electrical and Electronics Engineers Inc., India, pp. 1350—1356. https://doi.org/10.1109/ICIRCA51532.2021.9544795.

Kim, K.-S., Oh, S.J., Cho, H.B., Chung, M.J., 2021. One-class classifier for chest X-ray anomaly detection via contrastive patch-based percentile. IEEE Access 9, 168496—168510. https://doi.org/10.1109/access.2021.3136263.

Kim, K., Oh, S.J., Lee, J.H., Chung, M.J., 2024. 3D unsupervised anomaly detection through virtual multi-view projection and reconstruction: clinical validation on low-dose chest computed tomography. Expert Systems with Applications 236, 121165. https://doi.org/10.1016/j.eswa.2023.121165.

Kruk, M.E., Gage, A.D., Joseph, N.T., Danaei, G., García-Saisó, S., Salomon, J.A., 2018. Mortality due to low-quality health systems in the universal health coverage era: a systematic analysis of amenable deaths in 137 countries. The Lancet 392 (10160), 2203—2212. https://doi.org/10.1016/s0140-6736(18)31668-4.

Kukreja, H., 2016. An introduction to artificial neural network. International Journal of Advance Research and Innovative Ideas in Education 1, 27—30.

Manzoor, A., Pan, Q., Khan, H.J., Siddeeq, S., Bhatti, H.M.A., Wedagu, M.A., 2020. Analysis and detection of lung sounds anomalies based on NMA-RNN. In: Proceedings — 2020 IEEE International Conference on Bioinformatics and Biomedicine, BIBM 2020. Institute of Electrical and Electronics Engineers Inc., China, pp. 2498—2504. https://doi.org/10.1109/BIBM49941.2020.9313197.

Motai, Y., Ma, D., Yoshida, H., 2016. Smart anomaly prediction in nonstationary CT colonography screening. IEEE Transactions on Industrial Informatics 12 (6), 2292—2301. https://doi.org/10.1109/tii.2016.2595399.

Oladipupo, n.d. Types of Machine Learning Algorithms. New Advances in Machine Learning.

Park, S., Lee, K.H., Ko, B., Kim, N., 2023. Unsupervised anomaly detection with generative adversarial networks in mammography. Scientific Reports 13 (1). https://doi.org/10.1038/s41598-023-29521-z.

Pham, L., Phan, H., Palaniappan, R., Mertins, A., McLoughlin, I., 2021. CNN-MoE based framework for classification of respiratory anomalies and lung disease detection. IEEE Journal of Biomedical and Health Informatics 25 (8), 2938—2947. https://doi.org/10.1109/jbhi.2021.3064237.

Šabić, E., Keeley, D., Henderson, B., Nannemann, S., 2021. Healthcare and anomaly detection: using machine learning to predict anomalies in heart rate data. AI & Society 36 (1), 149—158. https://doi.org/10.1007/s00146-020-00985-1.

Salehinejad, H., Sankar, S., Barfett, J., Colak, E., Valaee, S., 2017. Recent Advances in Recurrent Neural Networks. arXiv, Canada. https://doi.org/10.48550/arxiv.1801.01078.

Samariya, D., Ma, J., Aryal, S., Zhao, X., 2023. Detection and explanation of anomalies in healthcare data. Health Information Science and Systems 11 (1). https://doi.org/10.1007/s13755-023-00221-2.

Sebastian, Y., Chew, J.T., Tiong, X.T., Raman, V., Fong, A.Y.Y., Then, P.H.H., 2018. Anomaly detection from diabetes similarity graphs using community detection and Bayesian techniques. In: ACM International Conference Proceeding Series. Association for Computing Machinery, Malaysia. https://doi.org/10.1145/3164541.3164643.

Tecuci, G., 2012. Artificial intelligence. WIREs Computational Statistics 4 (2), 168—180. https://doi.org/10.1002/wics.200.

The Top Anomaly Detection Techniques You Need to Know, 2023. Dataherose. https://dataheroes.ai/blog/anomaly-detection-techniques-you-need-to-know.

Ukil, A., Bandyoapdhyay, S., Puri, C., Pal, A., 2016. IoT healthcare analytics: the importance of anomaly detection. In: Proceedings — International Conference on Advanced Information Networking and Applications, AINA. 2016. Institute of Electrical and Electronics Engineers Inc., India, pp. 994–997. https://doi.org/10.1109/AINA.2016.158.

Woldaregay, A.Z., Launonen, I.K., Albers, D., Igual, J., Årsand, E., Hartvigsen, G., 2020. A novel approach for continuous health status monitoring and automatic detection of infection incidences in people with type 1 diabetes using machine learning algorithms (Part 2): A personalized digital infectious disease detection mechanism. Journal of Medical Internet Research 22 (8), e18912. https://doi.org/10.2196/18912.

Further reading

Mikayla, F., 2022. The Difference between AI, ML and DL. https://www.cengn.ca/information-centre/innovation/difference-between-ai-ml-and-dl/.

Wu, J., 2017, 23. Introduction to Convolutional Neural Networks. National Key Lab for Novel Software Technology. Nanjing University, China.

Yulia, G., 2021. What is Anomaly Detection in Machine Learning. https://serokell.io/blog/anomaly-detection-in-machine-learning.

Artificial intelligence—based computational intelligence solutions for robotic automation

Dasaradharami Reddy Kandati and Anusha Sirasanambeti

School of Computer Science Engineering & Information Systems, Vellore Institute of
Technology, Vellore, Tamil Nadu, India

1. Introduction

The integration of AI and CI techniques has led to significant advancements in the field of robotics in recent years. AI-based CI solutions have revolutionized robotic automation, enabling machines to perform complex tasks with precision and efficiency. The purpose of this section is to present a summary of how AI and CI contribute to robotic automation, emphasizing the advantages they offer and the various areas where they can be applied.

The field of computer science known as AI is dedicated to the creation of intelligent machines that can imitate human behavior, while CI encompasses a range of computational methods, including fuzzy logic, neural networks, and genetic algorithms, to tackle intricate problems (Said and Salem, 2015). By combining these two fields, researchers have been able to create powerful solutions that enhance the capabilities of robots, making them more adaptable, autonomous, and intelligent. AI is used to make robots smarter, and its intelligence is the capacity of a computer to accomplish any intellectual work in any situation that a person might perform (Collins et al., 2021), as demonstrated in Fig. 12.1.

One of the key advantages of AI-based CI solutions in robotic automation is their ability to learn and adapt to changing environments (Nandhini et al., 2023). Through machine learning (ML) algorithms, robots can acquire knowledge and improve their performance over time. This enables them to handle dynamic tasks and respond effectively to unforeseen situations, making them highly versatile in various industries (Soori et al., 2023).

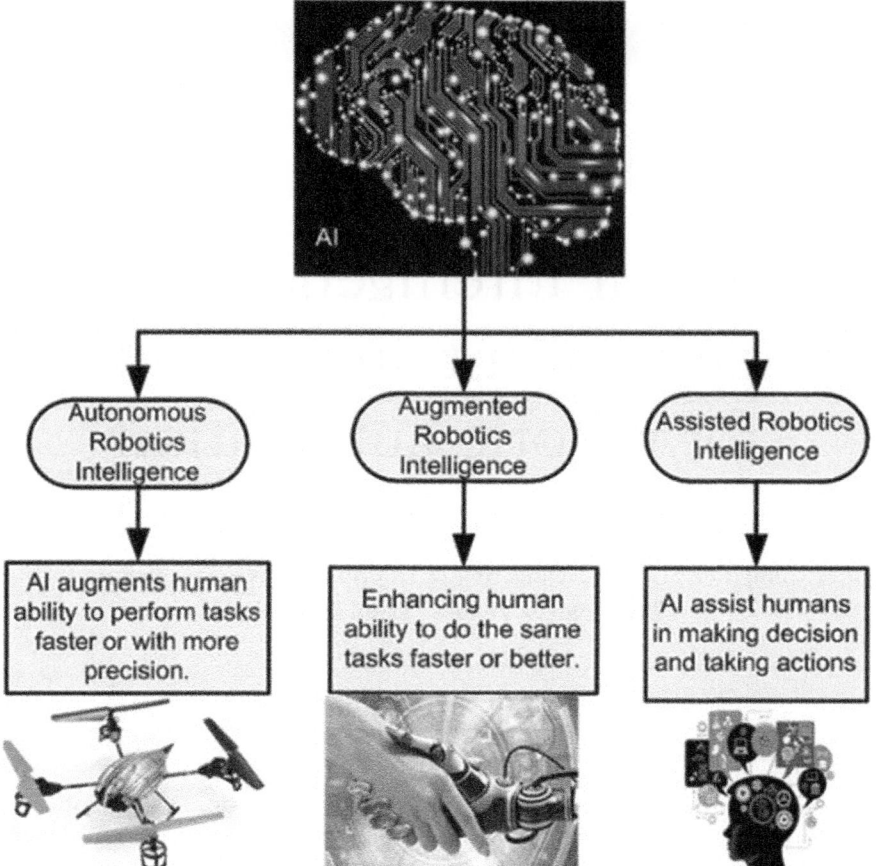

FIGURE 12.1 AI for robotic intelligence.

Moreover, AI-based CI solutions enable robots to perform tasks that are too dangerous or tedious for humans (Xu et al., 2021). For instance, in hazardous environments such as nuclear power plants or deep-sea exploration, robots equipped with AI and CI technologies can carry out inspections, repairs, and data collection without risking human lives. This not only increases safety but also reduces costs associated with human labor.

Furthermore, AI and CI techniques have found applications in areas such as manufacturing, healthcare, logistics, and agriculture. AI-enabled robots in the manufacturing industry have the potential to enhance overall efficiency, optimize production processes, and improve quality control (Gupta et al., 2022). In healthcare, AI-powered robots can assist in surgeries, monitor patients, and provide personalized care (Deo and Anjankar, 2023). In logistics, robots equipped with AI and CI can optimize warehouse operations, automate inventory management, and streamline supply chain processes. In agriculture, AI-based robots can perform tasks such as crop monitoring, harvesting, and pest control, leading to increased productivity and sustainability (Jha et al., 2019).

Furthermore, AI-based CI solutions have also addressed the challenge of human–robot interaction. Through natural language processing and computer vision techniques, robots can understand and respond to human commands, gestures, and expressions. This has opened up new possibilities for collaborative robotics, where humans and robots can work together seamlessly, sharing tasks and complementing each other's abilities (Li, 2023).

Another area where AI and CI have made significant contributions is in the field of perception and sensing. Robots equipped with advanced sensors and AI algorithms can perceive and interpret their surroundings, allowing them to navigate complex environments, avoid obstacles, and interact with objects (Soori et al., 2023). This has paved the way for applications such as autonomous vehicles, drones, and robotic assistants that can operate in diverse and dynamic settings.

Moreover, AI-based CI solutions have also improved the efficiency of robotic decision-making. Robots can utilize ML algorithms and analyze extensive data to make well-informed decisions in real-time, thereby optimizing their actions according to the present context. This has led to advancements in areas such as autonomous robots for industrial automation, smart home systems, and intelligent surveillance systems (Hasan et al., 2019).

However, it is important to address the ethical considerations associated with AI-based CI solutions in robotic automation. As robots gain more autonomy and decision-making capabilities, concerns arise about responsibility, accessibility, and the possible effects on employment (Vrontis et al., 2022). It is essential to maintain responsible development and deployment of these technologies with appropriate regulations and safety measures.

The utilization of AI-powered CI solutions has revolutionized robotic automation, empowering robots to execute intricate jobs, adjust to dynamic surroundings, and enhance their interaction capabilities with humans. These solutions have found applications in various industries, improving safety, efficiency, and productivity. To fully leverage the capabilities of AI and CI in robotic automation, it is crucial to find a harmonious equilibrium between technological progress and ethical concerns as the field continues to develop.

2. Role of AI in robotic automation

AI plays a crucial role in robotic automation by enabling robots to perceive and understand their environment through computer vision and sensor fusion techniques. Computer vision allows robots to interpret visual data from cameras and other sensors, enabling them to recognize objects, navigate their surroundings, and perform tasks with precision.

With AI-powered computer vision, robots can identify and classify objects, detect and track motion, and even understand complex scenes. This capability enables them to interact with their environment in a more intelligent and autonomous manner (Matsuzaka and Yashiro, 2023). For example, robots can use computer vision to locate and grasp objects, inspect and sort items, or navigate through dynamic environments.

Sensor fusion is another important technique that AI facilitates in robotic automation. By combining data from multiple sensors such as cameras, radar, and other environmental sensors, robots can create a more comprehensive and accurate understanding of their surroundings (Nubert et al., 2022). This fusion of sensor data allows robots to perceive depth, distance, and spatial relationships, enabling them to make informed decisions and adapt to changing conditions.

AI algorithms and ML techniques are used to process and analyze the data collected by robots' sensors. These algorithms can learn from experience and improve over time, enabling robots to continuously enhance their perception and understanding of the environment. This iterative learning process helps robots adapt to new situations, handle uncertainties, and perform tasks more efficiently.

Overall, AI empowers robots with the ability to perceive and understand their environment, making them more capable and autonomous in various applications such as manufacturing, logistics, healthcare, and even domestic settings. AI-powered robotic automation is causing a revolution in industries and reshaping our interactions with machines by utilizing computer vision and sensor fusion techniques.

AI algorithms such as DL and RL have revolutionized the field of robotics by enhancing their decision-making capabilities. DL is a branch of ML that focuses on training neural networks with multiple layers in order to effectively process and analyze large volumes of data. This enables robots to recognize patterns, make predictions, and perform complex tasks.

RL, on the other hand, involves training robots through trial and error. Their acquisition of knowledge is facilitated through the receipt of responses, which can be rewarding or punishing depending on their performance (Kormushev et al., 2013; Zhang and Mo, 2021). The iterative process allows robots to adapt and improve their decision-making abilities over time.

Overall, AI algorithms such as DL and RL play a crucial role in enhancing the decision-making capabilities of robots, making them more versatile and adaptable in various real-world scenarios. The AI-based applications sense, reason, acts, learn, and adapt, as shown in Fig. 12.2.

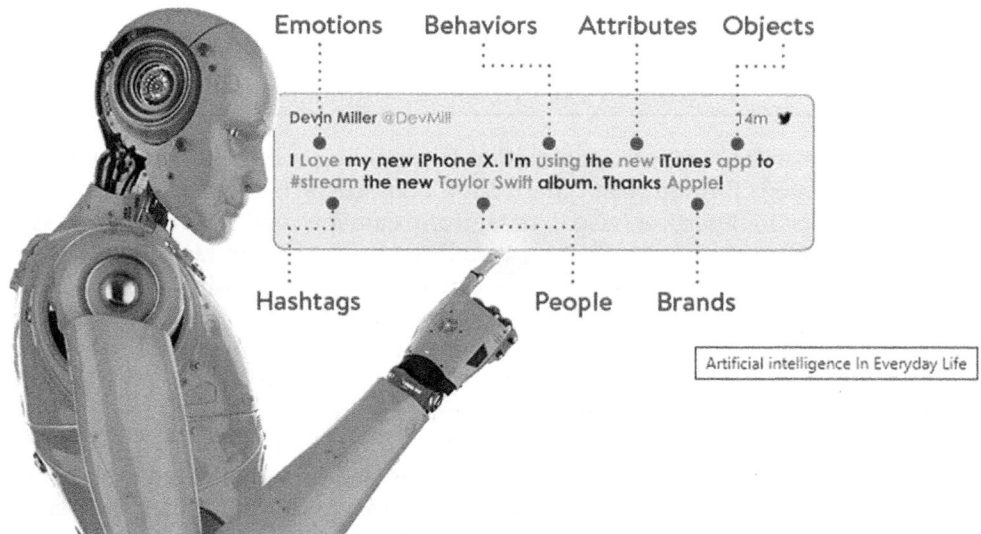

FIGURE 12.2   Role of artificial intelligence.

3. The integration of CI techniques

Indeed, the integration of CI techniques, such as fuzzy logic, genetic algorithms, and neural networks, into robotic automation systems has been a significant area of research. These techniques offer several advantages in enabling robots to handle uncertainty, optimize their performance, and learn from experience. The taxonomy of CI techniques is shown in Fig. 12.3.

Fuzzy logic allows robots to reason and make decisions based on imprecise or uncertain information. By using fuzzy sets and fuzzy rules, robots can handle vague or ambiguous inputs and produce appropriate outputs. This is particularly useful in situations where precise mathematical models are difficult to define.

Genetic algorithms enable robots to enhance their performance by emulating the principles of evolution. Through the iterative evolution of a number of potential solutions, robots can adapt and refine their behavior progressively (Katic and Vukobratovic, 2003). This is especially beneficial in complex and dynamic environments where traditional optimization methods may struggle.

Neural networks offer robots the ability to learn from experience and generalize from examples. By training on a dataset, robots can develop models that capture complex patterns and relationships. This allows them to make predictions, recognize objects, or perform other tasks based on previously unseen data.

Overall, the integration of AI and CI techniques in robotic automation systems provides a powerful framework for developing intelligent robots. These techniques enable robots to handle uncertainty, optimize their performance, and learn from experience, making them more adaptable and capable of handling various applications. As the research in this domain progresses further, it is possible to anticipate the emergence of increasingly advanced and intelligent automated methods in the forthcoming years.

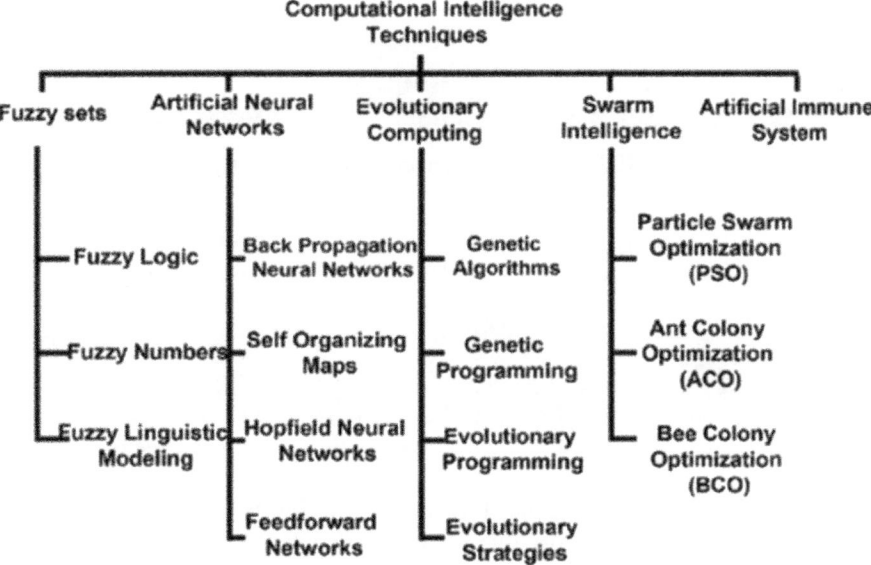

FIGURE 12.3 Taxonomy of CI techniques.

Furthermore, the integration of AI and CI techniques in robotic automation systems enables robots to adapt and evolve their behavior in real-time. This is particularly valuable in dynamic and unpredictable environments where robots need to continuously adjust their actions based on changing conditions.

4. AI-based CI solutions

AI-based CI solutions have indeed found applications in various domains, including manufacturing, healthcare, logistics, and agriculture. These solutions utilize the capabilities of artificial intelligence to examine vast quantities of data, make predictions, and automate processes, leading to improved productivity, quality, and safety.

In the manufacturing industry, AI-based solutions can optimize production processes, predict equipment failures, and enhance quality control (Ghahramani et al., 2020), as shown Fig. 12.4. For example, machine learning algorithms examine information collected by manufacturing machines to detect anomalies and identify maintenance requirements, minimizing production delays and improving overall efficiency.

AI-powered solutions in the healthcare field have the potential to aid in disease diagnosis, forecast patient outcomes (Patil and Shankar, 2023), and customize treatment plans, as depicted in Fig. 12.5. For example, deep learning algorithms can scrutinize medical images

FIGURE 12.4   AI in manufacturing industry.

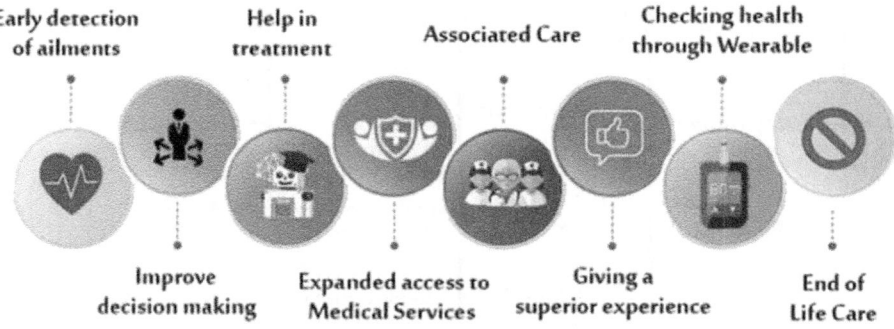

FIGURE 12.5   AI in healthcare.

to identify irregularities and support radiologists in making precise diagnoses. Additionally, AI can analyze patient data to detect patterns and anticipate the probability of readmission or adverse events, facilitating proactive interventions.

In logistics, AI-based solutions can optimize supply chain management, route planning, and inventory management. For instance, AI algorithms can be utilized by examining past data to forecast demand patterns and enhance inventory management, resulting in cost reduction and enhanced customer satisfaction. Additionally, AI can enhance route planning by utilizing traffic-related information, thereby reducing travel time and fuel consumption (Iyer, 2021).

In agriculture, AI-based solutions can improve crop yield, optimize resource allocation, and monitor plant health. For example, AI algorithms can analyze satellite imagery and sensor data to detect crop diseases, nutrient deficiencies, and water stress, enabling timely interventions. AI can also optimize irrigation schedules based on weather forecasts and soil moisture data, conserving water resources.

These real-world examples demonstrate the effectiveness of AI-based computational intelligence solutions in various domains. By harnessing the power of AI, organizations can achieve significant improvements in productivity, quality, and safety, leading to enhanced operational efficiency and better outcomes.

Additionally, AI-based computational intelligence solutions have the potential to revolutionize other domains as well. Let's explore a few more examples:

4.1 Finance

AI techniques can analyze huge quantities of financial information to identify trends, predict market trends, and automate trading decisions (Cohen, 2022). The result can be more accurate investment strategies, reduced risks, and improved financial decision-making.

4.2 Energy

AI can optimize energy consumption, predict energy demand, and enhance the efficiency of power grids (Sankarananth et al., 2023). For instance, machine learning algorithms can analyze historical energy usage data to identify patterns and optimize energy distribution, reducing costs and minimizing environmental impact.

4.3 Education

AI-based solutions can personalize learning experiences, provide intelligent tutoring, and automate administrative tasks (Kamalov et al., 2023). For example, adaptive learning platforms can analyze student performance data to tailor educational content and provide targeted feedback, improving learning outcomes.

4.4 Customer service

Chatbots and virtual assistants powered by AI have the ability to handle personalized recommendations, provide support, and handle customer inquiries (Adam et al., 2021).

Customer queries can be understood and responded to by natural language processing algorithms, which improves customer satisfaction and reduces response times.

4.5 Cyber security

AI can detect and respond to cyber threats in real-time, improving the security of digital systems (Kaur et al., 2023). ML algorithms can analyze network traffic patterns, identify anomalies, and proactively defend against cyber-attacks, safeguarding sensitive data and infrastructure.

Responsible implementation and ethical considerations are of utmost importance when utilizing AI-based CI solutions, despite their numerous benefits. Ensuring transparency, fairness, and accountability in AI systems is essential to build trust and mitigate potential risks.

The applications of AI-based CI solutions are vast and diverse. From manufacturing to healthcare, logistics to agriculture, finance to education, and beyond, AI has the potential to transform industries and improve various aspects of our lives. Continued research and development in this field will likely uncover even more innovative applications in the future.

5. Challenges and future directions

AI-based computational intelligence in robotic automation has made significant advancements in recent years, but it still faces several challenges and has exciting future directions. Here are some key challenges and future directions in this field:

1. **Limited Generalization:** AI models often struggle to generalize their knowledge to new and unseen scenarios. While they can perform well in controlled environments, they may struggle when faced with real-world complexities. Future research should focus on developing AI models that can adapt and generalize their knowledge effectively.
2. **Safety and Ethics:** As AI-based robotic automation becomes more prevalent; ensuring safety and ethical considerations becomes crucial (Torresen, 2018). It is necessary to create strong safety measures that can effectively prevent accidents and promote harmonious collaboration between humans and robots. Furthermore, it is crucial to establish ethical guidelines and regulations to address concerns regarding privacy, bias, and accountability.
3. **Explainability and Transparency:** The current lack of transparency in AI models poses a challenge in understanding their decision-making process, which can hinder trust and acceptance (Balasubramaniam et al., 2022). Future research should prioritize the development of explainable AI models that can offer clear explanations for their actions, empowering humans to comprehend and have confidence in their decisions.
4. **Human—Robot Interaction:** Effective interaction between humans and robots is essential for successful robotic automation (Hofmann et al., 2020). Future research should focus on developing natural language processing and understanding capabilities, as well as intuitive interfaces that enable seamless communication and collaboration between humans and robots.

5. **Lifelong Learning:** AI models should be able to continuously learn and adapt to new tasks and environments. Lifelong learning algorithms that can acquire new knowledge while retaining previously learned information are crucial for long-term robotic automation. This area requires further research to develop efficient and scalable lifelong learning techniques.

6. **Integration with Other Technologies:** AI-based computational intelligence in robotic automation can benefit from integration with other emerging technologies. For example, combining AI with augmented reality (AR) or virtual reality (VR) can enhance robot perception and enable more immersive human—robot interaction (Kaszuba et al., 2021). Exploring such integrations can open up new possibilities for robotic automation.

7. **Scalability and Deployment:** While AI models have shown promising results in research settings, deploying them at scale in real-world scenarios can be challenging. Future research should focus on developing scalable AI algorithms and architectures that can handle large-scale robotic automation tasks efficiently.

In summary, addressing the challenges of generalization, safety, explainability, human—robot interaction, lifelong learning, integration with other technologies, and scalability will shape the future directions of AI-based computational intelligence in robotic automation. Continued research and development in these areas will pave the way for more capable and intelligent robots that can effectively assist humans in various domains.

The future directions of AI-based computational intelligence in robotic automation are promising and hold great potential. Here are some key areas that are likely to see significant advancements:

1. **Enhanced Perception and Sensing:** AI algorithms will continue to improve the perception and sensing capabilities of robots. This includes advancements in computer vision, object recognition, and sensor fusion techniques, enabling robots to better understand and interact with their environment.

2. **Adaptive and Learning Robots:** AI will play a crucial role in developing robots that can adapt and learn from their experiences. Robots will be able to acquire new skills, optimize their performance, and adapt to changing tasks and environments through the use of reinforcement learning and other machine learning techniques.

3. **Human—Robot Collaboration:** AI will facilitate closer collaboration between humans and robots. This involves developing intelligent algorithms that allow robots to understand human intentions, work alongside humans in shared workspaces, and adapt their behavior to ensure safe and efficient collaboration.

4. **Explainable AI in Robotics:** The increasing complexity and autonomy of AI necessitate a greater emphasis on transparency and interpretability (Sanneman and Shah, 2020). Future studies will prioritize the development of techniques that make AI in robotics more explainable, allowing humans to comprehend and have confidence in the decisions made by robots powered by AI.

5. **Swarm Robotics:** AI will enable the coordination and cooperation of large groups of robots, known as swarm robotics. Swarm robotics has the potential to revolutionize various fields, such as search and rescue operations, environmental monitoring, and industrial automation (Bakhshipour et al., 2017).

6. **Ethical and Responsible AI:** As AI-powered robots become more prevalent, there will be an increased emphasis on ethical and responsible AI development (Müller, 2021). This includes addressing issues such as bias, fairness, responsibility, and confidentiality to guarantee AI-based robotic systems are created and used in a socially responsible manner.

7. **Autonomous Decision-Making:** AI algorithms will continue to advance in their ability to make autonomous decisions in complex and dynamic environments (Hagos and Rawat, 2022). This will enable robots to perform tasks with minimal human intervention, leading to increased efficiency and productivity in various industries.

Overall, the future of AI-based computational intelligence in robotic automation is likely to bring about significant advancements in perception, learning, collaboration, explainability, swarm robotics, ethics, and autonomous decision-making. These advancements will pave the way for more capable and intelligent robots that can effectively assist humans in various domains.

6. Conclusion

The integration of artificial intelligence (AI) and computational intelligence (CI) techniques has significantly enhanced robotic automation in various industries. AI enables robots to perceive and understand their environment through computer vision and sensor fusion, while algorithms such as deep learning and reinforcement learning enhance their decision-making capabilities. The incorporation of CI techniques such as fuzzy logic, genetic algorithms, and neural networks allows robots to handle uncertainty, optimize performance, and learn from experience. The combination of AI and CI provides a powerful framework for developing intelligent robotic systems.

The applications of AI-based computational intelligence solutions in domains such as manufacturing, healthcare, logistics, and agriculture have shown promising results in improving productivity, quality, and safety. Real-world examples have demonstrated the effectiveness of these solutions. However, there are still challenges to overcome and future directions to explore. These include developing more advanced robotic systems capable of autonomous decision-making, enabling human—robot collaboration, and incorporating learning from human feedback.

Overall, AI-based computational intelligence solutions have revolutionized robotic automation, and their continued development holds great potential for further advancements in the field. By addressing the challenges and exploring new possibilities, we can unlock the full capabilities of intelligent robotic systems in the future.

References

Adam, M., Wessel, M., Benlian, A., 2021. AI-based chatbots in customer service and their effects on user compliance. Electronic Markets 31 (2), 427—445. https://doi.org/10.1007/s12525-020-00414-7.

Bakhshipour, M., Jabbari Ghadi, M., Namdari, F., 2017. Swarm robotics search and rescue: a novel artificial intelligence-inspired optimization approach. Applied Soft Computing 57, 708—726. https://doi.org/10.1016/j.asoc.2017.02.028.

Balasubramaniam, N., Kauppinen, M., Hiekkanen, K., Kujala, S., 2022. Transparency and Explainability of AI Systems: Ethical Guidelines in Practice. Springer Science and Business Media LLC, pp. 3–18.

Cohen, G., 2022. Algorithmic trading and financial forecasting using advanced artificial intelligence methodologies. Mathematics 10 (18), 2227–7390. https://doi.org/10.3390/math10183302.

Collins, C., Dennehy, D., Conboy, K., Mikalef, P., 2021. Artificial intelligence in information systems research: a systematic literature review and research agenda. International Journal of Information Management 60, 02684012. https://doi.org/10.1016/j.ijinfomgt.2021.102383.

Deo, N., Anjankar, A., 2023. Artificial intelligence with robotics in healthcare: a narrative review of its viability in India. Cureus 2168–8184. https://doi.org/10.7759/cureus.39416.

Ghahramani, M., Qiao, Y., Zhou, M.C., O'Hagan, A., Sweeney, J., 2020. AI-based modeling and data-driven evaluation for smart manufacturing processes. IEEE/CAA Journal of Automatica Sinica. 7 (4), 1026–1037. https://doi.org/10.1109/jas.2020.1003114.

Gupta, S., Modgil, S., Lee, C.-K., Cho, M., Park, Y., 2022. Artificial intelligence enabled robots for stay experience in the hospitality industry in a smart city. Industrial Management & Data Systems 122 (10), 2331–2350. https://doi.org/10.1108/imds-10-2021-0621.

Hagos, D.H., Rawat, D.B., 2022. Recent advances in artificial intelligence and tactical autonomy: current status, challenges, and perspectives. Sensors 22 (24), 1424–8220. https://doi.org/10.3390/s22249916.

Hasan, R., Asif Hussain, S., Azeemuddin Nizamuddin, S., Mahmood, S., 2019. An autonomous robot for intelligent security systems. In: 2018 9th IEEE Control and System Graduate Research Colloquium, ICSGRC 2018 – Proceeding, vol. 3. Institute of Electrical and Electronics Engineers Inc., Oman, pp. 201–206. https://doi.org/10.1109/ICSGRC.2018.8657642.

Hofmann, P., Samp, C., Urbach, N., 2020. Robotic process automation. Electronic Markets 30 (1), 99–106. https://doi.org/10.1007/s12525-019-00365-8.

Iyer, L.S., 2021. AI enabled applications towards intelligent transportation. Transportation Engineering 5, 2666691X. https://doi.org/10.1016/j.treng.2021.100083.

Jha, K., Doshi, A., Patel, P., Shah, M., 2019. A comprehensive review on automation in agriculture using artificial intelligence. Artificial Intelligence in Agriculture 2, 1–12. https://doi.org/10.1016/j.aiia.2019.05.004.

Kamalov, F., Santandreu, C.D., Gurrib, I., 2023. New era of artificial intelligence in education: towards a sustainable multifaceted revolution. Sustainability 15 (16). https://doi.org/10.3390/su151612451.

Kaszuba, S., Leotta, F., Nardi, D., 2021. A Preliminary Study on Virtual Reality Tools in Human–Robot Interaction, vol. 12980. Springer Science and Business Media LLC, pp. 81–90. https://doi.org/10.1007/978-3-030-87595-4_7.

Katic, D., Vukobratovic, M., 2003. Genetic Algorithms in Robotics, vols. 426–4. Springer Nature, pp. 113–131. https://doi.org/10.1007/978-94-017-0317-8_4.

Kaur, R., Gabrijelčič, D., Klobučar, T., 2023. Artificial intelligence for cybersecurity: literature review and future research directions. Information Fusion 97, 15662535. https://doi.org/10.1016/j.inffus.2023.101804.

Kormushev, P., Calinon, S., Caldwell, D., 2013. Reinforcement learning in robotics: applications and real-world challenges. Robotics 2 (3), 122–148. https://doi.org/10.3390/robotics2030122.

Li, S., 2023. Proactive human–robot collaboration: mutual-cognitive, predictable, and self-organising perspectives. Robotics and Computer-Integrated Manufacturing 81, 102510.

Matsuzaka, Y., Yashiro, R., 2023. AI-based computer vision techniques and expert systems. AI 4 (1), 289–302. https://doi.org/10.3390/ai4010013.

Müller, V.C., 2021. Ethics of Artificial Intelligence. Informa UK Limited, pp. 122–137. https://doi.org/10.4324/9780429198533-9.

Nandhini, S.S., Karthiga, M., Goyal, S.B., 2023. Computational Intelligence in Robotics and Automation. CRC Press, 2023.

Nubert, J., Khattak, S., Hutter, M., 2022. Graph-based multi-sensor fusion for consistent localization of autonomous construction robots. In: Proceedings – IEEE International Conference on Robotics and Automation. Institute of Electrical and Electronics Engineers Inc., Switzerland, pp. 10048–10054. https://doi.org/10.1109/ICRA46639.2022.9812386.

Patil, S., Shankar, H., 2023. Transforming healthcare: harnessing the power of AI in the modern era. International Journal of Multidisciplinary Sciences and Arts 2, 60–70.

Said, H.M., Salem, A.-B.M., 2015. Exploiting computational intelligence paradigms in e-technologies and activities. Procedia Computer Science 65, 396–405. https://doi.org/10.1016/j.procs.2015.09.101.

Sankarananth, S., Karthiga, M.E., Suganya, S., Sountharrajan, B., Durga, P., 2023. AI-enabled metaheuristic optimization for predictive management of renewable energy production in smart grids. Energy Reports 10, 1299—1312. https://doi.org/10.1016/j.egyr.2023.08.005.

Sanneman, L., Shah, J.A., 2020. Trust Considerations for Explainable Robots: A Human Factors Perspective arXiv: United States. https://arxiv.org.

Soori, M., Arezoo, B., Dastres, R., 2023. Artificial intelligence, machine learning and deep learning in advanced robotics, a review. Cognitive Robotics. 3, 26672413. https://doi.org/10.1016/j.cogr.2023.04.001.

Torresen, J., 2018. A review of future and ethical perspectives of robotics and AI. Frontiers in Robotics and AI 4, 2296—9144. https://doi.org/10.3389/frobt.2017.00075.

Vrontis, D., Christofi, M., Pereira, V., Tarba, S., Makrides, A., Trichina, E., 2022. Artificial intelligence, robotics, advanced technologies and human resource management: a systematic review. International Journal of Human Resource Management 33 (6), 0958—5192. https://doi.org/10.1080/09585192.2020.1871398.

Xu, Y., Liu, X., Cao, X., Huang, C., Liu, E., Qian, S., Liu, X., Wu, Y., Dong, F., Qiu, C.-W., Qiu, J., Hua, K., Su, W., Wu, J., Xu, H., Han, Y., Fu, C., Yin, Z., Liu, M., Roepman, R., Dietmann, S., Virta, M., Kengara, F., Zhang, Z., Zhang, L., Zhao, T., Dai, J., Yang, J., Lan, L., Luo, M., Liu, Z., An, T., Zhang, B., He, X., Cong, S., Liu, X., Zhang, W., Lewis, J.P., Tiedje, J.M., Wang, Q., An, Z., Wang, F., Zhang, L., Huang, T., Lu, C., Cai, Z., Wang, F., Zhang, J., 2021. Artificial intelligence: a powerful paradigm for scientific research. The Innovation 2 (4), 26666758. https://doi.org/10.1016/j.xinn.2021.100179.

Zhang, T., Mo, H., 2021. Reinforcement learning for robot research: a comprehensive review and open issues. International Journal of Advanced Robotic Systems 18 (3), 1729—8814. https://doi.org/10.1177/17298814211007305.

Developing green computing awareness based on optimization techniques for environmental sustainability

A. Arivoli[1], B. Kalaavathi[1] and Chen Joy Iong-Zong[2]

[1]Vellore Institute of Technology, School of Computer Science and Engineering, Vellore, Tamil Nadu, India; [2]Dayeh University, Department of Electrical Engineering, Dacun, Taiwan

1. Introduction

1.1 Cloud computing

Cloud computing is a novel computing paradigm that incorporates "the principles of distributed computing, grid computing, virtualization, networking, service orientation, and market-oriented computing". The cloud is a scalable and elastic and computing solution that permits users to pay for resources based on their usage and access them across the Internet. These are commonly known as "infrastructure-as-a-service (IaaS), platform-as-a-service (PaaS), and software-as-a-service (SaaS)". Cloud services encompass various components, including computation and storage servers for infrastructure, operating systems for platforms, and application software. Technologically speaking, cloud computing offers a multitude of benefits. From a business perspective, these benefits encompass reduced capital and operational expenses, along with expedited timetables for the introduction of novel applications and new services. Prior to achieving widespread acceptance, there are several challenges that need to be tackled. These challenges include the implementation of flexible architectural solutions, efficient techniques for virtualizing CPU, storage, and network resources, optimizing performance through modeling, developing methods for Cloud-based systems and applications, modeling reliability, and establishing procedures and approaches to ensure security and privacy. Recent advancements in smart devices (such as phones and tablets) have brought attention to several areas that require focus. These include integrating cellular technologies and wireless into the Cloud paradigm enabling mobile with roaming access, developing dynamic applications, and creating "new virtualization, scheduling, and transport schemes to achieve energy savings and promote GC" as a crucial aspect of the Cloud.

The current cloud service architecture is highly centralized, enabling multiple service types to be operated from a single location referred to as a data center. As cloud computing advances rapidly, data centers are expanding. By 2030, the energy demand of data centers will increase to 2967 TW h, a significant rise from the 200 TW h recorded in 2016 (Koot and Wijnhoven, 2021). The worldwide market for Internet data centers is estimated to see a compound annual growth rate (CAGR) of 13.4% from 2020 to 2027. The market is forecast to increase from an estimated value of "US$59.3 billion in 2020 to US$143.4 billion by 2027" (Analysts, 2021).

The "Power Usage Effectiveness" (PUE) evaluations provide present and yearly measurements of the entire capacity power, specifically focusing on the power consumed by IT hardware. These assessments help quantify the power that is not utilized by IT equipment and is instead lost to non-IT equipment. Power consumption efficiency of a DC is quantified by this metric. The optimal PUE value is 1.0, which corresponds to 100% efficiency. Nevertheless, it is exceedingly improbable. In 2021, the average PUE of DC's was 1.57, which increased from 1.59 in 2020. This continues the pattern of PUE immobility observed over the past 5 years. Due to heat generation and high energy consumption, servers require cooling. The rapid growth of data collection and usage is driving the need for more data centers. Cloud computing offers on-demand services to multiple clients through a distributed network of computers and data centers, with payment based on actual usage. Servers, networking devices, cooling structures, and displays consume a significant amount of electricity and occupy a substantial area. Therefore, numerous corporate and government bodies give high importance to green technology toward promotion of environmentally friendly resources usage. Green IT encompasses a variety of programs aimed at addressing environmental concerns associated with information technology.

1.2 Green cloud computing

The utilization of computing resources in a sustainable and environmentally friendly manner is possible with GC computing. As shown in, GC computing helps to lower greenhouse gas emissions and is therefore a better strategy for the future of computing. The major goal in green cloud computing is to use sustainable energy sources that will benefit in the long run and contribute to the achievement of the sustainable development goal. Green cloud computing gives us the means to switch from using fossil fuels to power data centers to a more sustainable method that uses renewable energy.

Green cloud computing enables the utilization of computer resources in an ecologically sustainable and environmentally conscious manner. Fig. 13.1 demonstrates that green cloud computing effectively reduces greenhouse gas emissions, making it a superior approach for the next generation of computing. The key objective of GC computing is to utilize renewable energy sources that will yield long-term benefits and help to the attainment of the objective of

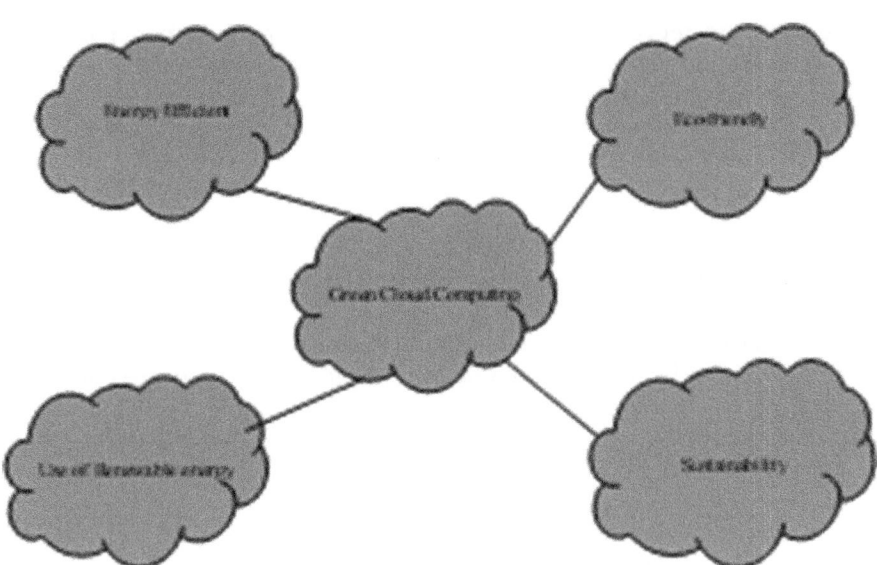

FIGURE 13.1   Green cloud computing.

sustainable development. Green cloud computing enables the transition from fossil fuel-based energy sources to a sustainable approach that relies on renewable energy for powering data centers.

As part of the advancement of green cloud computing, efforts are being made to construct environmentally friendly data centers that have lower carbon emissions and higher energy efficiency (Radu, 2017). Nevertheless, the idea is not limited just to the IT hardware within the DC. Entire other ecological components, including constructions, hardware, and cooling systems should strive to minimize energy consumption. Researchers have been increasingly dedicating their efforts since 2009 to finding energy-efficient solutions for decreasing the greenhouse gas emissions of data centers preserving quality of service (QoS) (Ardagna et al., 2014). There are two methods to attain GC computing: nontechnical methods and technological alternatives (Radu, 2017). Nontechnical approaches encompass the utilization of energy efficient cooling methods that do not squander electricity on cooling apparatus, as well as the adoption of sustainable energy sources. Several technological options "energy-aware virtual machine (VM) scheduling, dynamic voltage and frequency scaling (DVFS), thermal management strategies, virtualization, and VM consolidation with live migration".

1.2.1 Green cloud computing features
1. **Energy effectiveness:** Reducing data center and cloud infrastructure energy consumption is the goal of GC computing. It achieves this through having better optimized cooling infrastructure and hardware that uses less energy.
2. **Renewable energy resources:** Using renewable energy sources rather than fossil fuels is the main goal of green cloud computing. Resources for renewable energy include, but are not limited to, solar, wind, tidal, and bioenergy.

3. **Waste reduction and recycling:** To make computing more sustainable and environmentally benign, Green Cloud computing also seeks to reduce the amount of e-waste produced and increase recycling.
4. **Green hardware:** In this context, "green hardware" refers to hardware that is more energy-efficient and produced with consideration for the environment and sustainability.

1.2.2 Green cloud computing advantages
(1) **Improved performance:** Appropriate and optimal resource use will lead to improved performance and quicker reaction times.
(2) **Sustainable environment:** Since green cloud computing is centered on utilizing environmentally friendly and sustainable renewable energy resources, it contributes to the provision of a sustainable environment.
(3) **Economical:** Unlike electricity produced from fossil fuels, which is not even environmentally friendly, power produced from renewable energy resources will be more affordable.
(4) **Competitive advantages among providers:** As cloud service providers (CSPs) move toward eco-friendly cloud computing, green cloud computing will spur competition among them and encourage their adoption of green practices.

1.2.3 Green cloud computing limits
1. **Restricted availability of renewable energy resources:** Although the objective of GC computing is to utilize renewable energy resources, some geographical areas have little to no access to them.
2. **Upfront costs:** The implementation of green cloud computing will be prohibitively expensive for many enterprises due to the numerous infrastructure upgrades, renewable energy configurations, and other requirements.
3. **Complexity:** The deployment of GC computing infrastructure may be intricate, necessitating the use of specialist personnel and expertise.
4. **Difficulty in Adaptation:** Research on GC computing is a novel topic which is naturally evolving. As a result, it is challenging to adjust to how green technology and the natural world are changing.

1.3 Cloud computing and virtualization

Cloud computing refers to the progressive growth of "grid, parallel, and distributed computing". The framework is transformed into a "virtual resource pool" through the use of virtualization technologies. Computational, storage, and networking resources are distributed over multiple machines. Users no longer require to locally install server-clusters since cloud resources may be rented online on demand. Virtualization is a resource controlling system that operates on top of the hardware assets.

This technology enables the abstraction and segregation of the underlying "hardware resources, such as memory, CPU, network", etc., in order to enhance user experience. Virtualization is a vital element in cloud computing settings. It involves dividing the hardware resources of each host or (PM) among many execution contexts, known as VMs.

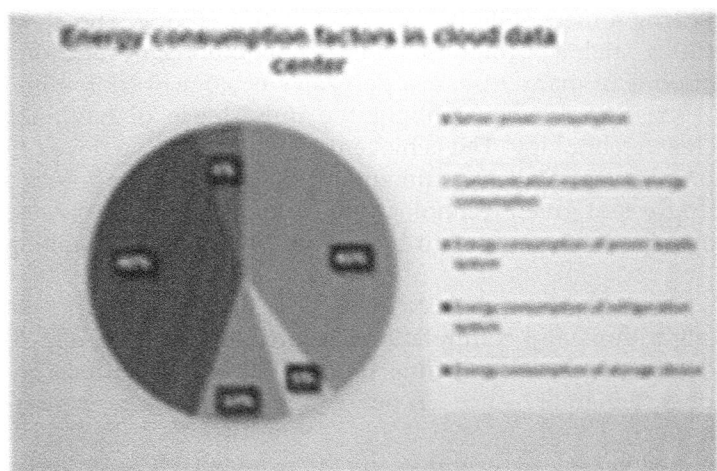

FIGURE 13.2 Energy consumption factors in cloud DCs.

Each VM has the capability to operate independently as a separate system for executing user programs, and is usually isolated from other VMs. The hypervisor or VM monitor (VMM) of every PM is accountable for supervising the deployed VMs on that PM. XEN and KVM are renowned open source hypervisors employed in cloud computing (DCs). Efficient management of the diverse virtual resources available in cloud data centers is essential for ensuring high-quality service to clients, maximizing PM utilization, achieving a good return on investment, and optimizing energy efficacy. Fig. 13.2 illustrates the prime factors that influence the energy consumption of cloud DC. It is evident that physical machines (PMs) account for the bulk of power usage in these data centers, as indicated by references (Ghobaei-Arani et al., 2017; Masdari et al., 2017). Efficient virtual machine (VM) administration is essential in cloud DC and can assist the "cloud service provider (CSP)" in attaining their company's objectives. VM placement, although an NP-hard problem, has been demonstrated to be solvable through a range of heuristic and metaheuristic strategies.

1.4 Green algorithms

GC focuses on attaining environmental sustainability by specifically "addressing the design, production, usage, and disposal of computers and related equipment in an environmentally benign manner". The green algorithm aims to promote the progress of computational science in a manner that is more environmentally sustainable. Green technologies comprise a wide range of methods and solutions that go beyond only environmental technologies. The term "green" refers to the integration of sustainability and environmental friendliness across various domains, "including the environment, economy, societies, and technologies" (Wu et al., 2016). The term "green in the field of communications and computing originated from the creation of the IEEE Technical Subcommittee on Green Communications and Computing (TSCGCC) in 2011".

Green technologies go beyond being a mere collection of separate technology. The globe is today facing formidable difficulties, such as profound climatic fluctuations, diminishing energy savings universal repercussions of many diseases, pervasive ecological contamination, and acute food shortages. These problems highlight the importance of having a wide range of transdisciplinary green technology. The achievement of sustainable development relies on the implementation of pragmatic environmentally friendly solutions across various sectors, promoting the progress of green technologies. Green technologies possess the capacity to foster new relationships between human beings, nature, and the changing environment.

Within the framework of the "Green algorithm," sustainable development refers to the development and application of algorithms and computational approaches that attempt to promote sustainability goals and minimize environmental impacts. These algorithms are tailored to tackle several aspects of sustainability, such as energy efficiency, resource conservation, and environmental monitoring.

Multiple examples demonstrate the vital role algorithms can play in advancing sustainable development.

- Energy-Efficient Algorithms: Energy-efficient algorithms are designed to decrease energy usage in computing systems, such as DCs or mobile devices. Their objective is to enhance energy efficiency by optimizing the allocation of resources, scheduling tasks, and managing power consumption, all while ensuring optimal performance.
- Environmental Modeling: Algorithms play a crucial role in simulating and modeling various environmental systems, including climate models, ecosystem simulations, and pollution dispersion models. Through these simulations, scientists and policymakers gain valuable insights into complex environmental changes. This information aids in making informed decisions related to environmental management and policy formulation.
- Smart Grid Optimization: In the realm of smart grids, algorithms are instrumental in optimizing energy distribution and enhancing overall efficiency. These algorithms contribute to reducing energy losses and effectively integrating renewable energy sources into the grid. By dynamically balancing the supply and demand of energy in real-time, smart grid algorithms promote a more sustainable and resilient energy infrastructure.

Algorithms play a crucial role in environmental monitoring by being used in satellite imagery and data analysis. They enable the monitoring of environmental conditions, rates of deforestation, and the evaluation of air and water quality. Moreover, algorithms play a role in identifying alterations in the surroundings as time progresses. Green and sustainable computing focuses on creating computer systems that are energy efficient, ecologically friendly, and socially responsible, with the aim of minimizing the adverse environmental effects of computing.

Algorithms play a crucial role in environmental monitoring by being used in satellite imagery and data analysis. They enable the monitoring of environmental conditions, rates of deforestation, and the evaluation of air and water quality. Moreover, algorithms play a role in identifying alterations in the surroundings as time progresses. Green and sustainable computing focuses on creating computer systems that are energy efficient, ecologically friendly, and socially responsible, with the aim of minimizing the adverse environmental effects of computing.

1.5 Optimization techniques

Algorithms play a crucial role in environmental monitoring by being used in satellite imagery and data analysis. They enable the monitoring of environmental conditions, rates of deforestation, and the evaluation of air and water quality. Moreover, algorithms play a role in identifying alterations in the surroundings as time progresses. Green and sustainable computing focuses on creating computer systems that are energy efficient, ecologically friendly, and socially responsible, with the aim of minimizing the adverse environmental effects of computing.

- Mathematical Programming: Consists of "nonlinear programming (NLP), integer programming (IP), and linear programming".
- Heuristic Methods: This group includes "particle swarm optimization, genetic algorithms, simulated annealing, and greedy algorithms". While these techniques yield fast, almost optimum solutions, they do not ensure global optimality.
- Metaheuristic Algorithms: Higher-level processes known as metaheuristic algorithms are responsible for coordinating heuristics and other lower-level optimization strategies. Ant colony optimization, simulated annealing, and genetic algorithms are a few examples.
- Dynamic programming: This technique divides complicated issues into more manageable, overlapping subproblems. It is especially helpful for problems with overlapping subproblems.

2. Virtual machine placement

2.1 Need for VMP

VMs are designed to allocate and utilize certain computer hardware such as central processing units (CPUs), input/output (I/O), bandwidth and memory. Cloud providers should allocate VMs to suitable (PMs) in order to optimize the goals of both the user and the provider. Cloud DC are derived from the use of DC infrastructure for providing cloud computing services. Cloud DC manage 94% of workloads (Index, 2018). DCs consume a substantial amount of electricity. With reference to Amazon's DC analysis, 42% of their operational expenses are allocated to energy costs (Index, 2018). The discourse surrounding climate change serves as an additional incentive to conserve energy. Servers contribute to 0.5% of the global carbon dioxide emissions (Speitkamp and Bichler, 2010). Consequently, the research on maintaining data center energy efficiency while ensuring high quality of service (QoS) is an expanding field.

DCs house several physical-servers. Around 60% of the energy usage in a DC is directly linked to the IT infrastructure, specifically the power modules. Virtualization is crucial for achieving energy savings and maximizing server utilization, since it allows for the installation of several VMs on a single physical-server. Therefore, usage of effective (VMP) approach can greatly decrease power consumption in DCs. The VMP optimization issue is proven to be NP-hard (Ullman, 1975). Virtualization enables the migration, isolation, and consolidation of VMs. Migration technology facilitates the transfer of VM from one PM to another. VM located on isolated hosts will migrate from their current host to a reduced number of hosts throughout the process of VM consolidation. This is done to reduce energy consumption by either switching off or hibernating the original host (Masdari et al., 2016). Insufficient or excessive usage of CPU and RAM leads to higher energy consumption. Efficient energy management in software requires the implementation of scheduling and virtualization techniques. Reduced resource use diminishes the energy efficiency of the system. Consolidating virtual machines enhances the efficiency of cloud resources. VM allocation is challenging in the process of VM consolidation. Identifying the most suitable performance monitoring solution for a VM decreases the number of PM in a DC.

2.2 Literature review

A key challenge for green cloud settings is distributing new VM requests across PMs to minimize energy consumption. Numerous research initiatives have explored this subject in terms of hardware, network, or software layers. Most methods focusing on the latter are preferred due to their flexibility and cost-effectiveness. However, centralized control mechanisms in most software approaches lead to single point of failure and limited scalability (Analysts, 2021). Emerging bio-inspired heuristics give decentralized control, reduce failures, and increase scalability for green cloud nodes.

2.2.1 ACO based VMP

Heuristics can be categorized into two types based on algorithm generalization. Meta-heuristics, such as the Ant Colony Optimization algorithm (ACO), are problem-independent heuristics that address various optimization issues. Meta-heuristics are used to improve candidate solutions based on a specific quality measure, such as objective function, starting with a simple or random answer. Metaheuristics are broad and easy to create and apply because they don't have to extensively adopt to a particular area (Farshin and Sharifian, 2019).

Dorigo (1992) proposed the ACO algorithm, a stochastic approach for identifying the most optimal route in a graph. Ant colonies, despite their lack of vision, can navigate toward a food supply by utilizing pheromone trails, as revealed by a comprehensive study undertaken by emulating biologists. The ants themselves emit a pheromone. An ant will determine its course at an intersection by evaluating the amount of pheromones in every direction. Higher pheromone concentration enhances the probability of being selected. The ant will emit the pheromone along the route it has recently traversed simultaneously. There is an inverse correlation between the length of the voyage and the concentration of pheromones. In addition, at intersections lacking pheromone trails, certain ants will arbitrarily select a path to follow, resulting in a constructive feedback loop. Once the ants have located the food, they will coming back to the source area, bringing the food, and continue to move between the two locations namely the food supply and the source. Consequently, ants traversing a brief route will exhibit increased frequency of back-and-forth movement, resulting in a naturally elevated concentration of pheromones that stimulates additional ants to trail along this path. Ant colonies exhibit a high degree of self-organization as per the ACO algorithm, where each ant functions as an independent entity, seeking its own path and communicating with other ants through the use of pheromones.

Al-Moalmi et al. (2019) presented an ACO-based metaheuristic algorithm for selecting adjacent PMs for data and VM placement. It introduced ACO-based metaheuristic method MinDistVMDataPlacement problem. The algorithm initialized a potential set of PMs based on proximity. The program evaluates the probability of selecting each PM based on distance and VM demand. The PM set receives vertices from the candidate set with the highest probability until their weights meet the need. This approach places selected PMs closer together, minimizing cross-network traffic and speeding calculation. This allocation approach executed jobs in allotted VMs better than others.

The study (Abdel-Basset et al., 2019) introduces a VMP method based on ACO, which utilizes artificial ants and global search data. Utilizing order interchange and migration local search approaches in combination. The optimization of this procedure has been achieved on a worldwide scale by assigning VMs to the fewest active PMs. This technique has been employed to address VMP concerns with different VM sizes in both homogeneous as well as heterogeneous cloud environments. The objective is to reduce the consumption of power and the wastage of cloud DC resources in cloud data centers. C++ effectively implements this method in the cloud data center and outperforms alternative VMP algorithms.

2.2.2 Other optimization

The authors (Saxena et al., 2022) introduced a system called "secure and multi-objective virtual machine placement (SM-VMP)" to tackle many difficulties in cloud data DCs, as well as resource inefficiency, power usage, and security vulnerabilities. The framework employs the "Whale Optimization Genetic Algorithm (WOGA)" to efficiently allocate physical resources among virtual machines (VMs), leading to decreased intercommunication delay and enhanced resource efficiency. The WOGA method combines the investigation and exploitation skills of the "Whale Optimization method (WOA)" with the pareto-optimal solution led by the nondominated sorting based Genetic Algorithm (GA). WOGA is a hybrid method that integrates whale generative optimization and nondominated ordering based GA. Its purpose is to discover the most optimal arrangement of VMS that fulfills the criteria of security, resource consumption, communication cost and power consumption. The program utilizes a random search strategy to examine the search space in various directions and enables random migration of virtual machines to update the allocations of VMs. The framework achieves "significant reductions in shared servers, intercommunication cost, power consumption, and execution time, while simultaneously enhancing resource usage, through the utilization of the WOGA algorithm".

The authors of the study (Singh et al., 2023) presented an innovative algorithm called "flower pollination-based nondominated sorting optimization (FP−NSO)". This algorithm focuses to optimize resource-utilization, while simultaneously minimizing energy utilization and carbon emissions in cloud DCs. The algorithm utilizes a combination of flower pollination optimization and a genetic algorithm called "nondominated sorting technique-based genetic algorithm (NSGA-II)" to efficiently perform (VMP) in a cloud environment. This approach leads to significant savings in power consumption, carbon emission, and execution time, while also enhancing resource utilization. The algorithm is assessed using the Google-cluster dataset, and performance indicators such as usage of resources, power expenditure and carbon release values are calculated for both static and dynamic set-ups. The statistical analysis, utilizing the Friedman test and Finner posthoc analysis, substantiates the superiority of the suggested Bio-VMP technique compared with other existing approaches. It reveals notable disparities in the outcomes, indicating substantial differences.

The primary objective of Mythrayee and Lavanya (2023) is to tackle the VMP problem in cloud environments, with the aim of enhancing (QoS) and minimizing computing cost. The suggested method, named "Improved Gray-wolf optimization (IGWO)", utilizes multilevel optimal VM placement to improve system performance and surpasses current methods. The IGWO algorithm has a unique "Hierarchical resolution assessment method". The IGWO model is stated as a multiple objective constraint and is solved using a bioinspired maximization approach. Efficiency of the proposed IGWO model is evaluated by an intensive simulation process conducted in a MATLAB environment. The objective of parameter analysis is to assess the efficacy of the IGWO algorithm and ascertain its sensitivity to various parameters.

The study (Ding et al., 2018) presented a highly effective VMP methodology for targeting IaaS side-channel assaults. The hit-rate and loss-rate indices are utilized for the analysis of system safety and the allocation of unauthorized user and selected tenant VMs to separate PMs for the purpose of physical isolation. The researchers assess the quality metrics of a cloud computing system and develop a model to analyze energy usage and resource loss. This strategy focuses on enhancing system security and efficiency by employing a "multi-object discrete firefly optimization technique to address the VMP problem".

Baalamurugan and Vijay Bhanu (2020) proposed a "multi-objective krill herd algorithm for efficient VMP in cloud computing, aiming to minimize resource wastage and power consumption". The suggested algorithm seeks nondominated solutions that reduce distributed cloud computing resource and power waste. Experiments compare the suggested method to multi-objective genetic, ant colony, First-Fit Decreasing, and Simplified-Ant Colony algorithms to assess its efficiency. Results relate that the proposed strategy is more competent than others. The scalability of the algorithm is tested by modifying the simulation environment of VM requests, demonstrating its effectiveness in placing a large number of VMs in data centers.

Fu et al. (2018), introduced an improved gray wolf optimizer for cloud DC VM allocation. As a minimization problem, VMP problem formulation optimizes and decreases active PMs. Another goal of this strategy is to manage cloud DC power and resource waste. The results from this discrete problem solution are better than others. Cloudsim tools are used for simulation.

2.2.3 PSO

The binary Particle Swarm Optimization (PSO) algorithm was proposed by Al-Moalmi et al. (2019) as a means to optimize VMP approaches. The goal is to minimize energy consumption by improving energy efficiency fitness. This approach necessitates a substantial number of repetitions and revisions. The proposed algorithm exhibits a greater reduction in energy consumption compared with four alternative algorithms. This approach emulates the VMP problem by utilizing Amazon EC2.

Abdel-Basset et al. (2019) proposed a cloud DC VMP paradigm using whale optimization. The objective of this strategy is to enhance and efficiently utilize cloud data center resources. This solution aims to enhance and minimize the of active DC machines in order to achieve the goal. This methodology considers the VMP as a variable sized bin packing problem, taking into account the bandwidth. The "improved Levy-based whale optimization technique" enhances the bandwidth of cloud data centers. This approach utilizes the Cloudsim toolkit to obtain precise results and do comparisons.

2.2.4 Hybrid ACO algorithm (HACOS)

The hybrid ACO algorithm is described in (Alharbe et al., 2022). Fig. 13.3 depicts the flowchart of the algorithm. The dashed rectangle in Fig. 13.3 indicates the areas where hybridization and improvement take place. The practice of simulated annealing enhances the pheromone update rule. Rather than upgrading the pheromone levels on all the roads that the ants have investigated, the algorithm chooses to update only a fraction of the trails. This strategy seeks to expedite the development of the algorithm and enhance the quality of the best possible outcomes obtained in each iteration. By doing so, it increases the likelihood of escaping from suboptimal solutions. The pheromone in this strategy refers to the rate at which virtual machines are deployed on physical nodes, and it is intentionally variable to prevent premature convergence. Heuristic information refers to the probability of allocating virtual machines to physical nodes. Greater indicates a higher level of achievement in deployment.

The quality of an algorithm is contingent upon the definition and updating rules of pheromones. Pheromones are generated to facilitate virtual machine placement, both among the VM and the physical-node (VM-Host) and among VMs (VM-VM). The authors (Alharbe et al., 2022; Alharbe et al., 2022) selected a specific procedure. (1) Let's assume that there are M physical nodes, N virtual computers, and K ants. Each each ant operates independently. Every virtual machine node possesses a queue of virtual machines awaiting deployment, and they are assigned randomly. Fig. 13.3 depicts the process of the HACOS algorithm.

2.2.4.1 Pheromone definition

The objective of optimization is to limit the amount of network traffic within the data center and maximize the use of links. Hence, the pheromone will evaluate both of these factors concurrently, and their influence on the maximization problem remains indefinite, hence the ant is making progress. The significance of network traffic and link utilization is denoted by a randomly generated number when the path is established is exchanged. Meanwhile absence of pheromone in the primary check, the ants chose a positioning path randomly.

2.2.4.2 Heuristic information

The heuristic information, denoted as η_{ij} in Eq. (13.1), represents the probability of allocating a VM V_i to a physical machine P_j. It minimizes DC network congestion and optimizes connection utilization without causing any negative impact. Therefore, when implementing a VM V_i, an ant must give preference to a physical-node P_i that has less network congestion.

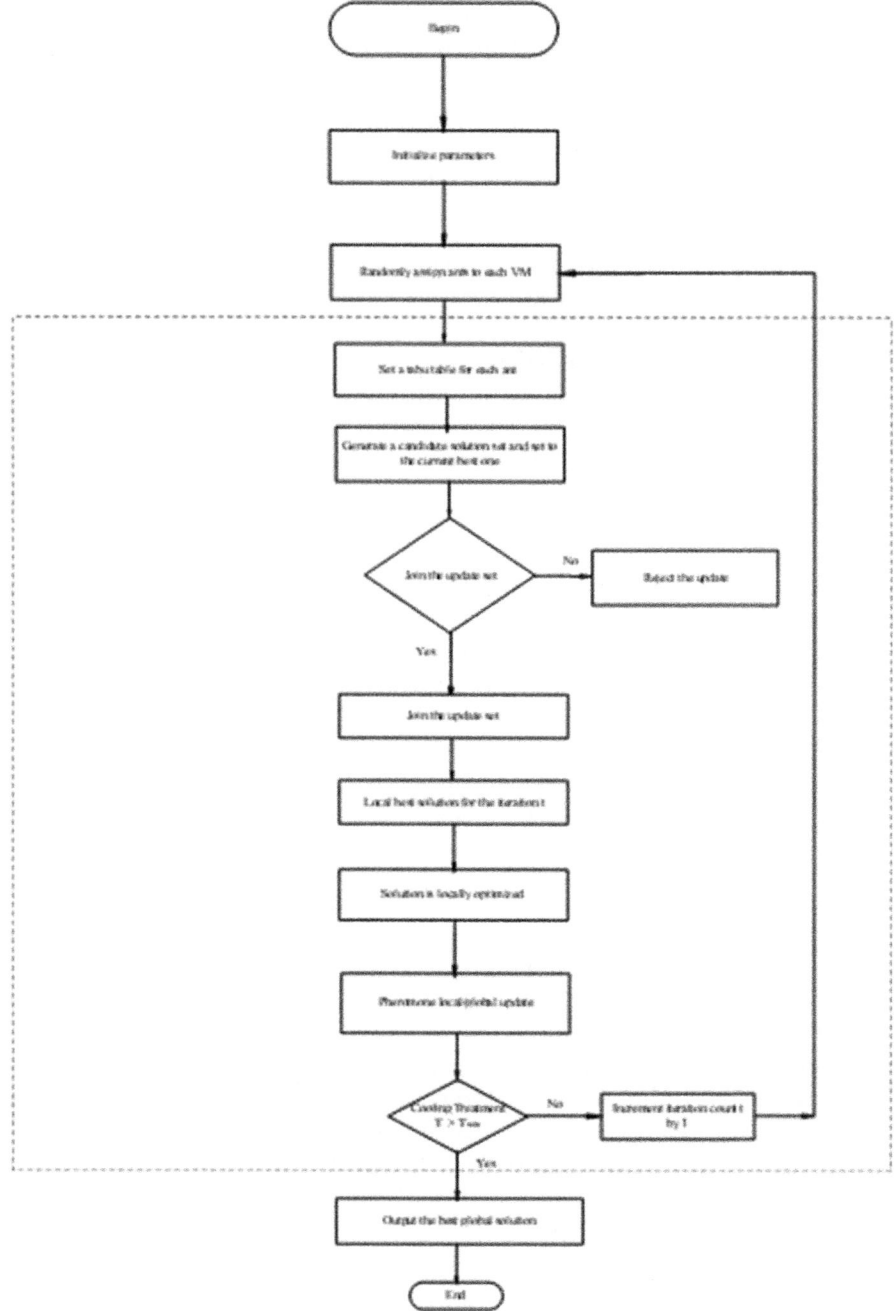

FIGURE 13.3   HACOS algorithm flow chart.

Heuristic data is derived from the combination of network flow and the distance of communication:

$$\eta_{ij} = \sum_{n=1}^{N} a_{in} b_{j\pi(n)} \qquad (13.1)$$

where N — number of VMs

a_{in} — network data flow between VMs, V_i and V_n

$b_{j\pi(n)}$ — number of route hops between two PMs, P_j and $P_{\pi(n)}$.

2.2.4.3 State transition rule

The ACO algorithm assigns task lists to ants k. When an ant deploys VM V_i to PM P_j, it adds P_j to the tabu list $tabu_k$. The succeeding search will exclusively target nodes which were not include in the tabu table. The ants sequentially allocate VMs to PMs, taking into account the probability of transfer. Eq. (13.2) displays the transition probability.

$$P_{ij}^k(t) = \begin{cases} \dfrac{\left(\tau_{ij}(t)\right)^{\alpha} \left(\eta_{ij}(t)\right)^{\beta}}{\displaystyle\sum_{i \notin tabu_k} \left(\tau_{ij}(t)\right)^{\alpha} \left(\eta_{ij}(t)\right)^{\beta}}; \text{if } j \in \text{allow}_k \\ 0; \text{Otherwise} \end{cases} \qquad (13.2)$$

$P_{ij}^k(t)$ — probability of ant k selecting physical node P_j to deploy VM V_i

allow_k — set of physical nodes available for ant k to choose from.

α — pheromone factor

β — heuristic factor.

2.2.4.4 Implementation of HACOS algorithm

HACOS algorithm is depicted in Fig. 13.3. The algorithm works as follows (Alharbe et al., 2022; Alharbe et al., 2022; Alharbe et al., 2022).

Input: Required virtual and physical machine resources respectively: R, H; G network topology, link capacity, expected virtual machine traffic; Algorithm parameters: K, α, β, ρ. Simulation of annealing parameters (T_{max}, T_{min}, Θ, φ).

Output: The mapping between virtual and real machines is π, and the objective function value is f.

Algorithm:

(1) Initialize parameters and randomly spawn VM deployment plans;

(2) Randomly put the ant k on virtual machine i; select a physical node according to (18); add the deployed virtual machine to $tabu_k$;

(3) In step 2, ants deploy virtual computers outside of $tabu_k$ until all are deployed, generating a deployment plan π_k.

(4) For all ants, repeat steps 2 and 3 to build a candidate set C, update f, and generate $f_{localbest}$;

(5) Optimise local best solution, then randomly exchange virtual machine pairs to generate $\text{ant}_{\text{new}}(S_{\text{new}}, f_{\text{new}})$, e.g., $f_{\text{new}} - f_{\text{localbest}} < 0$. Accept the solution, add it to update set U, and repeat random exchange h times (all restrictions must be met);

(6) Select the candidate set solution for the update set and update pheromone;

(7) Algorithm terminates with π, f if $T < T_{\text{min}}$. Otherwise, adjust T and repeat step 2;"

2.2.4.5 Experimental results

CloudSim is used to test HACOS. Experimentations are implemented in Java, and the Hardware configuration: CPU 3.2 GHz, RAM 8 GB.

The experiments have 100 VMs, 50 PMs, 1500 MIPS computing capacity, 8 GB memory.

C − Computing resource requests of the VMs − $C \in$ [200 MIPS; 500 MIPS; 700 MIPS]

M − memory resource requests − $M \in$ [512 ; 1024; 2560 MB]. According to the C and M, 150 VMs are randomly generated.

The experiments focused on optimizing the total link utilization rate and network traffic. The algorithm was evaluated in different network topology environments, including "Tree, Fat-Tree, and VL2 (Virtual Layer 2)". The outcomes showed that the current algorithm outperformed the fundamental ACO and Best fit Decreasing (BFD) algorithms in terms of and link utilization and total network traffic (Koot and Wijnhoven, 2021; Alharbe et al., 2022) also discusses the parameter selection for the algorithm, highlighting the importance of adjusting parameters to enhance global search performance. The testing findings conclusively showed that the HACOS algorithm is highly effective and efficient in reducing network traffic and maximizing link usage in a cloud computing environment.

3. Virtual machine migration

VM migration is the act of moving a running VM from one physical-server to another without interrupting the execution of the user's application. VM migration is the primary method employed in virtualized cloud systems to optimize energy consumption during runtime. The migration time is inversely proportional to the available network bandwidth and the size of the VM. However, the migration time can be reduced by consolidating VMs, which enhances resource efficiency. Efficient resource utilization, in turn, can lead to a drop in energy consumption. Moreover, the amount of energy used is directly influenced by the size of the virtual machine that is being moved. Interestingly, when considering a comparable group of VMs, the sequence in which the migrations occur affects the migration duration, which is dependent on the size of the VMs. Hence, it is imperative to examine the balance between the energy expenditure and the duration of VM transfer (Hossain et al., 2020).

3.1 Important issues with VM migration

- Transferring virtual machines (VMs) via shared bandwidth while upholding the Service Level Agreement (SLA) of an application is difficult due to the dynamic memory capacity of VMs, which can range from 1 GB to over 50 GB.

- WAN-based virtual machine migration encounters several challenges, including increased packet drop ratio, longer communication distances, diverse network architectural design, unexpected network behavior, higher latencies, and limited network bandwidth. These factors contribute to a higher risk of violating service level agreements (SLAs).
- The complexity of VM migration rises as WAN-based storage migration increases due to the requirement of both asynchronous and synchronous connection channels for transmitting storage blocks across CDCs.
- Furthermore, current VM migration strategies lack the capability to efficiently distribute workloads in a coordinated manner across many CDCs.
- Secure virtual machine migration on heterogeneous cloud data centers (CDCs) is a difficult undertaking because of the increased communication distances. Hijackers obtain hardware states, application-sensitive data, currently hosted apps, and OS kernel states for malicious purposes.

3.2 ACO based VMM

Increasing demand for cloud infrastructure has led to, operating costs, carbon emissions and high energy consumption. The chapter (Alvarez-Meaza et al., 2021) proposes a novel technique, known as Active & Idle Virtual Machine Migration (AIVMM), to tackle these concerns. AIVMM intends to condense virtual machines and promote environmentally friendly cloud computing. AIVMM underscores the constraints of current evolutionary computational algorithms and underscores the necessity for a novel method that not only tackles these limitations but also integrates VMs and guarantees environmentally friendly cloud computing. The efficacy of the AIVMM algorithm in attaining environmentally friendly cloud computing (Alvarez-Meaza et al., 2021) addresses the issues of high energy utilization and inappropriate resource consumption in cloud DC.

3.2.1 AIVMM algorithm

The AIVMM algorithm, when combined with the "Order Exchange Migration Ant Colony System (OEMACS)" algorithm, surpasses traditional approaches and provides substantial reductions in energy consumption and resource usage in data centers. The AIVMM algorithm, derived from the ACO algorithm, is designed to address the VMP problem. This solves both the issue of excessive energy usage and the effective utilization of physical resources in cloud architecture. The results illustrate the efficacy of the AIVMM algorithm.

The study in reference (Alvarez-Meaza et al., 2021) presents a power model and assesses the fitness rate of the generated solution to identify the better solution for the current iteration. In addition, it conducts local search and compares the CPU and storage usage of active and idle VMs. The chapter (Alvarez-Meaza et al., 2021) introduces a framework architecture for cloud computing and provides an analysis of relevant research in the subject.

AIVMM algorithm is a new evolutionary computational approach that addresses the problem of high energy consumption with improper resource consumption in cloud DC. The algorithm effectively migrates idle VMs from actively working servers to inactive servers, dipping power interruption for active machines. It is implemented with the OEMACS algorithm, resulting in a fusion algorithm that outperforms predictable methods and offers significant savings in DC resources and energy. The algorithm recognises the count of active idle VMs in a server. AIVMM algorithm exchanges idle VMs from a single server with dynamically working VMs from a nonoverloaded server, ensuring maximum active VMs in a single-server utmost of the time. This helps to minimize power consumption interruptions for actively working VMs.

The AIVMM algorithm utilizes a fitness function to evaluate the constructed solution and performs local pheromone updates on each solution. It also performs global pheromone updates on the best solution of each iteration. The algorithm continues iterating until a termination condition is met. The AIVMM algorithm is assured of the, OEM "(Order and Exchange Migration), ACS (Ant Colony System) algorithm and VMM Virtual Machine Migration". It ensures efficient utilization of physical resources and offers a solution for green cloud computing.

3.2.2 AIVMM results

The AIVMM algorithm along with OEMACS, overtakes basic methods also provides major savings in DC resources and energy. The evaluation of the AIVMM algorithm was conducted using the CloudSim simulator. The evaluation process included measuring the memory and time required for simulation instantiation, as well as comparing the time to migrate VMs between the existing algorithm OEMACS and the combination of OEMACS and AIVMM.

The findings indicated that the time needed for simulation instantiation grew exponentially as the number of VMs rose, although the memory required exhibited a linear correlation. The simulation was determined to be reasonably compatible with desktop computers equipped with typical processing capabilities. The comparison between the existing algorithm OEMACS and the partnership of OEMACS and AIVMM demonstrated the efficacy of the AIVMM algorithm in decreasing the duration required for VM migration. The AIVMM algorithm facilitated the migration of idle or shadow VMs from one server to another, ensuring that actively operating VMs did not experience any undesirable power supply lag.

4. Green cloud for sustainable development

4.1 Green cloud attributes

GC research places a significant focus on the growth of efficient cloud computing with eco-friendly attributes. This includes key aspects such as energy monitoring, virtualization, high-efficiency computing, load balancing, green data centers, extensibility, and recycling. The widespread use of millions of computers in cloud data centers to manage rapidly increasing data comes with a substantial environmental impact, particularly in terms of energy consumption. These servers demand an immense amount of power, comparable to that used by approximately 180,000 residences. To put this into perspective, consider the case of Amazon's data centers. Projected figures indicate that, over a 3-year repayment plan, these data centers account for around 53% of the total expenditure in terms of price and deployment purchase, with 42% allocated to energy-related expenses. Additionally, approximately 19% of the costs are associated with electricity usage and thermal architecture (Usvub et al., 2017). This underscores the critical need for GC to address the eco-friendly implications of cloud computing infrastructure.

4.2 RUAEE algorithm

Han et al. (2018) introduced several practises to improve the energy efficiency of cloud computing algorithm. Among these (Alvarez-Meaza et al., 2021), presented a "Resource Utilization Aware Energy saving server consolidation algorithm (RUAEE)". This algorithm is designed to optimize resource utilization while concurrently minimizing the sum of live migrations of virtual machines. Experimental outcomes demonstrate the effectiveness of RUAEE in achieving reductions in mitigating service-level agreement and energy consumption (SLA) violations within data centers. The primary goal is to enhance diminish power consumption and resource utilization in data centers. By proposing and executing RUAEE, the study contributes to the overarching objective of making cloud computing more energy-efficient while ensuring optimal resource allocation and performance in the cloud environment.

- A model for resource usage is suggested to direct the fine-tuning of unbalanced hosts and VM placements in order to enhance resource utilization and reduce active PMs count.
- An optimal strategy for consolidating unused hosts to reduce the number of live migrations of VMs.
- The algorithm is validated by running simulations using actual workload hints of Google cluster.

5. Green scheduling algorithm in cloud computing data centers

An essential process method in green manufacturing is green scheduling. Thus, scholarly research on green scheduling is aware in present years due to rush to develop manufacturing that lower the emission of contaminating gases and by promoting energy efficient systems. Among the green manufacturing tenets that seek to reduce energy waste and environmental harm is green scheduling. To discourse the issue of energy usage in GC, an innovative energy efficient load balancing global optimization algorithm known as the "Resource-aware load balancing clonal algorithm for task scheduling". This algorithm is based on the principles and concept of load balancing. To optimize both energy usage along with load balancing resource-aware scheduling algorithm are used (Lu and Sun, 2019).

5.1 DVFS

The Green Task Scheduling (GTS) technique was proposed in Sofia and Kumar (2017) as a scheduling technique to moderate the usage of cloud resources. This method also has the benefit of lowering hardware costs. A system known as DVFS is used to control the frequency and processor voltage without compromising its performance. When GTS is used with DVFS in cloud computing, positive outcomes are achieved.

A DVFS method for cloud data centers was proposed in Wu et al. (2014). A sizable cluster of servers connected by the Internet typically makes up a cloud data center. In a cloud data center, a job scheduler is required to plan out job executions. The cloud DC's resources must be effectively utilized by the job scheduler in order to complete tasks. The scheduling algorithm's performance problems are related to resource usage and execution time. A proficient scheduler can complete tasks faster and with less resources. Less energy is expended when fewer resources are used. One of the main obstacles to creating large-scale clouds is power usage.

The DVFS technology is employed to regulate voltage and frequency of servers in Cloud computing. This strategy can minimize the energy usage of a server during periods of inactivity or when it is handling low workloads. "According to a study in (Katal et al., 2023), the energy consumption of data centers is projected to rise from 200 TWh in 2016–2967 TWh in 2030". DCs' substantial power demands for service provision lead to increased carbon emissions (Katal et al., 2023) examines software based methodologies to establish environmentally friendly DCs with power control at the software level. Factors to examine include the environmental impact of data centers, particularly in relation to Electronic-waste. Additional aspects encompass energy efficiency and problem-solving methodologies employed to mitigate power usage in data centers. Novel technologies that can be used in individual software, including methods applied in operating systems, Virtualization, and applications. It presents a variety of techniques at each level to decrease energy consumption, which unquestionably enhances the ongoing environmental challenge of mitigating pollution.

5.2 CACO

The CACO task scheduling method, which minimizes energy usage in cloud systems, was proposed in Ari et al. (2017). It has numerous advantages that make GC possible. Service cloud providers' energy-aware work scheduling in the cloud has a significant impact on customers' optimal resource use and, consequently, cost effectiveness. For cloud computing, the conventional work scheduling techniques are insufficient. Tasks should be effectively planned in such an environment to minimize makespan. To enable GC, the author suggested biologically inspired scheduling method CACO (Cloud Ant Colony based Optimization), which is built on a modified version of the ant ACO and tries to minimize the makespan time while promising load balancing resources. This technique used a simulated environment that mimicked a homogeneous cloud computing setup.

5.3 Honeynet

In Pittman and Alaee (2023), a honeynet green scheduling technique was described, which primarily focuses on two scheduling objectives. The first subject pertains to the scheduling of cloud workloads in general. The second task involves customizing workload scheduling on IaaS platforms specifically for honeynets. To optimize the utilization of the cloud compute node for the honeynet, the algorithm takes into account the maximum resources available on each node. To limit energy consumption consumption and operating expenses through sustainable, green operations, power consumption and usage of honeynet nodes in addition to the resources available on cloud compute nodes. In order to solve the environmental issues related to cloud services, a successful algorithmic implementation may result in lower energy usage in cloud technologies.

5.3.1 Calculation of energy consumption

$$Ci = \{CPU.MEM, NET\} \tag{13.3}$$

Here Ci is the collection of CPU, RAM, and network resources for the cloud compute node. The implied relationship to virtualized computational resources allows the suggested approach to incorporate cloud hardware in (13.3). Furthermore, the honeynet is composed of virtual computer resources. In terms of cloud hardware resources, at the very least, the virtual compute resources are related 3-to-1 (13.4)

$$Ch = \{vCPU, vMEM, vNET\} \tag{13.4}$$

"The proposed algorithm must be aware of the maximum resources available on a given individual compute node $ci \in Ci$ such that the cloud compute node is optimally utilized for the honeynet" (13.5).

$$ci \epsilon Ci | \text{sum}(ch \epsilon Ch \leq \max(ci) \tag{13.5}$$

The method maintains a 3-to-1 relationship with the hardware cloud resources, while taking the honeynet's virtual machine resources into account.

5.4 EEDPFSP

In Li et al. (2021) "goal of reducing both total flow time and overall energy consumption, the study discusses the energy-efficient scheduling of the distributed permutation flowshop (EEDPFSP)". A novel operator, constructive heuristic methods, and problem-specific features are included into an enhanced NSGAII algorithm (INSGAII) to yield high quality primary solutions. Green scheduling has received little attention, especially for large-scale issues, and effective techniques such as meta-heuristic approaches are needed. The approach is intended to manage the EEDPFSP's distributed and multi-objective optimization complexity.

6. Green data center optimization

6.1 ABC based energy saving

In order to optimize energy usage in green data centers (GDCs), "an artificial bee colony (ABC)-based energy-saving technique" for Fat Tree switches is proposed in this research. The outcomes of the simulation show that the suggested algorithm is well-organized in cutting down on energy waste and speeding up completion times in scenarios with both high and low traffic. This model describes the phases required in the artificial bee colony algorithm optimization process, which minimizes the impartial function of energy usage in GDC. The simulation results confirm how well the suggested approach works to produce workable solutions that get close to the ideal answer (Zhou et al., 2019).

The algorithm uses roulette selection to choose honey sources and computes the likelihood. The stages of choosing honey sources, locally searching in their neighborhood, and updating the ideal fitness value are repeated throughout the optimization process. To minimizing energy consumption in GDC networks is established based on multi commodity network flow. To maximize energy usage in the switches, take into account variables including wear consumption, service demands, traffic load, and queuing delay (Patil and Duttagupta, 2014).

6.2 Server virtualization

Low environmental impact and energy efficiency are priorities in the construction of GDC. GDC aims to lower energy consumption and carbon impact without sacrificing or even improving availability and performance. Server virtualization is a prominent technology employed in GDC to decrease the demand for physical servers and enhance energy efficacy. Power management is an essential technology utilized in environmentally friendly data centers. Power management strategies are categorized as two types that can help data centers reduce energy consumption and minimize environmental harm are the application of renewable energy sources and the implementation of frequency scaling and dynamic voltage. The former involves utilizing sustainable sources of energy, while the latter adjusts the power usage of servers according to workload demands.

7. Conclusion

The capabilities of modern technology have been enhanced through research programs, leading to the creation of a vast array of tools, systems, and software, as well as the advancement of these technologies' ability to support GC. This chapter's objective was to analyze and investigate how sustainable technologies are applied in different GC sectors in order to arrive at ecologically responsible and sustainable development. The investigation found sustainable green approaches in the fields of green algorithms, green cloud data centers, green sustainable development, and green transportation. The research highlighted notable breakthroughs and achievements across multiple domains, including the development of integrated smart energy systems, green architectural frameworks, and algorithms. The integration of AI (Artificial Intelligence) and ML (Machine Learning) could lead to a sustainable growth of GC. Along with these challenges, the research identified a lack of standards, technical hurdles, and insufficient funding as further barriers to the development of GC. An outline of the latest progresses in GC is presented in the study along with a comparison to earlier studies. Examining the patterns in GC publications, it shows that research output increased steadily starting in 2010 and peaked in 2023. Some of the subjects covered are green data centers, green cloud computing, and green sustainable development.

References

Abdel-Basset, M., Abdle-Fatah, L., Sangaiah, A.K., 2019. An improved Lévy based whale optimization algorithm for bandwidth-efficient virtual machine placement in cloud computing environment. Cluster Computing 22, 8319–8334. https://doi.org/10.1007/s10586-018-1769-z.

Al-Moalmi, A., Luo, J., Salah, A., Li, K., 2019. Optimal virtual machine placement based on grey wolf optimization. Electronics 8 (3), 283. https://doi.org/10.3390/electronics8030283. MDPI AG.

Alharbe, N., Rakrouki, M.A., Aljohani, A., 2022. An improved ant colony algorithm for solving a virtual machine placement problem in a cloud computing environment. IEEE Access 10, 44869–44880. https://doi.org/10.1109/ACCESS.2022.3170103.

Alvarez-Meaza, I., Zarrabeitia-Bilbao, E., Rio-Belver, R.-M., Garechana-Anacabe, G., 2021. Green scheduling to achieve green manufacturing: pursuing a research agenda by mapping science. Technology in Society 67, 101758. https://doi.org/10.1016/j.techsoc.2021.101758.

Analysts, G.I., 2021. Internet Data Centres — Global Market Trajectory & Analytics.

Ardagna, D., Casale, G., Ciavotta, M., Pérez, J.F., Wang, W., 2014. Quality-of-service in cloud computing: modeling techniques and their applications. Journal of Internet Services and Applications 5 (1), 18690238. https://doi.org/10.1186/s13174-014-0011-3.

Ari, A.A.A., Damakoa, I., Titouna, C., Labraoui, N., Gueroui, A., 2017. Efficient and scalable ACO-based task scheduling for green cloud computing environment. Proceedings — 2nd IEEE International Conference on Smart Cloud, SmartCloud 2017 66–71. https://doi.org/10.1109/SmartCloud.2017.17.

Baalamurugan, K.M., Vijay Bhanu, S., 2020. A multi-objective krill herd algorithm for virtual machine placement in cloud computing. The Journal of Supercomputing 76 (6), 4525–4542. https://doi.org/10.1007/s11227-018-2516-1.

Ding, W., Gu, C., Luo, F., Chang, Y., Rugwiro, U., Li, X., Wen, G., 2018. 3, 1. DFA-VMP: an efficient and secure virtual machine placement strategy under cloud environment. Peer-to-Peer Networking and Applications 11 (2), 318–333. https://doi.org/10.1007/s12083-016-0502-z.

Dorigo, M., 1992. Optimization, Learning and Natural Algorithms.

Farshin, A., Sharifian, S., 2019. A modified knowledge-based ant colony algorithm for virtual machine placement and simultaneous routing of NFV in distributed cloud architecture. The Journal of Supercomputing 75 (8), 5520–5550. https://doi.org/10.1007/s11227-019-02804-x.

Fu, X., Zhao, Q., Wang, J., Zhang, L., Qiao, L., 2018. Energy-aware VM initial placement strategy based on BPSO in cloud computing. Scientific Programming 2018, 1–10. https://doi.org/10.1155/2018/9471356.

Ghobaei-Arani, M., Shamsi, M., Rahmanian, A.A., 2017. An efficient approach for improving virtual machine placement in cloud computing environment. Journal of Experimental & Theoretical Artificial Intelligence 29 (6), 1149–1171. https://doi.org/10.1080/0952813X.2017.1310308.

Han, G., Que, W., Jia, G., Zhang, W., 2018. Resource-utilization-aware energy efficient server consolidation algorithm for green computing in IIOT. Journal of Network and Computer Applications 103, 205–214. https://doi.org/10.1016/j.jnca.2017.07.011.

Hossain, M.K., Rahman, M., Hossain, A., Rahman, S.Y., Islam, M.M., 2020. Active Idle Virtual Machine Migration Algorithm-a new Ant Colony Optimization approach to consolidate Virtual Machines and ensure Green Cloud Computing. In: ETCCE 2020 — International Conference on Emerging Technology in Computing, Communication and Electronics, 12. Institute of Electrical and Electronics Engineers Inc, Bangladesh, p. 21. https://doi.org/10.1109/ETCCE51779.2020.9350915.

Index, C.G.C., 2018. Cisco Global Cloud Index: Forecast and Methodology.

Katal, A., Dahiya, S., Choudhury, T., 2023. Energy efficiency in cloud computing data centers: a survey on software technologies. Cluster Computing 26 (3), 1845–1875. https://doi.org/10.1007/s10586-022-03713-0.

Koot, M., Wijnhoven, F., 2021. Usage impact on data center electricity needs: a system dynamic forecasting model. Applied Energy 291, 116798. https://doi.org/10.1016/j.apenergy.2021.116798.

Li, Y.Z., Pan, Q.K., Gao, K.Z., Tasgetiren, M.F., Zhang, B., Li, J.Q., 2021. A green scheduling algorithm for the distributed flowshop problem. Applied Soft Computing 109, 15684946. https://doi.org/10.1016/j.asoc.2021.107526.

Lu, Y., Sun, N., 2019. An effective task scheduling algorithm based on dynamic energy management and efficient resource utilization in green cloud computing environment. Cluster Computing 22, 513–520. https://doi.org/10.1007/s10586-017-1272-y.

Masdari, M., Nabavi, S.S., Ahmadi, V., 2016. An overview of virtual machine placement schemes in cloud computing. Journal of Network and Computer Applications 66, 106–127. https://doi.org/10.1016/j.jnca.2016.01.011.

Masdari, M., Salehi, F., Jalali, M., Bidaki, M., 2017. A survey of PSO-based scheduling algorithms in cloud computing. Journal of Network and Systems Management 25 (1), 122–158. https://doi.org/10.1007/s10922-016-9385-9.

Mythrayee, D., Lavanya, V.S., 2023. An efficient wolf optimizer system for virtual machine placement in wireless network over the cloud environment. Wireless Personal Communications 129 (3), 2141–2156. https://doi.org/10.1007/s11277-023-10229-2.

Patil, T.C., Duttagupta, S.P., 2014. Hybrid self-sustainable green power generation system for powering green data center. In: International Conference on Control, Instrumentation, Energy and Communication, CIEC 2014. Institute of Electrical and Electronics Engineers Inc, India, pp. 331–334. https://doi.org/10.1109/CIEC.2014.6959104.

Pittman, J.M., Alaee, S., 2023. A green scheduling algorithm for cloud-based honeynets. Frontiers in Sustainability 3, 2673–4524. https://doi.org/10.3389/frsus.2022.1048606.

Radu, L.D., 2017. Green cloud computing: a literature survey. Symmetry 9 (12), 20738994. https://doi.org/10.3390/sym9120295.

Saxena, D., Gupta, I., Kumar, J., Singh, A.K., Wen, X., 2022. A secure and multiobjective virtual machine placement framework for cloud data center. IEEE Systems Journal 16 (2), 3163–3174. https://doi.org/10.1109/JSYST.2021.3092521.

Singh, A.K., Swain, S.R., Saxena, D., Lee, C.N., 2023. A bio-inspired virtual machine placement toward sustainable cloud resource management. IEEE Systems Journal 17 (3), 3894–3905. https://doi.org/10.1109/JSYST.2023.3248118.

Sofia, A.S., Kumar, P.G., 2017. Energy efficient task scheduling to implement green cloud. Asian Journal of Research in Social Sciences and Humanities 7 (2), 443–458. https://doi.org/10.5958/2249-7315.2017.00101.0.

Speitkamp, B., Bichler, M., 2010. A mathematical programming approach for server consolidation problems in virtualized data centers. IEEE Transactions on Services Computing 3 (4), 266–278. https://doi.org/10.1109/TSC.2010.25.

Ullman, J.D., 1975. NP-complete scheduling problems. Journal of Computer and System Sciences 10 (3), 384–393. https://doi.org/10.1016/S0022-0000(75)80008-0.

Usvub, K., Farooqi, A.M., Alam, M.A., 2017. Edge up green cloud in cloud data centres. International Journal of Advanced Research in Computer Science 8 (2).

Wu, C.M., Chang, R.S., Chan, H.Y., 2014. A green energy-efficient scheduling algorithm using the DVFS technique for cloud datacenters. Future Generation Computer Systems 37, 141–147. https://doi.org/10.1016/j.future.2013.06.009.

Wu, J., Guo, S., Li, J., Zeng, D., 2016. Big data meet green challenges: greening big data. IEEE Systems Journal 10 (3), 873–887. https://doi.org/10.1109/JSYST.2016.2550538.

Zhou, Q., Lou, J., Jiang, Y., 2019. Optimization of energy consumption of green data center in e-commerce. Sustainable Computing: Informatics and Systems 23, 103–110. https://doi.org/10.1016/j.suscom.2019.07.008.

Bio-inspired meta-heuristic algorithm for solving engineering optimization problems based on computational intelligence

S. Mohana Saranya, S. Mohanapriya and Dinesh Komarasamy

Department of Computer Science and Engineering, Kongu Engineering College, Erode, Tamil Nadu, India

1. Introduction to computational intelligence and engineering optimization

At the core of numerous industries, Engineering Optimization serves as a vital tool for solving intricate problems, whether it be designing efficient structures, enhancing energy systems, or streamlining complex scheduling tasks. The central objective of optimization is to uncover the best possible solution within the constraints and objectives at hand. However, as problems grow in complexity and dimension, traditional optimization methods often struggle to yield optimal results, often leading to less-than-ideal or even unattainable solutions.

1.1 Computational intelligence

Computational intelligence, at the crossroads of computer science, artificial intelligence, and cognitive psychology, equips computers and machines to emulate human-like intelligence, reasoning, and problem-solving. It bridges the gap between computational might and human-like learning, offering a versatile toolbox with methods such as neural networks, evolutionary algorithms, fuzzy logic, and expert systems. In a world filled with complex challenges, from data analysis to fine-tuning solutions, computational intelligence takes center stage, adapting to various tasks such as speech recognition, image analysis, financial predictions, and robotics, providing innovative solutions for a wide range of applications.

Computational Intelligence in Sustainable Computing and Optimization
https://doi.org/10.1016/B978-0-443-23724-9.00014-1

1.2 Bio-inspired algorithms in engineering optimization

This chapter, embark on an enlightening journey into the world of bio-inspired metaheuristic algorithms and their profound impact on engineering optimization. It delve into the principles of bio-inspired approaches, showcasing their ability to explore vast solution spaces, adapt to dynamic conditions, and avoid getting stuck in suboptimal solutions. Then explore various bio-inspired algorithms, including evolutionary algorithms, swarm intelligence, and others, highlighting their strengths and weaknesses for different optimization challenges through comparative analysis. Furthermore, this chapter delve into the future of bio-inspired optimization, where researchers are actively working on hybridizing these methods with other optimization techniques, employing parallelization strategies, and integrating machine learning to enhance their capabilities.

2. Bio-inspired approaches for solving optimization problems

In recent years, a paradigm-shifting solution has emerged in the form of bio-inspired metaheuristic algorithms. These ground-breaking algorithms draw their inspiration from the wonders of nature and employ computational intelligence to address these challenging optimization problems with remarkable effectiveness. These algorithms take inspiration from a variety of sources, including evolutionary biology, swarm behavior, and genetic inheritance. By doing so, they attempt to address complex problems with a level of adaptability and resilience that can surpass traditional optimization methods.

One of the notable advantages of bio-inspired optimization approaches is their ability to navigate vast solution spaces. Traditional optimization techniques often struggle when confronted with high-dimensional, nonlinear, or rugged problem landscapes. Bio-inspired methods, however, excel in exploring these intricate terrains by mimicking nature's evolution and adaptation. This allows them to find solutions that are not only more efficient but also robust in the face of uncertainty and change.

Within the realm of bio-inspired optimization, several key methodologies have gained prominence. These include evolutionary algorithms, which mimic the process of natural selection to iteratively refine potential solutions, and swarm intelligence, which harnesses collective behaviors observed in social organisms such as ants or birds to find optimal solutions. Ant Colony optimization, Bee colony optimization, particle swarm optimization and Genetic algorithms are just a few examples of bio-inspired techniques that have demonstrated their prowess in solving diverse optimization problems.

The strength of bio-inspired approaches lies in their adaptability and versatility. Different algorithms within this category can be tailored to specific problem types, and their parameters can be fine-tuned to address the unique characteristics of each problem (Dehghani and Trojovský, 2022). This adaptability has led to their widespread application in various domains, ranging from engineering and logistics to data mining and artificial intelligence. In practice, bio-inspired optimization has delivered impressive results (Wang et al., 2022; Dehghani et al., 2023; Braik et al., 2022). It has played a crucial role in optimizing complex scheduling tasks, designing structurally sound buildings, enhancing energy systems for sustainability, and extracting valuable insights from vast datasets.

3. Application of metaheuristic algorithms in engineering optimization

Metaheuristic algorithms have carved out a significant place in the field of engineering optimization, offering innovative solutions to a broad spectrum of complex problems. These algorithms, inspired by natural and artificial processes, have proven their role in enhancing the efficiency, robustness, and cost-effectiveness of various engineering applications. The key areas where metaheuristic algorithms find practical application are as follows.

3.1 Structural design and engineering

One of the standout applications is in structural design, where metaheuristic algorithms have been instrumental in creating efficient and cost-effective designs for buildings, bridges, and other infrastructure (Dehghani and Trojovský, 2022). They can optimize the placement and quantity of structural elements, resulting in sturdy yet resource-efficient constructions.

3.2 Scheduling and time management

Metaheuristic algorithms are invaluable for scheduling tasks and managing resources in engineering projects. They can optimize project timelines, allocate resources efficiently, and ensure that complex tasks are completed in the most time-effective manner, which is essential in industries such as construction, manufacturing, and project management.

3.3 Energy systems and resource allocation

Optimizing energy systems, such as power generation and distribution, is a critical application. Metaheuristic algorithms can help in finding optimal configurations for energy resources, minimizing waste, and maximizing efficiency. This has significant implications for sustainability and cost savings.

3.4 Mechanical design and product development

In the realm of mechanical engineering and product development, metaheuristic algorithms assist in optimizing the design of machinery and consumer products (Liu et al., 2023). By refining design parameters, these algorithms contribute to achieving the best performance, durability, and cost-effectiveness.

3.5 Network design and routing

Metaheuristic algorithms play a pivotal role in designing efficient communication and transportation networks. They optimize routing, reducing delays and costs, which is vital in applications such as telecommunications, logistics, and transportation systems.

3.6 Manufacturing and process optimization

Manufacturing processes are intricate and multifaceted. Metaheuristic algorithms assist in optimizing manufacturing processes, reducing waste, enhancing product quality, and minimizing production costs.

3.7 Aircraft and aerospace design

In the aerospace industry, where safety and efficiency are paramount, metaheuristic algorithms play a crucial role in aircraft design, trajectory optimization, and resource allocation.

3.8 Environmental engineering

Addressing environmental challenges requires optimizing solutions to minimize negative impacts. Metaheuristic algorithms can help in optimizing environmental engineering processes, such as waste management, pollution control, and renewable energy systems.

3.9 Data mining and decision support

In the age of big data, metaheuristic algorithms facilitate data mining and decision support by optimizing data analysis and pattern recognition. They help in extracting meaningful insights from vast datasets, improving decision-making processes in various engineering applications (Pourkhodabakhsh et al., 2023).

3.10 Robotics and automation

Engineering optimization is crucial for robotic systems and automation. Bio-inspired algorithms aid in improving robot control, motion planning, and sensor network optimization (Kiani et al., 2022).

These examples showcase the versatility of metaheuristic algorithms in engineering optimization. By exploring the solution space efficiently and effectively, they contribute to advancements in various engineering domains, offering better, more cost-effective, and sustainable solutions to complex challenges. The applications of these algorithms are ever-expanding, making them a valuable asset for engineers and researchers seeking innovative ways to tackle real-world problems.

4. Evolutionary algorithms: Mimicking natural selection for optimization

Evolutionary algorithms are a fascinating category of optimization methods inspired by the principles of natural selection. Much like the process of evolution in the natural world, these algorithms seek to improve solutions iteratively over generations. Here, we delve into the core concepts and mechanisms behind evolutionary algorithms.

At the heart of evolutionary algorithms is the concept of a population, which represents a group of potential solutions to a problem. Just as in nature, these algorithms operate on a principle of survival of the fittest as below. This algorithm works on the below fashion.

1. The Population is initialized with solutions generated randomly
2. Each solution in the population is assessed using an objective function to measure their quality with respect to the problem's goals.
3. Repeat for a specified count of generations or until reaching an end condition:

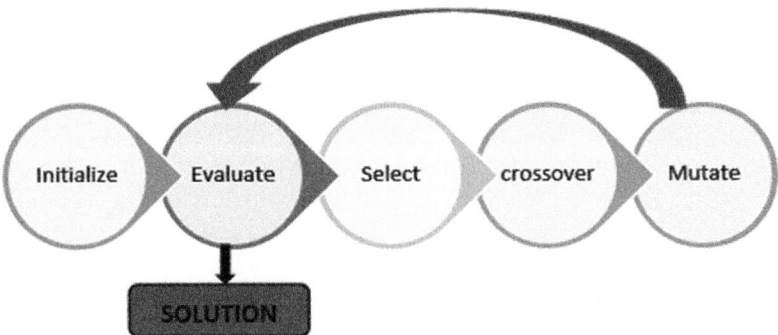

FIGURE 14.1   Evolutionary algorithm steps.

a. Select solutions from the population to form a mating pool, favoring solutions with better performance.
b. Apply crossover (recombination) and mutation operators to create new offspring solutions from the mating pool:
   - Crossover: To create offspring, Combine the genetic makeup of two or more parent solutions
   - Mutation: Introduce small, random changes into the genetic makeup of solutions.
c. The fitness is assessed using the objective function for the newly produced offspring
4. The better doing solutions are selected and newly created offspring to create following generation
5. Steps 3 and 4 are continued for some number of generations or end condition is reached
6. Output of the algorithm is the optimal solution created in the final population

Evolutionary algorithms excel in exploring complex and multi-dimensional solution spaces, making them valuable for solving a wide range of optimization problems (Deng et al., 2022). One of the advantages of evolutionary algorithms is their adaptability. They can be fine-tuned to suit specific problem types and objectives. Steps in Evolutionary Algorithm is shown in Fig. 14.1.

By adjusting parameters such as the population size, mutation rate, and crossover methods, the algorithm's performance can be tailored to address the unique characteristics of each problem. In practice, evolutionary algorithms have found applications in diverse fields, such as engineering, finance, biology, and artificial intelligence. They are particularly well-suited to problems with many variables and complex, nonlinear relationships.

5. Swarm intelligence: Harnessing collective behavior for problem solving

Swarm Intelligence is a collective problem-solving approach inspired by the decentralized and self-organized behavior of natural systems, particularly those found in birds and animals such as ants, bees, and termites. This approach leverages the intelligence of a group of simple, often autonomous agents (individuals) to achieve complex tasks or make decisions. These

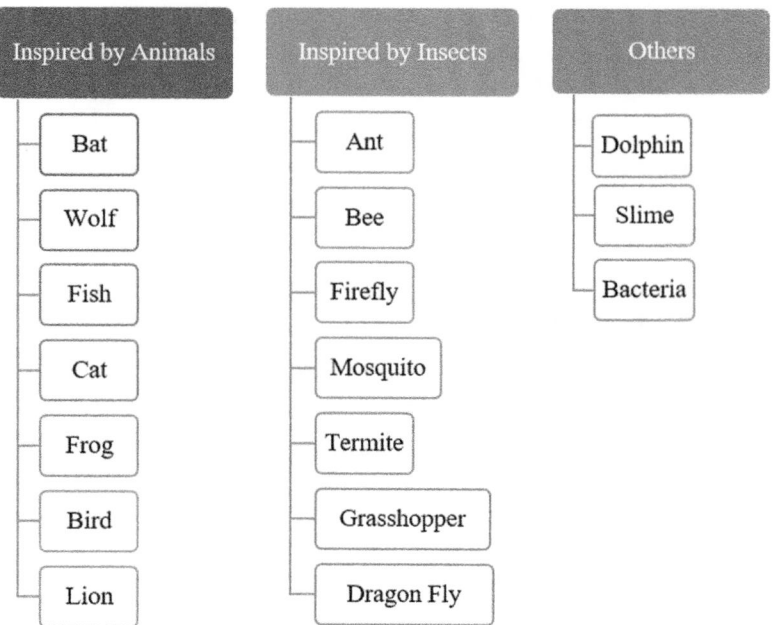

FIGURE 14.2   Categories of swarm intelligence algorithm.

agents work collaboratively, adapt to their environment, and often use local interactions and simple rules to accomplish their goals. Swarm intelligence algorithms are particularly useful in solving complex problems, where a centralized approach might be impractical or inefficient. They are often used in optimization problems, routing, scheduling, and other tasks where finding the best solution involves exploring a large solution space. The simplicity, scalability, and adaptability of swarm intelligence algorithms make them well-suited for a huge number of applications in various fields, including engineering, logistics, finance, and biology. Categories of swarm intelligence algorithm is shown in Fig. 14.2.

Following are some of the prominent Swarm Intelligence Algorithms.

5.1 Ant colony optimization (ACO)

One of the important types of optimization algorithm is the Ant Colony Optimization (ACO) which mimics the action of ant in locating and moving toward food (Dorigo and Stützle, 2019; Rezvanian et al., 2023). This algorithm was introduced by Marco Dorigo in 1992. Ants are prime examples of eusocial insects, which exhibit a high degree of social organization and cooperation within their colonies. Ants indeed communicate with each other using a variety of methods, including sound, touch, and chemical substance (pheromones). Pheromones are chemical compounds that are secreted by ants that trigger another ant. Pheromones are actually perfumes dropped by ants in wherever place they go. These perfumes are sensed by other ants and their behavior is changed based on it. Ants start to move in search for food in random paths from their nests. This random search helps to obtain multiple

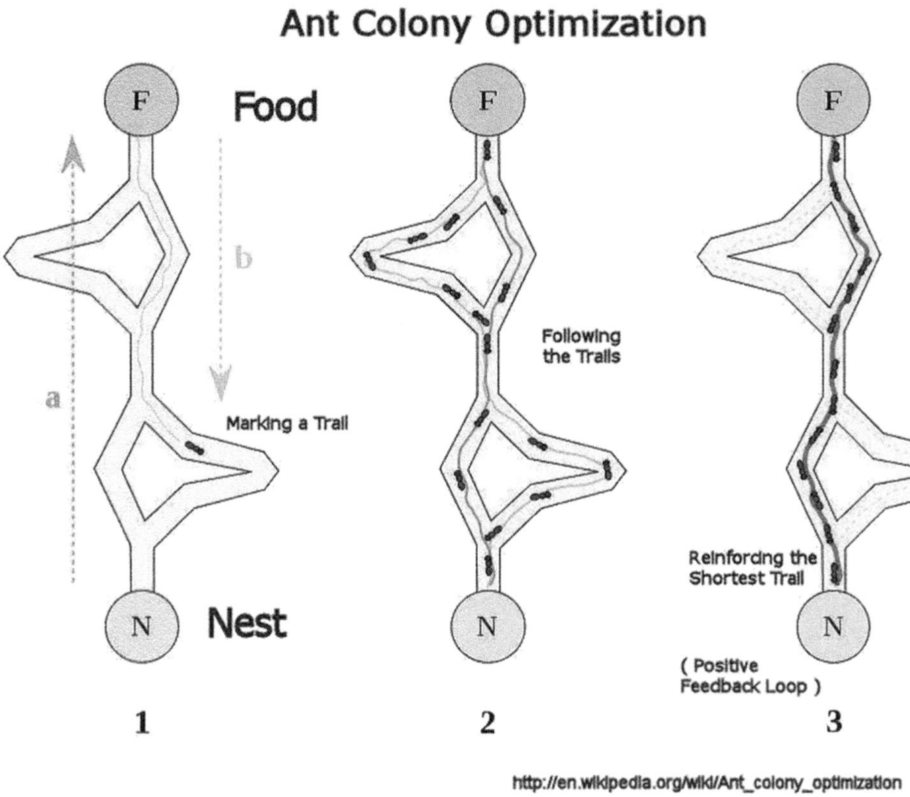

FIGURE 14.3   Movement of ants in search of food.

food path. Ants in return carry a portion of food back to their nests by leaving a pheromone on the path they found. Based on this pheromone, the selection of path is made by the following ants. The probability of selection of path by the following ants is based on the concentration and evaporation rate of pheromone. The length of each path can be easily calculated based on the evaporation rate of the pheromone. This procedure makes the following ants to choose the path to their food source which is shortest. Fig. 14.3 the Movements of Ants in search of food in the shortest path.

5.1.1 *Working of ACO algorithm*

The problem is represented in the form of a graph, where nodes represent possible solutions, and edges between nodes represent the feasibility of transitioning from one solution to another. A population of artificial ants is created, and each ant is placed on a random node in the graph. Ants start building solutions by moving from one node to another based on a set of probabilistic rules. The rules are influenced by two factors: Pheromone Levels, Heuristic Information. The probabilistic rules are calculated based on the Equation (Dehghani and Trojovský, 2022)

$$P_{ij} = \frac{\left(T_{ij}{}^{\alpha} * N_{ij}{}^{\beta}\right)}{\sum\left(\left(T_{ij}{}^{\alpha} * N_{ij}{}^{\beta}\right)\right)}$$

where,

P_{ij}: Probability if selecting path (i, j)

T_{ij}: The concentration level of pheromone on path (i, j).

N_{ij}: A heuristic value related to the quality of the solution component.

α and β: factors to control the impact of pheromone and the heuristic value, respectively.

Σ: Sum taken over all available paths

Begin

Initialize

While not termination condition do

Each ant is positioned in a starting node

Repeat

For every ant do

Use probabilistic rule to choose the next node

Update the pheromone

End for

Until solution is built by the ant

Update solution which is optimal

Do pheromone update

End While

End

Ants construct complete solutions, and each solution's quality is evaluated based on the optimization problem's objective function. Once solutions are built by every ant, pheromone levels on the edges are updated. Edges used in high-quality solutions receive more pheromone, while those used in low-quality solutions receive less. Pheromone levels also decay over time to prevent convergence to suboptimal solutions. Until stopping condition is not met or for a certain iteration of loop the process is continued. Throughout the iterations, the best solution found by the ants is recorded. This best solution improves over time as ants explore and update pheromone levels. Once optimal solution is found or final condition is met or number of repetitions have been reached the algorithm stops.

5.2 Particle swarm optimization

Particle Swarm Optimization (PSO) (Jain et al., 2018; Wang et al., 2018) is a community-based algorithm inspired by the manners of birds and fish, particularly the coordinated movement seen in group of birds or group of fish. This algorithm was introduced in the year 1995 by Dr. Eberhart and Dr. Kennedy.

The basic idea behind PSO is to model potential solutions to a problem as individual particles within a swarm. The algorithm starts by creating initial particles and velocities are assigned to them. Each particle represents a possible solution in the search space. The particles go through the search space, and its positions and velocities are adjusted based on their own experiences and the experiences of their neighboring particles. The algorithm iteratively refines the solutions by adjusting these positions and velocities, with the goal of converging toward an optimal solution.

5.2.1 Algorithm

Initialize the population of particles:
For every particle in the population:
Particle's position (c_pos) is randomly set up in search space.
Initialize particle's velocity in the search space randomly.
Fitness of the current position of the particle is evaluated.
Particle's Personal best position is updated (p_best_pos) to its current position (c_pos)
Update the best fitness value to the evaluated fitness.
Particle having good personal best fitness is set as the global best particle.
Set the algorithm parameters.
Repeat for certain iterations or until convergence are met:
For each particle do
velocity = velocity + c1*rand*(p_best_pos − c_pos) + c2*rand*(global_best_position − c_pos)
position = position + velocity.
Estimate new position fitness.
If new position > personal best:
Personal best position and fitness is updated.
Global best particle is updated based on particle having best personal best fitness.
End of loop.
Optimal solution found is returned.

In the above algorithm,

- $c1$ and $c2$ are cognitive parameter and social parameter respectively
- rand generates a random number between 0 1st 1

The algorithm iteratively updates the values of each particles, trying to discover the optimal solution within the search space.

PSO's potential to find the solution space effectively and find best solutions makes it a valuable tool for resolving optimization problems in various domains. It is often used when the problem is complex, nonlinear, and lacks a known analytical solution. Many variants of PSO have come into picture which have been explored in detail in this chapter (Nayak et al., 2023). It can also be combined with other algorithms thereby making a hybrid approach.

5.3 Bee colony optimization (BCO)

Bee Colony Optimization (BCO) (Agarwal et al., 2016) is an optimization algorithm inspired by the food searching nature of honeybee colonies. It falls under the category of swarm intelligence algorithms, which are motivated by the collaborative nature of social insects and other animals to solve complex problems. BCO was developed by Dervis Karaboga in the early 2000s and is mainly used for solving optimization and search problems.

The BCO algorithm is working on the basis of food searching manner of honeybees, where individual bees (agents) search for the best food sources (solutions) in their environment. Fig. 14.4 shows the behavior of Bees in collecting Nectar. Various optimization problems,

FIGURE 14.4   Behavior of bees in collecting nectar.

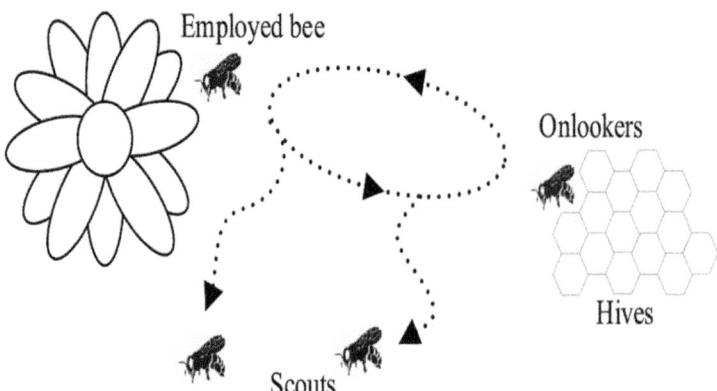

including continuous, combinatorial, and discrete problems can be solved using BCO (Kaya et al., 2022).

Key elements of the Bee Colony Optimization algorithm.

Population of Bees: In BCO, in order to research the solution space, a population of artificial bees is generated. Possible solutions to the optimization problem are represented by these bees. Fig. 14.1 shows the behavior of Bees in collecting Nectar.

Employed Bees: A portion of the population consists of employed bees. Employed bees are responsible for searching the solution space and improving their solutions. They exchange information with each other through a process known as neighborhood search.

Onlooker Bees: Based on the caliber of solutions found by employed bees, Onlooker bees pick the solutions. High-quality solutions are more likely to be chosen.

Scout Bees: Scout bees are responsible for exploring new, unvisited solutions. They are introduced to avoid the algorithm from getting glued in local optima. When a scout bee exhausts its search, it generates a new solution randomly.

Flowchart of BCO Algorithm is shown in Fig. 14.5 Working of this algorithm is explained below:

1. Begin by Initializing a population of candidate solutions.
2. Each employed bee chooses a solution from the current available population and modifies it to generate a new one. This modification can involve local search or other optimization techniques. Employed bees update their solutions if they find an improvement.

Based on the caliber of the solutions found by employed bees, Onlooker bees pick solutions to explore. Solutions with good quality will be chosen mostly.

1. Scout bees are responsible for diversifying the search by exploring new, unvisited solutions. An employed or onlooker bee becomes a scout bee if it cannot find optimal solution after a certain number of iterations. Scout bees generate new solutions randomly, often without considering the current population.
2. The algorithm proceeds through a predefined number of iterations, or it terminates when a convergence criterion is met.
3. Once the algorithm terminates, it returns the best solution found throughout the process.

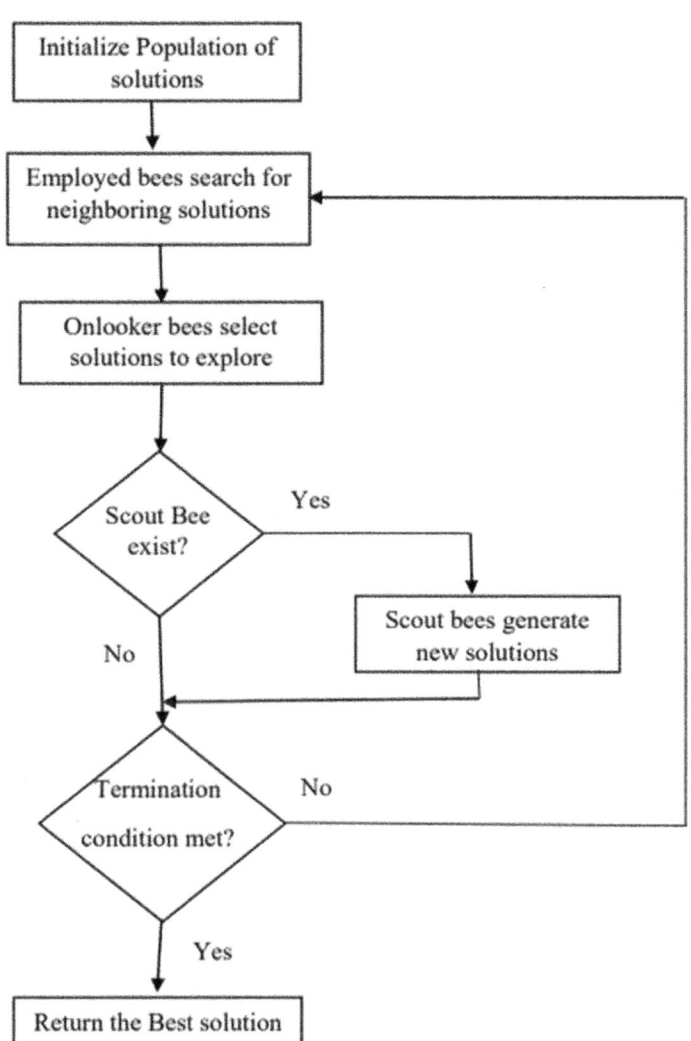

FIGURE 14.5 Flowchart of BCO algorithm.

BCO has been used for numerous optimization problems, including the traveling salesman problem (TSP), job scheduling, function optimization, and more. It has the advantage of being easy to implement and able to handle both discrete and continuous optimization problems.

5.4 Firefly algorithm

One of the nature-inspired optimization algorithm (Kumar and Kumar, 2021) that is based on the flashing light behavior of fireflies at night is the Firefly Algorithm. Every Firefly has one important feature called flash light. This flash light can be used to impress other fireflies and to alert about threats or enemies. It was introduced by Xin-She Yang in 2008. Numerous optimization problems can be solved with the help of Firefly Algorithm. The FA is particularly effective in handling complex, multimodal, and nonlinear optimization tasks. Firefly algorithm have come in different versions which have been discussed in this chapter (Tilahun et al., 2019). The working of firefly algorithm is shown in Fig. 14.6.

The steps in Firefly algorithm are as follows:

1. *Population of fireflies is created. Each of the firefly is regarded as the problem's solution.*
2. *The objective function is defined. This objective function should take a potential solution*
3. *The fitness value of the solution the firefly represents is used to determine the light intensity of the firefly. For the brighter firefly the quality of the solution is good*
4. *Fireflies are attracted to brighter fireflies, and they move toward the brighter ones in the solution space. The attractiveness of one firefly to another depends on the distance between them and the brightness of the other firefly. The attractiveness variation β on distance 'r' is calculated using Equation* (Wang et al., 2022)

$$\beta = \beta_0 e^{-\gamma r^2}$$

where β_0 denotes the attractiveness

1. *The movement of fireflies is calculated. Assume there are two firefly x_i and x_j. Movement of ith firefly toward jth firefly because of more brightness is defines as using Equation* (Dehghani et al., 2023)

$$x_i^{t+1} = x_i^t + \beta_0 e^{-\gamma r_{ij}^2}\left(x_j^t - x_i^t\right) + \alpha_t \in_i^t$$

FIGURE 14.6 Working of firefly algorithm.

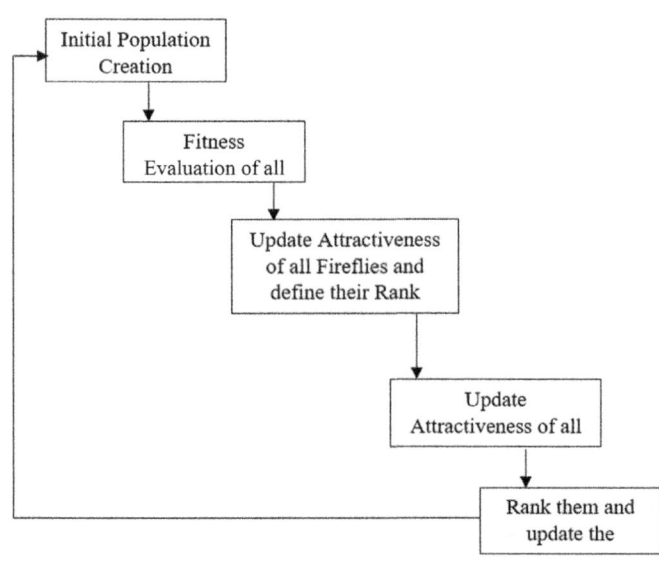

where $\in_i^t$ denotes the Random number vectors with time 't'

1. *The process is continued for certain number of iterations and stops if end condition is encountered or after specified number of loop*

5.5 Cuckoo search algorithm

The algorithm is drawn by the brood parasitism behavior of cuckoos, in which the eggs are laid by the cuckoos in the breeding ground of other birds. In the context of optimization, cuckoos represent potential solutions to a problem, and the nests represent locations or candidate solutions. The motive of Cuckoo Search is to create the optimal solution by iteratively improving the quality of these candidate solutions.

Working of Cuckoo search is as follows:

Initialization: A population of cuckoos (candidate solutions) is generated. These cuckoos represent potential solutions to the optimization problem.

Objective Function: Objective Function is used evaluate the caliber of each cuckoo.

Levy Flights: Levy flights has been incorporated by this Cuckoo Search algorithm, which are a type of random walk with heavy-tailed step-length distributions. In the context of Cuckoo Search, Lévy flights represent the movement of cuckoos between nests or candidate solutions.

Replace Eggs: Some of the cuckoos may lay eggs (generate new solutions) and replace the eggs in nests with lower quality solutions. This emulates the idea of survival of the fittest.

Abandon Eggs: In some cases, cuckoos may abandon their eggs (solutions) if they do not perform well. This is similar to discarding poor solutions and favoring better ones.

Selection: The best solution is chosen by this algorithm from the current population and retains them for the next iteration.

Termination: The algorithm stops when termination condition is met or after specified number of loop.

Some of the key issues such as parameter tuning, numerical and functional optimization can be resolved with the help of this Cuckoo search algorithm. This algorithm can be used in numerous areas (Adegboye and Deniz Ülker, 2023) such as machine learning, computer science and engineering and proved its power in finding high-quality results for a wide range of problems.

6. Genetic algorithms: Evolutionary principles for problem solving

Genetic Algorithm (GA) (Katoch et al., 2021) is one of the evolutionary algorithms mostly used for searching or optimization. It generally works similar to biological evolution. It also a meta-heuristic algorithm used to find the appropriate solutions for searching and optimizing problems (Alhijawi and Awajan, 2023). The traditional algorithms are struggled to find an appropriate solution especially for complex or nonlinear problems. But, Genetic Algorithms effectively solve those problems which works optimally based seven key principals or compounds. They are Initialization, Fitness Evaluation, Selection, crossover, mutation, replacement and termination criteria.

6.1 Initialization

The initial stage of genetic algorithm is to create or initialize the population. The population is randomly generated from the problems. It provides a set of solutions for the problems. Therefore, each solution to the problem is treated as a chromosome. Each chromosome is represented as string of binary value or real values. After initialization, the population is passes through the next pass of Genetic Algorithm called fitness Evaluation. There are several techniques to initialize the chromosome such as random initialization, initialization with heuristic methods, initialize with priori knowledge and so on. Among these, random generation is most common way to initialize the populations. Each gene has assigned a random value based on the range of the chosen problem. In random generation, the value of each chromosome is assigned using Equation (Braik et al., 2022)

$$G_i = \text{random}(i, j)$$

where i, j varies based on problem space
i denotes the minimum number and j represents the maximum range through which the number is randomly generated.

6.2 Fitness evaluation

The fitness of each chromosome to the given problem is evaluated using fitness function. The fitness function validates the each and every chromosome by ensuring how each chromosome effectively solves the problem. Therefore, the higher fitness function is required to choose a better solution for solving the problem. Consider $f(X)$ is a fitness function and then applied to solve individual chromosome. Thus, the fitness function $f(X)$ is applied on each chromosome and find out the optimal one for solving the problem.

The mathematical representation of fitness function is shown in Equation (Lagaros et al., 2022). Initially, the genetic algorithm has an initial population given as a set.

$$X = \left\{ g_1, g_2, ..., g_n \right\}$$

where 'n' denotes the number of chromosome or gene in initial population. After that, the fitness value is computed using fitness function using Equation (Liu et al., 2023)

$$Fitness\ (X) = f(X)$$

where the $f(X)$ changes based on the chosen algorithms. There are several algorithms used for choosing fitness value such as maximization problem, minimization problem, multi-objective optimization and constraint satisfaction.

6.3 Selection

The fitness function is applied on each chromosome to find out the fitness score. The output of the fitness function is a probability value which varies from 0 to 1. Therefore, the selection of chromosome directly depends on the fitness value of each chromosome. There are several algorithms to choose the best fitness value. Here, some of the algorithms are proportional selection, rank-based selection, tournament selection, stochastic universal sampling and so on. Among these, proportional selection is most commonly used selection algorithm. Here, the probability value of each selection is directly proportional to the each fitness score and is calculated using Equation (Pourkhodabakhsh et al., 2023)

$$P(x) = \frac{\text{Fitness}(x)}{\sum\limits_{y=0}^{n} \text{Fitness}(y)}$$

The fitness of each chromosome and initial population size is considered to calculate the probability value.

6.4 Crossover

The crossover is process of combining the existing chromosome to produce the one or more offspring. The main purpose of crossover is to combine a two individual chromosome to inherit good one from the parents. There are several algorithms for doing crossover and single point crossover is most commonly used method.

6.5 Mutation

Mutation will do small random modification in individual chromosome. The gene can be changed to get a new one. There are several algorithms in mutation such as mutation operation, mutation rate and so on.

6.6 Replacement and termination

After crossover and mutation, the new chromosome is generated from the old one to give a new Offspring. The new Offspring can be used for swap the existing chromosome based on the replacement algorithm and the process is repeated based on the termination condition. The termination conditions contain information about the number of iterations, achieving fitness level and so on.

7. Differential evolution: Efficient optimization through mutation and recombination

Differential evolution is one the popular method and gives a best solution based on the optimization criteria. The differential evolution contains subset of mutation, crossover and selection phases in genetic algorithm. Differential evolution is particularly effective for real valued optimization problem. Initially, the differential evolution is a subset of genetic algorithm. Thus, the first phase of differential evolution is initial population. The population is initialized based on candidate solution. Each chromosome is represented as a vector. The vector composes of decision variable.

7.1 Mutation in differential evolution

Mutation is the process of creating one or more offspring form the existing one. Thus, the individual chromosomes are combined to produce new one. The chromosomes are stored in a vector (V). After mutation, new variant is created by adding the weighted difference of two chromosomes X_1 and X_2 which are randomly picked and scaling factor (Kiani et al., 2022).

$$V = X_1 + D * (X_i - X_j)$$

where 'i' and 'j' denotes the two different donor vector and D denotes differential weight.

7.2 Crossover

The crossover is used to combine the mutant vector (V) with vector (X) to generate new one which will be stored in 'U'. Crossover is used to choose which component will choose from V and X. The crossover can been taken place binomial orexponential distribution.

7.3 Selection

The selection process is used to compare the trail vector U with another vector(X). The selection process will choose the best one based on the objective function. Thus, the selected vector becomes the target vector for the next generation.

7.4 Population update

After selecting a specified chromosome, the selecting vector U restores the target vector X only if it has better fitness value than selecting vector. Thus, differential evolution uses the above steps iteratively for multiple generations to produce the optimal solution. The key parameters in differential evolution are the population size, scaling factor, mutation, crossover and selection.

8. Harmony search: Finding optimal solutions through musical harmony

Harmony search (Dubey et al., 2021) is one of the optimized algorithms which is inspired by the improvisation process of searching for musical harmony. It is the process of musicians improvising to find best harmony by considering a set of musical parameters and assign to design variables. The harmony search will give a optimal or near-optimal combination of parameters based on the fitness function. Harmony search algorithm initially defines the problem. After defining the problem, the harmony algorithms define a set of decision variables, their range of values and define fitness function. New Harmony is created by combining the elements of existing harmony using the improvisation rules. HM contains a repository of set of solutions obtained for optimization process. Initially, its randomly generated by a improvisation rules. Some of the improvisation rules or steps are random selection of harmonies from HM, combine elements of selected harmonies, apply pitch adjustment, evaluation of new harmony. The new generated harmony will be determined using value of fitness function. After obtaining a best fitness value, the New Harmony with best fitness value will replace the harmony with worst value in a repository. The number of iteration of generating new harmony is based on the maximum number of iteration. Harmony algorithm is widely applied in several optimization algorithms.

9. Hybrid metaheuristic algorithms: Combining multiple approaches for improved performance

Generally, the heuristic algorithms will verify apply conditions to choose a optimal solutions. Thus, meta heuristic algorithms is obtained by combining two or more heuristic algorithms to produce a best optimization algorithm. Hybrid heuristic algorithm is used to enhance the performance and effectiveness of analyzing and solving complex problems. Here, some of the approaches to combine to produce met heuristic algorithms are Genetic Algorithms (GA), Ant Colony Optimization (ACO), Differential Evolution (DE), Simulated Annealing (SA), Particle Swarm Optimization (PSO), Tabu Search and so on.

9.1 Combining strategies

The hybrid meta heuristics algorithms are generated by combining one or more heuristic algorithms. There are different approaches to merge different heuristics algorithms. There are sequential combination, parallel combination, interweaved combination and component combination. Among these, sequential combination will combine one meta heuristic with another. The output of one meta heuristic algorithm is consider as input for the other. Later, parallel combination will concurrently process multiple meta heuristic algorithm by combining the results to process the optimal one. Further, Interweaved combination will combine the meta heuristic by alternating different heuristic during optimization process based on the specified rules. The component combination will combine different components of various meta heuristic algorithms to create a new one.

9.2 Advantage of hybridization

The meta heuristic algorithms are combined together to form a hybrid meta heuristic algorithms. The hybrid meta heuristic algorithms will have several advantages such as

- Enhanced Exploration and Exploitation:
- Robustness and Flexibility:
- Speeding up Convergence
- Overcoming Local Optima

Thus, designing hybrid meta heuristics algorithms play a vital role in choosing specific selection and integrating the appropriate meta heuristics. It will be framed by combining meta heuristics algorithms by determining the combination strategy, fine tuning parameters, effective mechanism for communication and cooperation among the meta heuristics approaches. Thus, the hybrid meta-heuristic algorithm has a great potential to solve large complex real—world problems effectively by integrating two or more meta heuristic algorithms.

10. Case studies of bio-inspired metaheuristic algorithms

Few case studies of Bio-Inspired Metaheuristic Algorithms in Engineering Optimization are:

10.1 Optimizing structural design with bio-inspired algorithms

One compelling application of bio-inspired metaheuristic algorithms in engineering lies in structural design optimization. By mimicking natural processes such as genetic evolution and swarm behavior, engineers have harnessed the power of computational intelligence to create efficient and robust structural designs. In a case study involving the construction of a large-scale bridge, a combination of genetic algorithms and finite element analysis was employed to find the optimal arrangement of materials while ensuring the structure's stability and cost-effectiveness. The results not only exceeded the capabilities of traditional optimization methods but also highlighted the potential of bio-inspired algorithms in revolutionizing the field of civil engineering.

10.2 Enhancing energy systems efficiency

The energy sector faces complex challenges in optimizing the generation, distribution, and consumption of energy. Bio-inspired metaheuristic algorithms, such as particle swarm optimization, have been applied to address these challenges effectively. In a case study focused on a smart grid system, this algorithm was used to determine the most efficient allocation of energy resources, minimizing waste and maximizing sustainability. The results showcased significant improvements in energy utilization and a reduction in environmental impact. Such applications demonstrate the transformative potential of bio-inspired approaches in shaping the future of energy systems.

10.3 Revolutionizing data mining with ant colony optimization

Bio-inspired algorithms are not confined to physical engineering challenges; they also excel in the domain of data mining and analytics. Ant colony optimization, inspired by the foraging behavior of ants, has proven to be a valuable tool in uncovering hidden patterns within vast datasets. In a case study involving customer behavior analysis for a retail company, this algorithm was applied to optimize marketing strategies. By efficiently exploring the solution space, it identified customer segments and product recommendations, leading to a substantial increase in sales and customer satisfaction. This case underscores the adaptability of bio-inspired metaheuristic algorithms in solving complex data-driven problems.

10.4 Solving complex scheduling tasks with swarm intelligence

Scheduling complex tasks in industries such as manufacturing, transportation, and healthcare can be a formidable challenge. Bio-inspired algorithms, specifically those based on swarm intelligence principles, have demonstrated remarkable capabilities in this regard. In a case study involving the optimization of nurse scheduling in a busy hospital, a particle swarm optimization algorithm was employed. It successfully allocated shifts and duties, considering various constraints and preferences, leading to improved staff satisfaction and operational efficiency. This example illustrates how bio-inspired metaheuristic algorithms are revolutionizing scheduling and resource allocation in diverse sectors.

These case studies underscore the real-world impact of bio-inspired metaheuristic algorithms in engineering optimization problems. By leveraging the power of computational intelligence and drawing inspiration from nature, these algorithms offer innovative solutions to some of the most intricate and challenging problems across different domains, pushing the boundaries of what is achievable in engineering and beyond.

11. Challenges and future directions in engineering optimization with computational intelligence

As bio-inspired metaheuristic algorithms continue to redefine the landscape of engineering optimization, several challenges and promising future directions emerge on the horizon. Which is seen in Fig. 14.7.

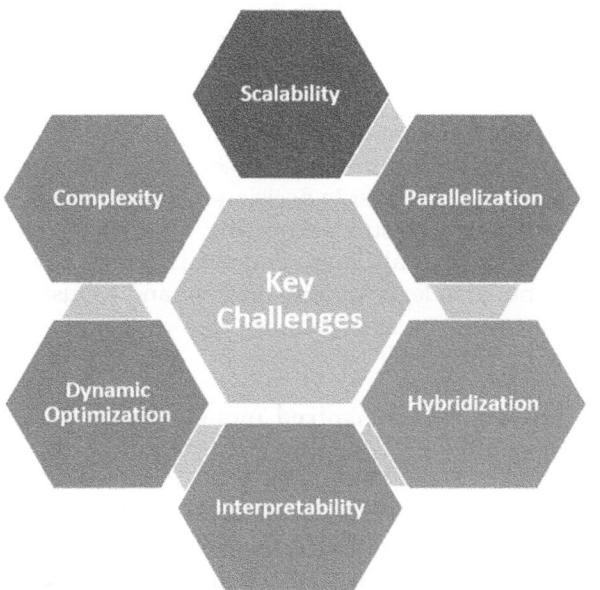

FIGURE 14.7 Key challenges in engineering optimization.

Scalability and Complexity: One of the primary challenges is dealing with increasingly complex and large-scale optimization problems. Many real-world engineering challenges involve high dimensionality and intricate constraints. Future directions must focus on developing algorithms that can efficiently handle these complexities, ensuring that they remain practical and effective.

Hybridization and Adaptation: The integration of bio-inspired algorithms with other optimization methods and machine learning techniques holds tremendous potential. Hybridization can lead to the development of more robust and versatile optimization approaches. The challenge is in identifying the most effective combinations and adapting them to specific problem domains.

Parallelization and High-Performance Computing: The computational demands of solving complex optimization problems are substantial. Leveraging the power of parallel and distributed computing can significantly enhance the efficiency and speed of bio-inspired algorithms. Future directions should explore how to harness high-performance computing resources effectively.

Incorporating Domain Knowledge: To further enhance the performance of bio-inspired algorithms, it's crucial to incorporate domain-specific knowledge. This can help guide the search process, particularly in engineering optimization, where specialized insights can significantly impact the quality of solutions.

Real-time and Dynamic Optimization: Many engineering systems are dynamic and change over time. Future directions must focus on developing bio-inspired algorithms that can adapt in real-time to shifting conditions, ensuring optimal performance in dynamic environments.

Benchmarking and Standardization: Establishing benchmarks and standardized testing protocols for bio-inspired algorithms is essential for comparing their performance accurately. Future research should focus on creating comprehensive benchmark suites and evaluating algorithm performance under various scenarios.

Ethical and Responsible AI: As computational intelligence techniques continue to advance, there's a growing need to consider the ethical implications of their application, especially in engineering optimization. Future directions should explore how to ensure responsible and ethical use of these powerful tools.

Explainability and Interpretability: The black-box nature of some bio-inspired algorithms can be a barrier to their adoption in critical applications. Future research should aim to make these algorithms more explainable and interpretable to gain trust in decision-making processes.

To fully unlock the potential of Bio-inspired metaheuristic, addressing these challenges and pursuing the outlined future directions will be pivotal. These endeavors will not only advance the field of computational intelligence but also lead to more efficient and robust solutions, ultimately benefiting a wide range of engineering applications.

12. Conclusion: Harnessing the power of bio-inspired metaheuristic algorithms for engineering optimization

Bio-inspired metaheuristic algorithms are powerful tools for engineering optimization. Drawing inspiration from nature, they exhibit adaptability and resilience in solving complex problems, often surpassing traditional methods. Evolutionary algorithms, swarm intelligence, and genetic algorithms each bring unique strengths to various optimization challenges, with real-world applications spanning structural design, scheduling, energy systems, and data mining. The future of bio-inspired optimization looks promising, with ongoing explorations of hybridization, parallelization, and integration with machine and deep learning techniques. Engineers and researchers are encouraged to embrace these innovative approaches, leveraging computational intelligence for more efficient solutions and brighter prospects in the ever-evolving field of engineering design and decision-making.

References

Adegboye, O.R., Deniz Ülker, E., 2023. Hybrid artificial electric field employing cuckoo search algorithm with refraction learning for engineering optimization problems. Scientific Reports 13 (1), 2045–2322. https://doi.org/10.1038/s41598-023-31081-1.

Agarwal, D., Gupta, A., Singh, P.K., 2016. A systematic review on artificial bee colony optimization technique. International Journal of Control Theory and Applications 9 (11), 5487–5500.

Alhijawi, B., Awajan, A., 2023. Genetic algorithms: theory, genetic operators, solutions, and applications. Evolutionary Intelligence 1864–5909. https://doi.org/10.1007/s12065-023-00822-6.

Braik, M., Hammouri, A., Atwan, J., Al-Betar, M.A., Awadallah, M.A., 2022. White Shark Optimizer: a novel bio-inspired meta-heuristic algorithm for global optimization problems. Knowledge-Based Systems. https://doi.org/10.1016/j.knosys.2022.108457.

Dehghani, M., Trojovský, P., 2022. Serval optimization algorithm: a new bio-inspired approach for solving optimization problems. Biomimetics 7 (4), 2313–7673. https://doi.org/10.3390/biomimetics7040204.

Dehghani, M., Montazeri, Z., Trojovská, E., Trojovský, P., 2023. Coati optimization algorithm: a new bio-inspired metaheuristic algorithm for solving optimization problems. Knowledge-Based Systems. https://doi.org/10.1016/j.knosys.2022.110011.

Deng, W., Ni, H., Liu, Y., Chen, H., Zhao, H., 2022. An adaptive differential evolution algorithm based on belief space and generalized opposition-based learning for resource allocation. Applied Soft Computing 127. https://doi.org/10.1016/j.asoc.2022.109419.

Dorigo, M., Stützle, T., 2019. Ant colony optimization: overview and recent advances. International Series in Operations Research and Management Science 311–351. https://doi.org/10.1007/978-3-319-91086-4_10.

Dubey, M., Kumar, V., Kaur, M., Dao, T.-P., Gritli, H., 2021. A systematic review on harmony search algorithm: theory, literature, and applications. Mathematical Problems in Engineering 2021, 1–22. https://doi.org/10.1155/2021/5594267.

Jain, N.K., Nangia, U., Jain, J., 2018. A review of particle swarm optimization. Journal of the Institution of Engineers (India): Series B 99 (4), 407–411. https://doi.org/10.1007/s40031-018-0323-y.

Katoch, S., Chauhan, S.S., Kumar, V., 2021. A review on genetic algorithm: past, present, and future. Multimedia Tools and Applications 80 (5), 8091–8126. https://doi.org/10.1007/s11042-020-10139-6.

Kaya, E., Gorkemli, B., Akay, B., Karaboga, D., 2022. A review on the studies employing artificial bee colony algorithm to solve combinatorial optimization problems. Engineering Applications of Artificial Intelligence 115. https://doi.org/10.1016/j.engappai.2022.105311.

Kiani, F., Seyyedabbasi, A., Nematzadeh, S., Candan, F., Çevik, T., Anka, F.A., Randazzo, G., Lanza, S., Muzirafuti, A., 2022. daptive metaheuristic-based methods for autonomous robot path planning: sustainable agricultural applications. Applied Sciences 12 (3). https://doi.org/10.3390/app12030943.

Kumar, V., Kumar, D., 2021. A systematic review on firefly algorithm: past, present, and future. Archives of Computational Methods in Engineering 28 (4), 3269–3291. https://doi.org/10.1007/s11831-020-09498-y.

Lagaros, N.D., Plevris, V., Kallioras, N.A., 2022. The mosaic of metaheuristic algorithms in structural optimization. Archives of Computational Methods in Engineering 29 (7), 5457–5492. https://doi.org/10.1007/s11831-022-09773-0.

Liu, K., Zheng, J., Dong, S., Xie, W., Zhang, X., 2023. Mixture optimization of mechanical, economical, and environmental objectives for sustainable recycled aggregate concrete based on machine learning and metaheuristic algorithms. Journal of Building Engineering 63.

Nayak, J., Swapnarekha, H., Naik, B., Dhiman, G., Vimal, S., 2023. 25 Years of particle swarm optimization: flourishing voyage of two decades. Archives of Computational Methods in Engineering 30 (3), 1663–1725. https://doi.org/10.1007/s11831-022-09849-x.

Pourkhodabakhsh, N., Mamoudan, M.M., Bozorgi-Amiri, A., 2023. Effective machine learning, meta-heuristic algorithms and multi-criteria decision making to minimizing human resource turnover. Applied Intelligence 53 (12), 16309–16331. https://doi.org/10.1007/s10489-022-04294-6.

Rezvanian, A., Mehdi Vahidipour, S., Sadollah, A., 2023. An Overview of Ant Colony Optimization Algorithms for Dynamic Optimization Problems. IntechOpen. https://doi.org/10.5772/intechopen.111839.

Tilahun, S.L., Ngnotchouye, J.M.T., Hamadneh, N.N., 2019. Continuous versions of firefly algorithm: a review. Artificial Intelligence Review 51 (3), 445–492. https://doi.org/10.1007/s10462-017-9568-0.

Wang, D., Tan, D., Liu, L., 2018. Particle swarm optimization algorithm: an overview. Soft Computing 22 (2), 387–408. https://doi.org/10.1007/s00500-016-2474-6.

Wang, L., Cao, Q., Zhang, Z., Mirjalili, S., Zhao, W., 2022. Artificial rabbits optimization: a new bio-inspired meta-heuristic algorithm for solving engineering optimization problems. Engineering Applications of Artificial Intelligence. https://doi.org/10.1016/j.engappai.2022.105082.

Secure sharing of health records stored in cloud using cryptographic secret sharing schemes through computational intelligence: A review

Sameera Mahammad and K. Usha Rani

Department of Computer Science, Sri Padmavati Mahila Visvavidyalayam (Women's University), Tirupati, Andhra Pradesh, India

1. Introduction

Cloud Computing is a cost saving and adaptable technology in today's world. It provides a massive data center to manage vast amount of data. Organizations gain from cloud computing, because they don't have to invest to manage their own infrastructure. It provides computer services such as internet based networking, software, data storage and servers. There are three major classification of services depending on the level of abstraction offering to the customers. These services include IaaS (Infrastructure as a Service), PaaS (Platform as a Service) and SaaS (Software as a Service). Companies and individuals can get these services on demand from cloud service providers. They can use different cloud deployment models based on the location of cloud infrastructure and access level of resources. There are four main types of cloud delivery models such as Public, Private, Hybrid and Community clouds (Sameera and Rani, 2023).

This paradigm of cloud enables healthcare environments to shift their EHRs (Electronic Health Records) to clouds instead of building and maintain dedicated data centers. EHRs systematically stores and manages each patient's health information record, personal information, medical information such as medical/family history, drug reaction, health status, medical examination and admission/discharge records in database format. However, the adoption of cloud computing in healthcare systems may also raise many security challenges associated with authentication, identity management, access control, trust management, and so on (Wu et al., 2012). To address these security challenges there are many techniques, among that one essential aspect for secure communication is Cryptographic Secret Sharing Schemes through Computational Intelligence which is the focus of this chapter.

1.1 Cryptographic secret sharing schemes

Cryptography is foundation for information security for keeping data secure from untrusted entities. It resolves security issues by converting readable form of data into non readable form. It is comprised of three types of algorithms as shown in Fig. 15.1. They are.

The primary goal of cryptography is to ensure the confidentiality, integrity, and authenticity of data (Kessler, 2003).

Secret Sharing Schemes is an important tool in cryptography which is main build box in many secure protocols. In this cryptographic method a dealer distributes a secret among a group of participants. The secret can only be revealed when certain predefined subsets of participants (specified in set Γ) combine their shares. This security feature ensures that any group of participants not included in Γ will have no access to the secret or any related information (Beimel, 2011).

There are several secret sharing techniques used in cryptography and information security, each with its own characteristics and applications. The following are most notable secret sharing schemes techniques.

1.1.1 Shamir secret sharing (SSS)

This algorithm is created by Adi Shamir a famous Israeli cryptographer, who also contributed to the invention of RSA algorithm. SSS algorithm divides secret that needs to be encrypted into various distinct parts as depicted in Fig. 15.2. After dividing it, a number K is chosen by user to decrypt parts and find the original secret. The number K is chosen in such a way that, if we know the parts lesser than K we will not be able find the secret. If we know K or more parts, then only we can reconstruct the secret. SSS is based on the mathematical concept of polynomial interpolation which can be applied to secret sharing scheme by embedding the secret into polynomial (Brickell, 1989).

1.1.2 Blakely's secret sharing (BSS)

In this covert distribution method, the initiator distributes the key or confidential information among a group of n members, with each distribution referred to as a "share." These shares are created by defining hyperplanes. The crucial aspect of this approach is that the key will only be revealed if t or more of these shares are combined. This scheme relies on a geometric approach, assuming that the secret is represented as a point in a t-dimensional space. To reconstruct the secret key, the hyperplanes must intersect at a point within this t-dimensional space. Similar to SSS, this scheme is also a linear threshold scheme. To gain access to the secret, an unauthorized party would need to compromise at least t shareholders to acquire their respective shares (Shamsoshoara, 2019).

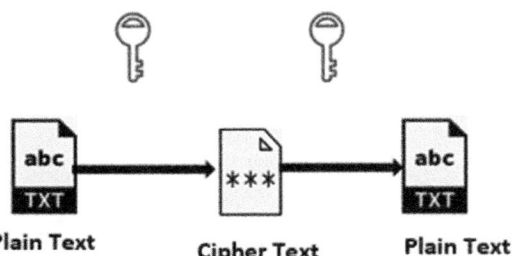

Secret Key (Symmetric) Cryptography

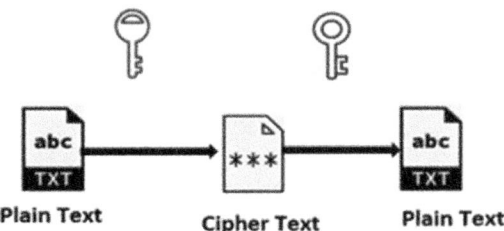

Public Key (Asymmetric) Cryptography

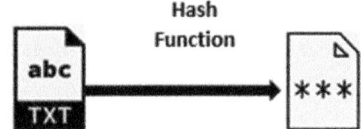

Hash Function (One Way Cryptography)

FIGURE 15.1   Three types of cryptography: symmetric key, public key and hash function.

- Symmetric Key Cryptography: uses same key for both encryption and decryption
- Public Key Cryptography: uses two different keys for encryption and decryption
- Hash Functions: It uses mathematical function for information security

1.1.3 Verifiable secret sharing (VSS)

It is fundamental primitive in several distributed computing tasks such as Byzantine agreement and secure multi party computation (Chandramouli et al., 2022). It allows dealers to share a secret among multiple parties. Even if an adversary controls some of the parties, the dealer can share secret. This scheme is verifiable because it includes auxiliary information which allows players to verify the shares they receive are consistent with the secret or not. It ensures that even if the dealer is malicious the well-defined secret makes the players to reconstruct it later (Chandramouli et al., 2022).

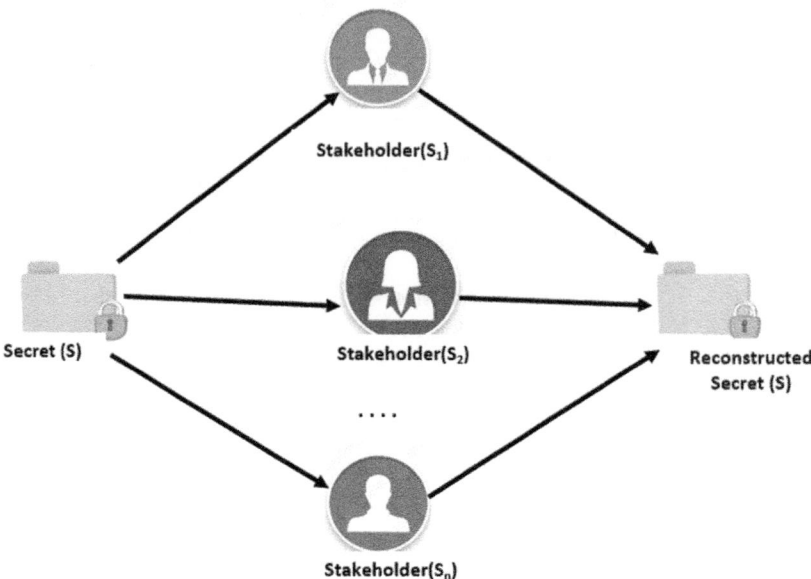

FIGURE 15.2   Secret sharing between 'n' participants. This figure shows sharing of secret between 'n' stakeholders and its reconstruction.

1.1.4 Quantum secret sharing (QSS)

QSS is a quantum cryptographic scheme that uses quantum mechanics and the no-cloning theorem to secure communications. In QSS message is divided into several copies and only authorized participants can recover the secret, while unauthorized participants cannot (Li and Li, 2023).

1.1.5 Visual cryptography

This secret sharing scheme decomposes secret images into multiple shares. These shares need to be digitally or physically overlapped to recover the original image, without any mathematical operations or additional hardware as illustrated in Fig. 15.3 (Ibrahim et al., 2021).

1.1.6 Homomorphic secret sharing (HSS)

HSS is designed to send a confidential message to multiple recipients simultaneously. The procedure entails taking a secret message, applying a homomorphic transformation to it, dividing the transformed secret into multiple segments, with each recipient receiving one of these segments. In HSS its functions are evaluated on shared secrets by processing input shares to generate output shares locally. The resulting output shares are of short length (Boyle et al., 2018).

1.1.7 Attribute based secret sharing (ABSS)

In ABSS, access control is integrated into the secret sharing scheme. ABSS enables the reconstruction of the secret if and only if participants with the relevant attributes collaborate, effectively enforcing access control policies based on the attributes (Lakshmi and Nitya, 2015).

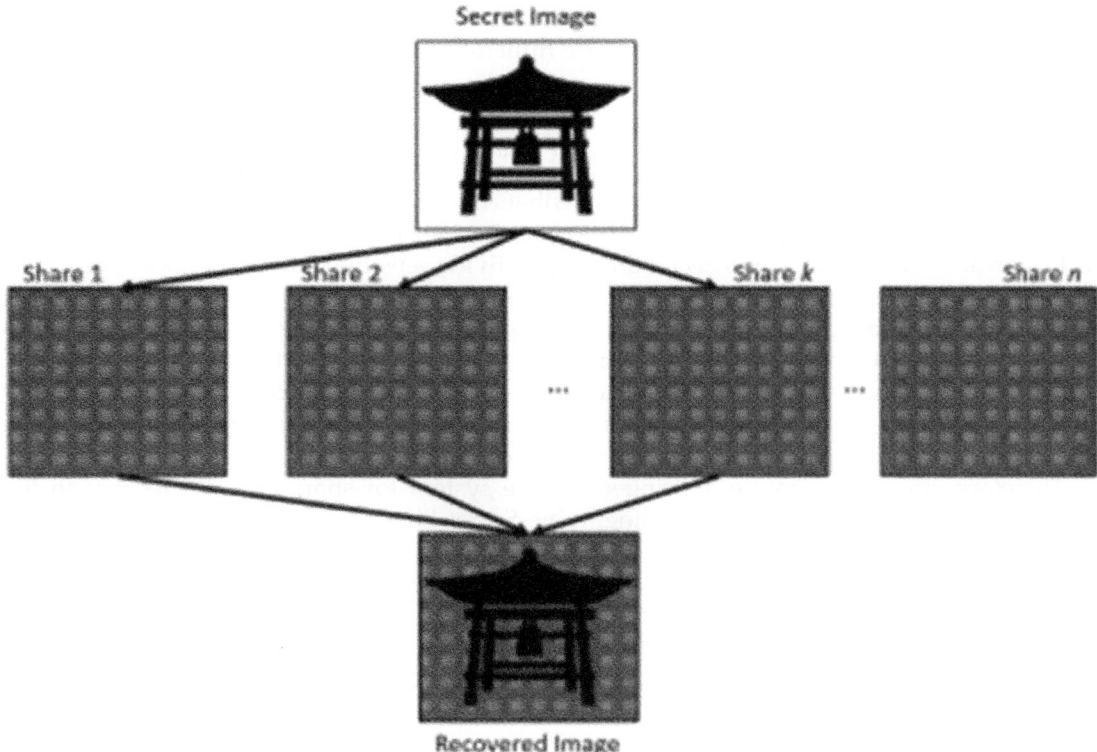

FIGURE 15.3   Visual cryptography (Ibrahim et al., 2021). This figure shows decomposing of secret image into multiple images and its reconstruction.

1.1.8 Ramp secret sharing (RSS)

The RSS (Ramp Secret Sharing) method transforms a secret, denoted as 'S,' into multiple shares, namely s1, s2, …, sn. These individual shares, taken by themselves, contain no information about the original secret 'S.' However, the reconstruction of secret 'S' becomes possible by collecting several of these shares (Çalkavur and Solé, 2023). RSS can be divided into two categories: strong and weak schemes. The strong RSS schemes are particularly valuable because they do not reveal any portion of the secret, even if some information regarding the secret leaks from a nonqualified set of shares. As a result, they are considered more desirable than weak RSS schemes (Iwamoto and Yamamoto, 2006).

1.1.9 Dynamic secret sharing (DSS)

In some scenarios, the number of participants or the threshold required for reconstruction may change over time. DSS scheme allow for flexibility in such situations (Yuan and Li, 2019).

1.2 Computational intelligence

Computational Intelligence is a branch of Artificial Intelligence (AI) that focuses on developing algorithms and models inspired by human intelligence and cognitive process to solve complex problems. It includes variety of techniques and methods such as Machine Learning,

Neural Networks, Fuzzy logic, Image recognition, Evolutionary computation Robotics and so on. CI is used to study the problems that are difficult to solve using conventional computational algorithms (Poole et al., 1998).

1.3 Cryptographic secret sharing schemes with computational intelligence

Cryptographic secret sharing schemes, enhanced through computational intelligence, represent a sophisticated approach to secure data sharing. These schemes employ mathematical algorithms such as Shamir's Secret Sharing to divide sensitive information into encrypted fragments, distributed among authorized parties. Computational intelligence techniques, such as machine learning and artificial intelligence, are then utilized for adaptive access control, anomaly detection, and dynamic threshold adjustments. This fusion of cryptography and computational intelligence ensures not only robust data protection but also the ability to adapt to evolving threats and user behaviors, making it a powerful solution for safeguarding sensitive data, particularly in healthcare and other critical domains.

Therefore, for implementing a stringent secure system for sharing health records in the cloud requires a combination of Cryptographic Secret Sharing, Computational Intelligence and strict security measures.

The layout of this chapter is structured as follows: In Section 2, thorough review is conducted on existing literature concerning Cryptographic Secret Sharing Schemes. Moving to Section 3, comparative analysis is performed on previous research efforts. Section 4 delves into the intersection of Computational Intelligence and Secure Data Sharing. Finally, in Section 5, entire chapter is concluded.

2. Literature review

The below Table 15.1 illustrates research pertaining to Cryptographic Secret Sharing techniques designed to ensure the secure sharing of EHRs stored in the cloud.

3. Comparative analysis

In this section a comparison on characteristics of various cryptographic secret sharing schemes used in the above literature are highlighted in Tables 15.2 and 15.3 and security analysis presented in Table 15.4.

After a thorough examination of the features, security assessments, and constraints within the current body of literature regarding cryptographic methods for securely sharing health records stored in the cloud, it becomes evident that incorporating computational intelligence could significantly enhance the secure sharing of cloud-stored health records.

4. Computational intelligence and secure sharing

This section delves into the computational intelligence techniques and their significance in ensuring the secure sharing of EHRs.

TABLE 15.1 Review of the literature on secret sharing schemes for EHRs.

Ref.	Proposed scheme/ technique	Brief description	Limitations/future scope
Li et al. (2010)	Multi Authority attribute based Encryption (MA-ABE)	Authors proposed a novel framework for access control to personal health Records (PHRs) stored in cloud. They utilized MA-ABE to reduce the complexity of key management brought by multiple PHR owners and users. Therefore, patients can allow access not only to the personal users, but also various users form different public domains with different professional qualifications and affiliations.	Enhancing MA-ABE scheme to accommodate more versatile access policies defined by the owner.
Ulutas et al. (2011)	Shamir's secret sharing scheme	This chapter introduces a secret sharing scheme with parameters (k, n) designed for distributing medical images among a team of 'n' healthcare professionals. The objective is to ensure that a minimum of 'k' clinicians must collaborate to access and diagnose the medical image, thereby addressing various security concerns. To achieve this, the authors employ Shamir's secret sharing scheme, which serves as a comprehensive solution to tackle multiple security aspects. The proposed approach not only enables the storage of extended EPR (electronic patient record) strings but also enhances the authenticity and confidentiality features. These outcomes are demonstrated through the results, illustrating the scheme's ability to meet all specified requirements effectively.	—
Parakh and Kak (2011)	K-threshold computational secret sharing technique	Authors proposed a computational secret sharing method with a K-threshold, which divides a secret 'S 'in to shares, each of size approximately $|S|/(K-1)$, where $|S|$ represents the size of the secret. This share size closely approaches the space-optimal limit of $|S|/K$ when the secret needs to be reconstructed from k shares. Therefore, this technique can be considered as an innovative information dispersal approach that offers nearly optimal space efficiency. It has promising applications in securely dispersing information on the internet and within the sensor networks.	—

(*Continued*)

TABLE 15.1 Review of the literature on secret sharing schemes for EHRs.—cont'd

Ref.	Proposed scheme/ technique	Brief description	Limitations/future scope
Askari et al. (2012)	Visual secret sharing scheme	Authors introduce an innovative visual secret Sharing (VSS) scheme that overcomes the challenges of traditional VSS such as losing image resolution and contrast, as well as expanding the image size, which requires additional storage space.	Some noise is identified in the recovered image.
Ermakova and Fabian (2013)	Multi- provider cloud architecture using both SSS scheme and Robin's Dispersal algorithm.	As the transfer of EHRs to cloud implies security and privacy risks, authors proposed a novel secure multi provider cloud architecture which satisfies many of the requirements by providing increased availability, confidentiality and integrity of medical records stored in cloud. This architecture features secret sharing as important measure to share the health records as chunks to different cloud services. Authors performed several experiments to measure the execution time with both SSS scheme and Robin's algorithm. The outcome of experiment is Robin's algorithm is low overhead and feasible.	Full implementation of multi provider cloud architecture with Robin's dispersal algorithm is the future work.
Jani Anbarasi and Anandha Mala (2014)	SSS scheme combined with DNA cryptography	The approach proposed by the authors enhance the security of medical images and electronic patient Records (EPRs) by changing them into shadow images. In this process EPRs are concealed in medical images using DNA hiding techniques. The secret images encoded in DNA is compressed with Huffman encoding and securely divided into the shadow images using SSS scheme. These shadow images are concealed within host image using stenographic techniques. During reconstruction of medical images and EPRs, a minimum of 't 'shadow images are combined and the result is decoded using Huffman decoding to unveil the DNA-encoded secret image. Finally, the original medical images and EPRs are separated using reverse DNA hiding technique.	—

TABLE 15.1 Review of the literature on secret sharing schemes for EHRs.—cont'd

Ref.	Proposed scheme/ technique	Brief description	Limitations/future scope
Fabian et al. (2015)	Attribute based encryption and secret sharing across multi-clouds	An innovative architecture and its practical implementation is designed to facilitate the secure and private sharing of patient data across different organizations. This architecture employs Attribute-based encryption to grant selective access authorization and cryptographic secret sharing to distribute data across multiple clouds to mitigate the potential risks posed by inquisitive cloud providers.	Future work will focus on addressing key management and role based access policy management in the context of inter organizations and need to conduct usability studies and various improvements for the multi-cloud proxy.
Rezaeibagha and Mu (2016)	Threshold secret sharing and role based access Control (RBAC)	An efficient architecture for an EHR system was suggested. This system was categorized into direct and indirect access areas and implemented with threshold secret sharing and RBAC to protect the patient's privacy.	Future work is to implement this system in multiple domain systems including clinic and hospital where multiple physicians and other medical staff are required to access electronic patient records.
Au et al. (2017)	A general framework for secure sharing PHRs with linear secret sharing scheme.	The proposed system enables patients to securely store and share their PHR in cloud to their care takers. Through this system the doctors can further refer patients medical record to specialists for research purposes, by keeping the patient's personal information private. It also supports cross domain operations.	As a future work, the same system can be implemented to data stored in cloud other than PHRs.
Luo et al. (2018)	Privacy Protector with Slepian-Wolf coding based secret sharing scheme (SW-SSS)	This chapter introduced a new framework called privacy Protector to safeguard patient's personal data. Additionally, a secret sharing scheme named SW-SSS is presented to optimize size of the secret share and enable exact share repairs. SW-SSS divides patient data and stores it across multiple cloud servers to ensure privacy even if two or more servers are compromised. Performance analysis also demonstrates that privacy Protector frame work is resilient to various attacks.	—
Marwan et al. (2019)	SSS scheme and multi-cloud architecture.	The core idea of this framework is to divide a digital image into multiple portions in a way that no valuable information can be obtained from a	Incorporating watermarking methods to safeguard ownership rights, validate the integrity of medical images and integrating

TABLE 15.1   Review of the literature on secret sharing schemes for EHRs.—cont'd

Ref.	Proposed scheme/ technique	Brief description	Limitations/future scope
		single share. To enhance the data security, DepSky system is used as distributed storage solution to prevent accidental disclosure of medical information.	attribute based access control within proposed frame work to further enhance data protection
Tang et al. (2019)	SSS scheme with Boneh-Goh-Nissim Crypto system	Authors proposed a privacy Preserving health data Aggregation scheme that securely collects data from multiple sources while ensuring fair incentives for participating patients by utilizing signature techniques. Furthermore, they combine the SSS scheme with Boneh-Goh-Nissim Crypto system to ensure data obfuscation security and fault tolerance.	—
Ismail et al. (2020)	SSSs with RSA	Proposed a framework known as hybrid secure data sharing Architecture (HSDSA). In this framework, they employed a multi-cloud environment to store medical records by encrypting with RSA method and then, they implement SSS approach to distribute data shares across different independent cloud providers. In the retrieval phase, (t, n) strategy is used to reconstruct the data. A Schnorr-based technique is applied to prove data possession and to verify the requester identity. A Diffie-Hellman algorithm is used to securely exchange decryption key.	Adding governmental organization as data requester
Sarosh et al. (2021)	Computational secret sharing with RC6	In this chapter the residual image problem is addressed by applying RC6 encryption to PHRs, the encrypted image is transformed into secret shares using computational secret sharing approach and the secret key is distributed through Perfect secret sharing.	Reducing computational complexity, enabling fast secret recovery and generating noise resistant shares associated with CSIS scheme is future research effort.
Peter Čuřík, Ploszek and Zajac (2022)	SSS scheme	Datachest is a proof-of-concept smartphone application that enhances the security of users' private data. It achieves this by combining symmetric encryption with Shamir's (2, 3)-threshold secret sharing scheme. In this	—

TABLE 15.1 Review of the literature on secret sharing schemes for EHRs.—cont'd

Ref.	Proposed scheme/technique	Brief description	Limitations/future scope
		setup, randomly generated symmetric keys are shared using the secret sharing scheme, and these shares are stored in three separate cloud locations. Importantly, even if one of these cloud storage services is compromised or experiences connectivity issues due to internal problems or authentication attacks, the user's private data remains secure. This is because the encryption key can still be reconstructed using the remaining two key shares, ensuring the data's confidentiality and integrity.	

TABLE 15.2 Characteristics of various schemes and frameworks (part-1).

Ref.	Scheme/technique	Single/multiple	Proactive/threshold/verifiable
Ulutas et al. (2011)	SSSs	Multiple	Threshold and verifiable
Parakh and Kak (2011)	k-threshold computational secret sharing	Multiple	Threshold
Askari et al. (2012)	Novel visual secret sharing	Single	Threshold
Ermakova and Fabian (2013) and Fabian et al. (2015)	SSSs and Robin's information Dispersal algorithm	Multiple	Threshold
Jani Anbarasi and Anandha Mala (2014)	SSSs with DNA cryptography	Multiple	Threshold
Rezaeibagha and Mu (2016)	Secret sharing with role based access control	Multiple	Threshold and proactive
Au et al. (2017)	Attribute based encryption and Proxy re-encryption	Multiple	—
Luo et al. (2018)	SW-SSS	Multiple	Threshold
Marwan et al. (2019)	SSSs and multi cloud	Multiple	Threshold
Tang et al. (2019)	SSSs with Boneh-Goh-Nissim Crypto system	Multiple	Threshold
Ismail et al. (2020)	SSSs with RSA	Multiple	Threshold and verifiable
Sarosh et al. (2021)	Computational secret sharing with RC6	Single	Threshold
Peter Čuřík et al. (2022)	Symmetric encryption with Shamir's(2,3)-threshold secret sharing scheme	Single/multiple	Threshold

TABLE 15.3 Characteristics of various schemes and frameworks (part-2).

Ref.	Scheme/technique	Secret sharing	Secret reconstruction	Key creation
Ulutas et al. (2011)	SSSs	Polynomial interpolation	Lagrange interpolation polynomial	Steganography is in use, therefore no key creation
Parakh and Kak (2011)	k-threshold computational secret sharing	Repetitive polynomial interpolation	Repetitive polynomial interpolation	Lacks clarity on key generation
Askari et al. (2012)	Novel visual secret sharing	By splitting image into same size share images of black and white	Superimposing shares with XOR operation	No cryptographic computations
Ermakova and Fabian (2013) and Fabian et al. (2015)	SSSs and Robin's information Dispersal algorithm	Polynomial interpolation	Lagrange interpolation polynomial	Random
Jani Anbarasi and Anandha Mala (2014)	SSSs with DNA cryptography	Polynomial interpolation	Lagrange interpolation, secret Shadow and participants key	Random
Rezaeibagha and Mu (2016)	Secret sharing with role based access control	Attribute based encryption and Proxy re-encryption	To access data, a party must possess a key associated with specific authorized attributes	Trusted Authority is responsible for key creation
Au et al. (2017)	Attribute based encryption and Proxy re-encryption	Attribute based encryption and Proxy re-encryption	To access data, a party must possess a key associated with specific authorized attributes	The Central Authority is incharge of creating keys, while attribute Authority assigns attributes to those keys
Luo et al. (2018)	SW-SSS	XOR operation	XOR operation	Random (Predeployed)
Marwan et al. (2019)	SSSs and multi cloud	Polynomial interpolation	Lagrange interpolation	Random
Tang et al. (2019)	SSSs with Boneh-Goh-Nissim Crypto system	Polynomial interpolation	Discrete Logarithm	Random
Ismail et al. (2020)	SSSs with RSA	Polynomial interpolation	Lagrange interpolation	Involves RSA, SSS and Diffie-Helman
Sarosh et al. (2021)	Computational secret sharing with RC6	Polynomial based	Linear Equations	—
Peter Čuřík et al. (2022)	Symmetric encryption with Shamir's (2,3)-threshold secret sharing scheme	Polynomial interpolation	Lagrange interpolation	Random

TABLE 15.4 Security analysis of various schemes and frameworks.

Ref.	Scheme/technique	Security analysis
Ulutas et al. (2011)	SSSs	It provides data authenticity and data confidentiality
Parakh and Kak (2011)	k-threshold computational secret sharing	• Security is highest when using Shamir's scheme directly without dividing the secret, resulting in shares of equal size to the secret. • The least security is achieved when the shares are only of 1/kth size of the secret • It also mentions to choose the level of security by dividing the secret into different number of pieces 'm' based on the desired security level. • Security level is determined by the probability of correctly guessing the kth share with only the knowledge of $k - 1$ shares
Askari et al. (2012)	Novel-visual secret sharing scheme	• It focuses on high quality of the recovered image without pixel expansion. • This technique can be applied on both binary and halftone images
Ermakova and Fabian (2013) and Fabian et al. (2015)	SSSs and Robin's information Dispersal algorithm	• It provides additional privacy protection by distributing EHRs as fragments in multi cloud providers. • This architecture is very useful in the case of key compromise or if encryption algorithms are broken. • Adding secret sharing approach to multi cloud scenario results low computational overhead.
Jani Anbarasi and Anandha Mala (2014)	SSSs with DNA cryptography	• Simulation results prove that this method hides longer electronic patient records. • It provides authenticity and confidentiality of patient records. • Better PSNR (Peak to Signal Noise Ration) is achieved • It has ability to resist various attacks
Rezaeibagha and Mu (2016)	Secret sharing with role based access control	• When comparing execution times, the computational overhead is highest for smallest samples of 10 KB. • ABE involves an initial computational overhead to higher throughput on larger documents. • Share creation, secret recovery, signature creation, and verification have consistent throughput. • PUT requests' execution time is less critical because pending jobs can be queued. • GET requests' execution time is more crucial to minimize user waiting time.

(Continued)

TABLE 15.4 Security analysis of various schemes and frameworks.—cont'd

Ref.	Scheme/technique	Security analysis
		• Bandwidth and network connectivity are limitations, especially for large files, which are not specific to the architecture. • CPU processing power and memory resources are bottlenecks since all operations are performed in-memory. • Disk I/O may be required on mobile devices, increasing execution time.
Au et al. (2017)	Attribute based encryption and Proxy re-encryption	• It achieves data confidentiality. • Selectively secure against chosen Ciphertext attack for both initial and re-encrypted cipher text • It allows straightforward user revocation • It simplifies emergency access with 'ER dept' (Emergency Room Department) policy • This scheme can be further strengthened by using adaptive security model to composite order bilinear groups and dual system encryption
Luo et al. (2018)	SW-SSS in privacy Protector	• Additional layer of security is added, by encrypting patient data with public key provided by the health care provider, so that only health care provider can decrypt with their private key • To ensure robustness of the security measures, random Initialization Vector numbers are generated using SHA-3 algorithm. • These numbers are combined with keys and data to create unique encryption keys making it more challenging for potential attackers to decipher the information
Marwan et al. (2019)	SSSs and multi-cloud	• The technique 'Histograms of encrypted images' is used to prove the data protection against untrusted cloud providers. • It is proved that proposed solution is efficient for protecting digital records • It ensures high fault tolerance, high availability, data confidentiality and anonymity.
Tang et al. (2019)	SSSs with Boneh-Goh-Nissim Crypto system	• This scheme resist differential attacks, tolerate health center failures and keep fair incentives for the patients • Performance evaluations demonstrate cost-efficient computation, communication and storage overhead.
Ismail et al. (2020)	SSSs with RSA	It protects privacy by allowing patient to get total control over generation and management of the decryption keys.

TABLE 15.4 Security analysis of various schemes and frameworks.—cont'd

Ref.	Scheme/technique	Security analysis
Sarosh et al. (2021)	Computational secret sharing with RC6	• The proposed technique has advantage of fault tolerance and reduced share size. • The residual image problem is mitigated using RC6 encryption method. • The proposed framework employs lossless computational secret sharing by making use of prime number equal to 257.
Čuřík et al. (2022)	Symmetric encryption with Shamir's (2,3)-threshold secret sharing scheme	• It provides privacy and integrity of the data • It minimizes probability of DoS attacks

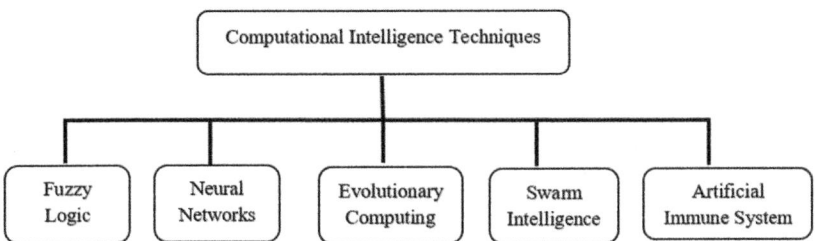

FIGURE 15.4 Classification of computational intelligence techniques. This figure shows five types of computational intelligence techniques.

4.1 Computational intelligence techniques

Computational Intelligence is a broad field within AI that concentrates on developing algorithms and systems capable of mimicking and simulating human-like intelligence and cognitive processes. It includes various techniques that enable computers and machines to perform tasks that typically require human intelligence such as learning data, reasoning, problem solving, decision making, and adapting to changing environments. CI is applied in wide range of fields, including healthcare, finance, robotics and many more. The techniques that elaborate on computational intelligence are shown in Fig. 15.4.

4.1.1 Fuzzy logic

Fuzzy logic is a mathematical framework that deals with uncertainty and imprecision. It allows for the representation and processing of data that are not strictly binary (true or false), making it suitable for applications where decisions are based on degrees of truth or membership values as depicted in Fig. 15.5.

4.1.2 Neural networks

Neural networks are computational models inspired by the structure and function of the human brain. They consist of interconnected nodes (neurons) organized in layers as shown in Fig. 15.6 and are used for tasks such as pattern recognition, image and speech processing, and

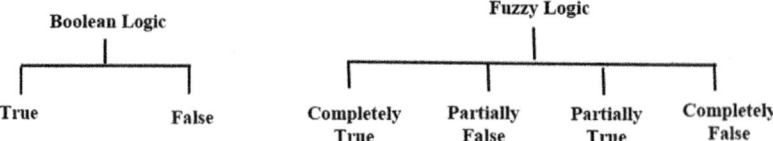

FIGURE 15.5 Boolean logic vs fuzzy logic. This figure shows boolean logic (true or false) and fuzzy logic (completely true, completely false, partially true and partially false).

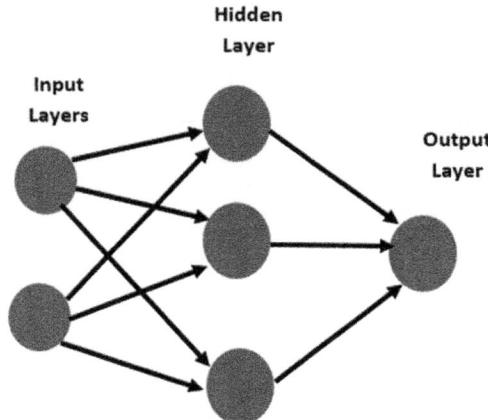

FIGURE 15.6 A simple neural network. This figure shows simple neural network consists of interconnected nodes organized in layers.

natural language understanding. Deep learning, a subset of neural networks, has been particularly influential in recent years.

4.1.3 Evolutionary computing algorithms

Evolutionary algorithms, such as genetic algorithms and evolutionary strategies, are optimization techniques inspired by the process of natural selection. The workflow of these algorithms are as illustrated in Fig. 15.7. They are used to find solutions to complex problems by evolving a population of potential solutions over successive generations.

4.1.4 Swarm intelligence

Swarm intelligence models are inspired by the collective behavior of social insects and animals, such as ants and birds. These models use decentralized and self-organized systems to solve problems, such as optimization and routing.

4.1.5 Artificial immune system

This technique is inspired by the human immune system's principles of pattern recognition, memory, and self/non-self-discrimination. It excels in identifying irregularities and safeguarding against threats leveraging the robust mechanisms of the natural immune system.

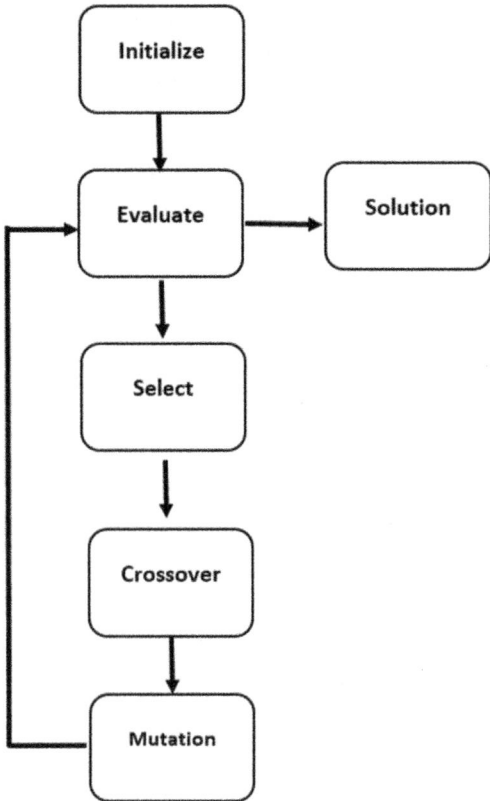

FIGURE 15.7 Workflow of evolutionary algorithm. This figure shows stages of evolutionary computing algorithms.

4.2 Enhancing EHRs security with computational intelligence

Computational intelligence plays a significant role in improving in secure sharing of health records stored in cloud environment. The following are the ways in which computational intelligence can be applied:

1. **Anomaly Detection:** The unusual patterns or behaviors in data access and data sharing process can be identified using machine learning algorithms. If there are any unauthorized or suspicious activities, the system can automatically trigger alerts or take corrective actions.
2. **Access Control:** Computational intelligence can be used to create adaptive access control mechanisms. Machine learning models can learn user behavior and adapt access permissions accordingly.
3. **Privacy-Preserving Techniques:** Techniques such as homomorphic encryption and federated learning of computational intelligence allow data to be processed and analyzed without exposing the raw data. This ensures that sensitive health records remain confidential while still enabling useful insights.

4. **User Authentication:** Authentication methods such as biometrics (e.g., fingerprint or facial recognition) coupled with machine learning, can provide strong user authentication, making it more difficult for unauthorized users to access data.

5. **Behavioral Analysis:** Machine learning can be used to analyze the behavior of users and devices accessing the health records. If there are deviations from expected behavior, it can trigger additional security measures.

6. **Predictive Security:** AI and machine learning can predict potential security threats based on historical data and emerging patterns. This proactive approach allows for the prevention of security breaches.

7. **Dynamic Key Management:** Computational intelligence can help in managing cryptographic keys dynamically, making it harder for attackers to gain access through key compromise.

8. **Natural Language Processing (NLP):** NLP algorithms can be used to analyze text embedded in health records to identify and redact sensitive information automatically. This can help in maintaining privacy of patient data.

9. **Continuous Monitoring:** AI-powered systems can continuously monitor the security posture of the cloud environment, identifying vulnerabilities and recommending corrective actions.

10. **Threat Intelligence:** Thread Intelligence data and real-time updates on potential threats to the health data can be analyzed using Machine learning algorithms.

The following chapters show how the security of EHRs stored in cloud enhanced by the combination of Computational Intelligence with Cryptographic secret sharing schemes.

Sarkar et al. (2019) have presented a new, highly secured method for generating secret keys by utilizing synchronized mechanism called Two Tier Neural Networks (TTNN). This approach enhances security by increasing the depth of synaptic connections within the TTNN architecture. Unlike conventional methods where a patient and cardiologist need to transmit a shred secret key over public channel, this novel approach depends on their unique and indistinguishable weight values as the shared secret key for encryption and decryption. Therefore, the risk of attack on this approach is significantly lowered when compared with other secret key exchange algorithms.

Rehman et al. (2021) introduce an energy-efficient IoT e-health system that incorporates Artificial Intelligence and Homomorphic Secret Sharing to enhance the reliability and energy efficiency of data transmission in medical applications. This system determines the most cost-effective rout for transmitting medical data using AI heuristics. This approach helps to identify and prevent malicious machine interactions, ensuring the security of medical data. It establishes trust among medical nodes through the distribution of secret pairs using homomorphic encryption, ensuring data privacy and authorized access. Moreover, this system employs multi-hop hashing mapping to make it difficult for intruders to compromise data integrity. RSA cryptosystem is also used to secure the medical records from network edges to sink node without imposing computational demands on medical sensors. However, the experiments have shown an increased packet drop rate when the network is heavily loaded.

Ghazal et al. (2022) suggested a private block chain-based encryption framework using computational intelligence approach for the sake of encryption. CI is beneficial in obtaining similar attributes in the gathered data. This framework provides better results in terms of 0.93 in the training phases and 0.91 in the validation accuracy.

Alabi et al. (2022) propose a multi layered encryption approach to ensure the security and privacy EHRs management. Advanced Encryption Standard algorithm is used for encrypting local hospitals' electronic medical records and Elliptic Curve Cryptography for cloud based HER systems. They also integrate artificial intelligence model to measure the trustworthiness of medical personnel. The effectiveness of the privacy model is evaluated in terms encryption/decryption speed and system output size. Their system outperformed in terms of security and reputation when compared with existing systems.

Marwan et al. (2018) have introduced an innovative solution to address main challenge associated with existing frameworks which is high computational cost linked with the image processing tasks and protecting medical data. This novel approach utilizes machine learning techniques such as Support Vector Machine (SVM) and Fuzzy C-means Clustering (FCM) for efficient classification of image pixels. Moreover, they introduce an additional layer, the CloudSec module to reduce the risk of medical data exposure. The simulation results of this architecture indicates that SVM is most suitable choice for segmentation of images and ensuring data security in the adoption of cloud services within the healthcare industry.

5. Conclusion

In the dynamic landscape of healthcare and its intersection with information technology, the secure sharing of EHRs stored in the cloud is of paramount importance. This review thoroughly explores the various cryptographic methods used to ensure the secure sharing of EHRs, offering a comparative analysis based on their characteristics and security considerations. Moreover, it investigates the fusion of Computational Intelligence with cryptographic secret sharing techniques to bolster the safeguarding of confidential health data in the cloud. This integration empowers organizations to establish stronger and more adaptable security systems capable of effectively shielding sensitive data from diverse array of threats and vulnerabilities.

References

Alabi, O., Gabriel, A.J., Thompson, A., Alese, B.K., 2022. Privacy and Trust Models for Cloud-Based EHRs Using Multilevel Cryptography and Artificial Intelligence. Springer Science and Business Media LLC, pp. 91–113. https://doi.org/10.1007/978-3-030-80821-1_5.

Askari, N., Moloney, C., Heys, H.M., 2012. A novel visual secret sharing scheme without image size expansion. In: 25th IEEE Canadian Conference on Electrical and Computer Engineering: Vision for a Greener Future, CCECE 2012. Canada. https://doi.org/10.1109/CCECE.2012.6334888.

Au, M.H., Yuen, T.H., Liu, J.K., Susilo, W., Huang, X., Xiang, Y., Jiang, Z.L., 2017. A general framework for secure sharing of personal health records in cloud system. Journal of Computer and System Sciences 90, 46–62. https://doi.org/10.1016/j.jcss.2017.03.002.

Beimel, A., 2011. International Conference on Coding and Cryptology. Springer.

Boyle, E., Gilboa, N., Ishai, Y., Lin, H., Tessaro, S., 2018. Foundations of homomorphic secret sharing. Leibniz International Proceedings in Informatics, LIPIcs 94. https://doi.org/10.4230/LIPIcs.ITCS.2018.21. http://drops.dagstuhl.de/opus/institut_lipics.php?fakultaet=04.

Brickell, E.F., 1989. Workshop on the Theory and Application of of Cryptographic Techniques. Springer.

Çalkavur, S., Solé, P., 2023. A ramp secret sharing scheme from finite fields. In: International Conference on Pioneer and Innovative Studies, 1, pp. 107–110. https://doi.org/10.59287/icpis.813.

Chandramouli, A., Choudhury, A., Patra, A., 2022. A survey on perfectly secure verifiable secret-sharing. ACM Computing Surveys 54 (11s), 1–36. https://doi.org/10.1145/3512344.

Čuřík, P., Ploszek, R., Zajac, P., 2022. Practical use of secret sharing for enhancing privacy in clouds. Electronics 11 (17). https://doi.org/10.3390/electronics11172758.

Ermakova, T., Fabian, B., 2013. Secret sharing for health data in multi-provider clouds. In: Proceedings – 2013 IEEE International Conference on Business Informatics, IEEE CBI 2013. IEEE Computer Society, Germany, pp. 93–100. https://doi.org/10.1109/CBI.2013.22.

Fabian, B., Ermakova, T., Junghanns, P., 2015. Collaborative and secure sharing of healthcare data in multi-clouds. Information Systems 48, 132–150. https://doi.org/10.1016/j.is.2014.05.004.

Ghazal, T.M., Hasan, M.K., Abdullah, S.N.H.S., Bakar, K.A.A., Al Hamadi, H., 2022. Private blockchain-based encryption framework using computational intelligence approach. Egyptian Informatics Journal 23 (4), 69–75. https://doi.org/10.1016/j.eij.2022.06.007.

Ibrahim, D.R., Teh, J.S., Abdullah, R., 2021. An overview of visual cryptography techniques. Multimedia Tools and Applications 80 (21–23), 31927–31952. https://doi.org/10.1007/s11042-021-11229-9.

Ismail, T., Touati, H., Hajlaoui, N., Hamdi, H., 2020. Hybrid and secure E-health data sharing architecture in multi-clouds environment. Lecture Notes in Computer Science 12157, 249–258. https://doi.org/10.1007/978-3-030-51517-1_21.

Iwamoto, M., Yamamoto, H., 2006. Strongly secure ramp secret sharing schemes for general access structures. Information Processing Letters 97 (2), 52–57. https://doi.org/10.1016/j.ipl.2005.09.012.

Jani Anbarasi, L., Anandha Mala, G.S., 2014. EPR hidden medical image secret sharing using DNA cryptography. International Journal of Engineering and Technology 6 (3), 1346–1356.

Kessler, G.C., 2003. An Overview of Cryptography.

Lakshmi, R., Nitya, 2015. Analysis of attribute based encryption schemes. International Journal of Computer Science and Engineering Communications 3, 1076–1081.

Li, L., Li, Z., 2023. An efficient quantum secret sharing scheme based on restricted threshold access structure. Entropy 25 (2). https://doi.org/10.3390/e25020265.

Li, M., Yu, S., Ren, K., Lou, W., 2010. Securing Personal Health Records in Cloud Computing: Patient-Centric and Fine-Grained Data Access Control in Multi-Owner Settings. Springer Science and Business Media LLC, pp. 89–106. https://doi.org/10.1007/978-3-642-16161-2_6.

Luo, E., Bhuiyan, Md Z.A., Wang, G., Rahman, Md A., Wu, J., Atiquzzaman, M., 2018. PrivacyProtector: privacy-protected patient data collection in IoT-based healthcare systems. IEEE Communications Magazine 56 (2), 163–168. https://doi.org/10.1109/MCOM.2018.1700364.

Marwan, M., Kartit, A., Ouahmane, H., 2018. Security enhancement in healthcare cloud using machine learning. Procedia Computer Science 127, 388–397. https://doi.org/10.1016/j.procs.2018.01.136.

Marwan, M., Alshahwan, F., Sifou, F., Kartit, A., Ouahmane, H., 2019. Improving the security of cloud-based medical image storage. Engineering Letters 27 (1), 175–193.

Parakh, A., Kak, S., 2011. Space efficient secret sharing for implicit data security. Information Sciences 181 (2), 335–341. https://doi.org/10.1016/j.ins.2010.09.013.

Poole, D.I., Randy, G., Goebel, A.K., Mackworth, 1998. Computational Intelligence, 1. Oxford University Press.

Rehman, A., Saba, T., Haseeb, K., Larabi Marie-Sainte, S., Lloret, J., 2021. Energy-efficient IoT e-health using artificial intelligence model with homomorphic secret sharing. Energies 14 (19). https://doi.org/10.3390/en14196414.

Rezaeibagha, F., Mu, Y., 2016. Distributed clinical data sharing via dynamic access-control policy transformation. International Journal of Medical Informatics 89, 25–31. https://doi.org/10.1016/j.ijmedinf.2016.02.002.

Sameera, M., Rani, K.U., 2023. Enhancement of cloud data protection using attribute based encryption with multiple keys: a survey. Mathematical Statistician and Engineering Applications 72, 1952–1967.

Sarkar, A., Dey, J., Bhowmik, A., Mandal, J.K., Karforma, S., 2019. Computational intelligence based neural session key generation on E-health system for ischemic heart disease information sharing. Advances in Intelligent Systems and Computing 812, 23–30. https://doi.org/10.1007/978-981-13-1540-4_3.

Sarosh, P., Parah, S.A., Bhat, G.M., Heidari, A.A., Muhammad, K., 2021. Secret sharing-based personal health records management for the internet of health things. Sustainable Cities and Society 74. https://doi.org/10.1016/j.scs.2021.103129.

Shamsoshoara, A., 2019. Overview of Blakley's Secret Sharing Scheme.

Tang, W., Ren, J., Deng, K., Zhang, Y., 2019. Secure data aggregation of lightweight E-healthcare IoT devices with fair incentives. IEEE Internet of Things Journal 6 (5), 8714–8726. https://doi.org/10.1109/jiot.2019.2923261.

Ulutas, M., Ulutas, G., Nabiyev, V.V., 2011. Medical image security and EPR hiding using Shamir's secret sharing scheme. Journal of Systems and Software 84 (3), 341–353. https://doi.org/10.1016/j.jss.2010.11.928.

Wu, R., Ahn, G.J., Hu, H., 2012. Secure sharing of electronic health records in clouds. In: CollaborateCom 2012 – Proceedings of the 8th International Conference on Collaborative Computing: Networking, Applications and Worksharing, pp. 711–718. https://doi.org/10.4108/icst.collaboratecom.2012.250497.

Yuan, J., Li, L., 2019. A fully dynamic secret sharing scheme. Information Sciences 496, 42–52. https://doi.org/10.1016/j.ins.2019.04.061.

16

Private blockchain-based encryption framework using Computational Intelligence approach

T. Sarath[1], K. Brindha[1], Rajesh Kumar Dhanaraj[2] and Balamurugan Balusamy[3]

[1]School of Computer Science Engineering and Information Systems, Vellore Institute of Technology, Vellore, Tamil Nadu, India; [2]Symbiosis Institute of Computer Studies and Research (SICSR), Symbiosis International (Deemed University), Pune, Maharashtra, India; [3]Shiv Ndar University, Delhi, India

1. Introduction

The blockchain makes use of cryptography, asymmetric key methods, and hash functions. With the use of hash roles, all blockchain participants can see the same information. SHA-256 is widely used as the hash function in blockchains. Security features including data privacy and access control are monitored using encryption techniques. However, securing access switch presents a substantial challenge. Bettencourt was pioneer in using the Cipher Text Policy Attribute-Based Encryption (CP-ABE) technique, which was developed in 2007. In CP-ABE, the private user's keys are generated from attributes, and the cipher text is linked to an open structure (Sharma et al., 2022).

Malware detection is only one example of an IoT (Internet of Things) security solution that makes use of Computational Intelligence techniques. Identifying cyber threats, tracking suspicious behavior, spotting intrusions, and recognizing cyberattacks. In order to strengthen its cybersecurity skills and ensure the safety of IoT apps and users, the IoT may adopt CI strategies. Communication among the database provider service, user and owner must be protected, and this requires a secure and computationally intelligent solution (Majhi et al., 2022).

There is no easy way to reconcile decentralized nature of blockchain technology with traditional security measures. A lot of progress needs to be made in this environment. While many scholars have studied blockchain-based security systems in recent years, few have provided concrete designs for them. As a result, this chapter offers a concise summary of several chapters that have advanced the area. First and foremost, we hope to gain insight into the potential advantages of combining CI with blockchain-based encryption in order to uncover novel means of bolstering computer security.

2. Technologies of blockchain

2.1 Block structure

Each block in a blockchain records information about a specific transaction. A cryptocurrency trade or other sort of data transfer could be implied here. Each block has a header and a body that make up its logical structure. Each block can be conceptually broken down into its head and its body. The previous block's hash value is one of several fields in each block's header. As can be seen in Fig. 16.1, the blocks remain connected in a fashion analogous to that of a linked slope. The block's body is where transactions are recorded. In blockchain terminology, the initial block is referred to as the "genesis" block. A cryptographic hash is used to determine the block's hash value, and each block includes its predecessor's hash to ensure that the blockchain's data remains unaltered. The initial hash value will be rendered useless if hackers modify the data in preceding blocks. Because to the domino effect, succeeding blocks' hash values will become invalid. Because of this, hackers who wish to alter a block's content must alter the hash charge in the shot of all subsequent blocks and type comparable changes in the majority of nodes in order toward achieve an agreement (Aparna, 2021; Kaur and Ali, 2021).

The consensus algorithm (Gupta et al., 2021) includes the operation in the bulge confirmation block. The hash value of the data representing each business in the block serves as the ID for that transaction. Create a hash tree of transaction identifiers in pairs and insert it into the block body (see Fig. 16.2). In conclusion, the block header is where the hash tree's starting point is kept. Therefore, the transaction can be verified without first verifying the local duplicate transaction by using Merkle tree branch that contains the transaction. The hash value on

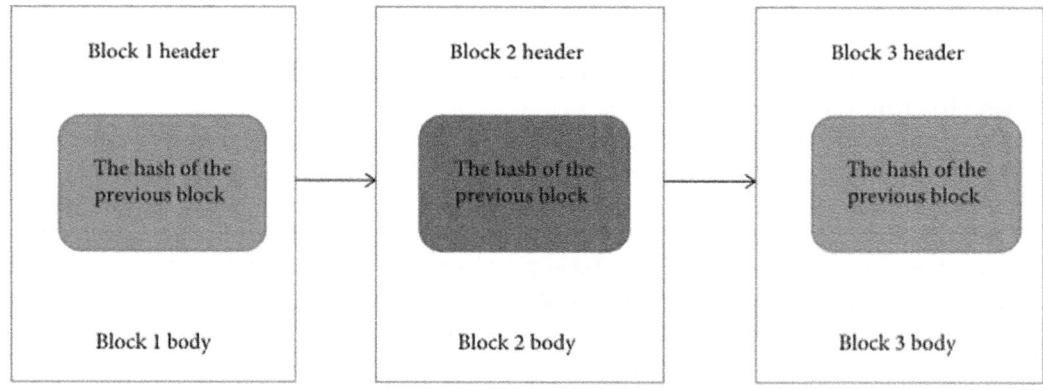

FIGURE 16.1   Building of chain.

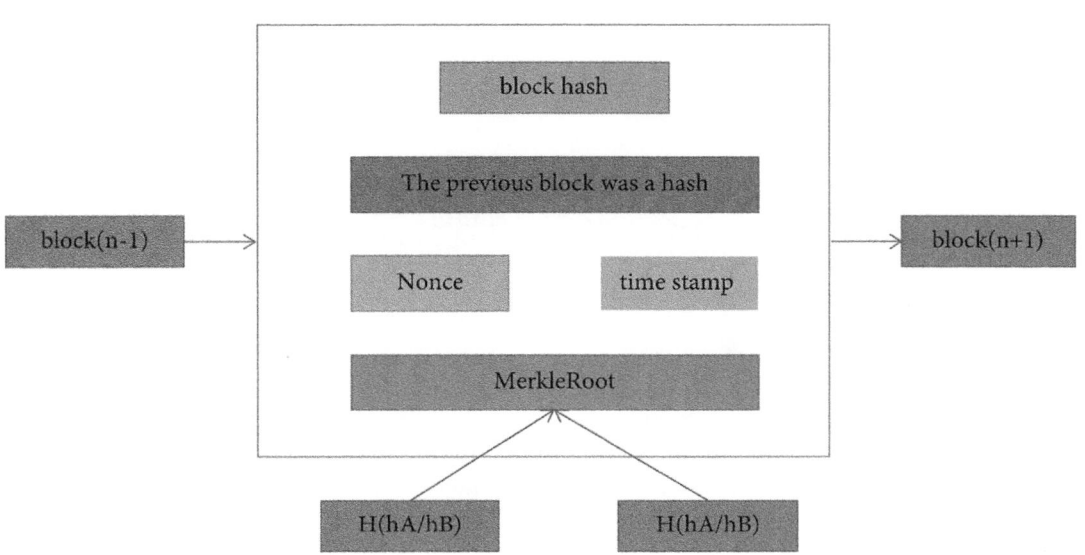

FIGURE 16.2   Block body and block head.

the branch where the altered transaction occurred will change, allowing for easy detection of the manipulated transaction.

2.2 Algorithm of consensus

Blockchain relies heavily on consensus algorithm. The goal is to apprize the system allocation state in a secure manner. If a blockchain is built on a "copy of state machine" then the algorithm of consensus must guarantee all copies of state machine always reflect the same state. Safety, practicality, and error-tolerance are the backbones of every good consensus method. Where total number of nodes represents n and f represents the maximum number of nodes that can tolerate faults, protection (n, f) denotes the degree of resistance to a system failure like a system that is resistant to force failure. Consensus algorithm activity ensures that all properly engaged nodes can endure to attain distributed consensus in the presence of f failing nodes. Consensus networks are vulnerable to attacks from nodes that suddenly stop contributing. Two main types of consensus algorithms that are rely on evidence and those rely on majority vote. In what follows, we'll look at several examples of each category. Fig. 16.3 also provides a classification of the many consensus algorithms.

2.3 Types of blockchain networks

Based on who has access to add new blocks to the blockchain. These networks are typically divided into two types: public and private (Gupta et al., 2021; Manikandan and Sakthi, 2020). There is a different approach that uses read-write access (Yoo et al., 2021). Based on the permissions required to add new blocks to the blockchain, this concept divides blockchains into open and closed categories. Therefore, there are two additional types of private blockchain networks.

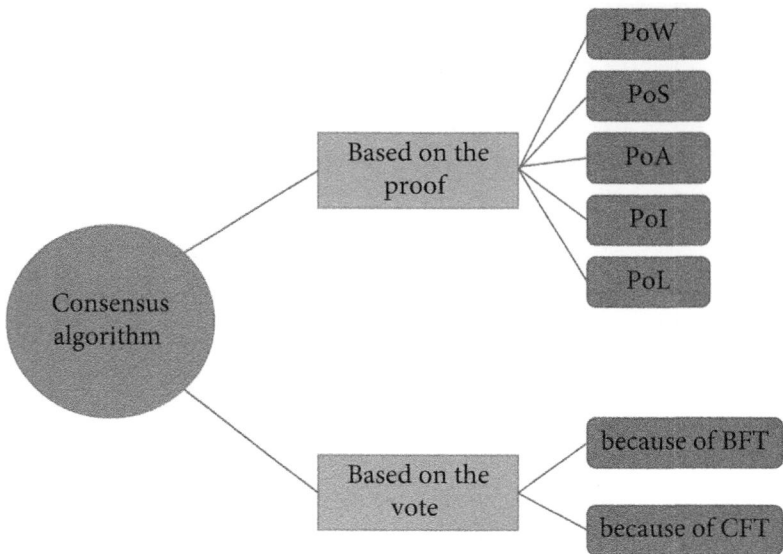

FIGURE 16.3   Consensus algorithm.

1) Open distributed ledgers
   This refers to blockchains where both reading and adding data to the blockchain are open to anybody (Yoo et al., 2021). Anyone can join a public blockchain network and begin doing transactions immediately, without needing the permission of any other node in the network. Information stored in such blockchains can be viewed by anyone; participation in the network is not required. In public blockchains, all nodes are created equal and have the ability to validate blockchain transactions through mining. blockchains are used by digital currencies such as Bitcoin and Ethereum. Such blockchains allow users to remain anonymous because they don't require registration with personal information and blockchain addresses aren't tied to real-world identities.

2) Internal blockchains
   Since only authorized users are allowed to view or write to a private blockchain, the data stored there is protected from prying eyes. In terms of who gets access to the blockchain's information, there are two types of private blockchains. However, in either case, the information is kept secret while the individuals' identities remain known.

   a) Blockchains, both public and private
      Once data has been written to this blockchain, however, it may be accessed by anybody. However, only a select group of people has authority to actually add new information to the blockchain. Permissioned blockchains require users to be known and validated before they can add information to blockchain. The release of public financial records by a government is one possible use case for this kind of blockchain network. Although everyone can access the blockchain, only authorized government

personnel can upload new entries. Since private blockchains lack the anonymity of public blockchains, their transactions must be verified on the basis of trust (Xiao and Chen, 2021). There is less incentive for bad behavior because the individuality of the user publishing the information is known in private blockchains.

b) Blockchains that are not open to the public

In this blockchain network configuration, only authorized users can add new blocks of data to the blockchain, and only those users can read the blocks that have already been added. Since on a private and closed blockchain, only authorized users have access to data, it is common for private blockchains to include identity verification tools (Manikandan and Sakthi, 2020). After confirming the user's identity, Users can be assigned varying degrees of access. The filing of tax returns is a possible use case for this blockchain design. Companies and individuals are verified as users before they are granted permission to add data to the blockchain, and only a select group of people are granted read-only access.

2.3.1 Coin: Ethereum

Ethereum is a decentralized application platform that is both open-source and built on the blockchain (Xiao and Chen, 2021). The Ethereum network as a whole relies on a distributed system of nodes all over the world. Volunteer nodes are responsible for upholding the entire system's consensus rules through transaction validation in exchange for a portion of the Ether coin. By transmitting and receiving the ether money and mining transaction blocks previously they are additional to the blockchain nodes container conduct business on the Blockchain Ethereum network. Smart contracts can be created and distributed by nodes, and nodes can also access the functionality of smart contracts that have been deployed elsewhere in the network. The main Ethereum network, Ethereum test networks, and private blockchain networks are three distinct varieties of Ethereum networks.

1) The core network of Ethereum

Mainnet, short for "main Ethereum network," is the community blockchain implementation of Ethereum wherever ether has actual value (Heister and Yuthas, 2020). In other words, a node on the main network can only acquire Ether through either mining or direct transfers from other nodes. Although there is a cost related with deploying a smart convention to the main network, once the transaction price has paid, the convention remains permanently placed on the Ethereum Blockchain network with no further costs. A smart contract can no longer be modified after it has been released to the main network. That means a new smart convention must be organized to the network if the previous smart contract has error, as it cannot be erased or altered.

2) The Ethereum testnet

An Ethereum test network is a blockchain network that mimics the main Ethereum network's behavior and features without actually exchanging any ether. This enables nodes to deploy and test smart contracts, as well as send and receive phoney ether, at no cost. Because smart contracts can be updated in a test network for no cost while still retaining the same properties as the main network, test networks are ideally suited to the testing of such contracts. At now, there are three primary test networks where users can put their Ethereum smart contracts through their paces. The first is Ropsten, which is a blockchain network based on proof-of-work similar to Ethereum itself. Nodes have the ability to mine for phoney ether and perform transactions. Kovan and Rinkeby, the other two test networks, also employ the proof-of-authority consensus technique. Nodes can appeal either on the appropriate web applications in order to conduct transactions and launch smart contracts using the counterfeit cryptocurrency.

3) A nonpublic blockchain based on Ethereum

Private networks built on the Ethereum protocol are permissioned Blockchains. In terms of transaction processing and the ability to host smart contracts, it is quite similar to the core Ethereum network. It's the best possible option for businesses.

3. Privacy regulations

In many parts of the world, the guarantee of personal solitude is among the most fundamental guarantees of citizenship. Individuals may have a right to privacy in regards to their own data. Careful protection of this right is warranted because of the far-reaching effects that control over one's own data can have on one's reputation, social life, and even sense of self (Heister and Yuthas, 2020).

Personal information security laws are now being drafted at a quick pace. In 2016, the European Union passed General Data Protection Regulation (GDPR), which sets a new standard for data privacy legislation. Organizations showing business with EU member states are essential by law to defend the private information of EU citizens.

The GDPR (Grimes, 2019) is reflected in some aspects of the California Privacy Rights Act (CPRA), which is a development of the California Consumer Privacy Act (CCPA) of 2018. The California Consumer Privacy Act (CCPR) is meant to give Californians the following rights:

The right to know what personal information is being composed.

The right to know if and to whom such data is being sold or shared.

To opt out of having their private data sold.

To view their private information.

To request that an organization erase one's personal information.

Right to privacy and the right to not be punished for exercising it (Uribe, 2020)

Democrat senator Maria Cantwell introduced the Consumer Online Privacy Rights Act (COPRA) in December 2019. Data privacy legislation at the level of federal has failed before, but lawmakers are still trying to adopt new, more stringent regulations (Yoon, 2019). Businesses that deal with customer information will need to change their processes and put in place new technology to comply with privacy rules. Companies can use blockchain and DLT to comply with current and future regulation for the protection of individuals' property rights and personal information.

3.1 Blockchain and privacy

Organizations may now share data in ways that were not possible before implementing blockchain technology, which can lead to more opportunities for teamwork, streamlined processes, and increased earnings. Since the data are deposited on shared ledgers that may be nearby by multiple blockchain contributors, queries about how to guarantee privacy over the data become more pressing in these settings.

When discussing the safety of public blockchains, the blockchain technology solutions provider ConsenSys claims, "In reality, privacy is not a property of any blockchain." Instead, each blockchain can consume additional levels of anonymity added to it (Consensys, n.d.). Designers need to think about who can read and write transactions, who can validate and store them, and how they are broadcasted. Permissions and security actions, including how they are enforced and updated are additional factors to think about. Discussions of privacy are further complicated by decisions concerning data ownership and permissible data uses by organizations and computer applications (Consensys, n.d.).

3.1.1 Decentralized identity

A common belief among blockchain advocates is that people should be able to make their own decisions about what parts of their identities to make public and how. DID is a self-sovereign identity realized through Blockchain technology, which greatly enhances the confidentiality and security of sensitive information.

DID is the concept of an individual having control of their own digital data that relates to many aspects of their identity. Microsoft's participation in the development of DID standards reflects the company's commitment to a user-first approach. Right now, our digital personas and the interactions we have with them are possessed and measured by third parties, some of whom we may not flush know about (Microsoft, n.d.). Organizations that would normally be accountable for securing data can gain from giving that responsibility back to the people whose information it is.

Blockchain technology allows for DID and stretches users a way to keep personal data separate from databases of organizations with which trade. Users can verify the veracity of claims made about their personal information by storing references to this data on the blockchain, which they own and manage. A user might, for instance obtain a driver's licence from the DMV and keep it in their own secure location. To prove their licensing status, users can show their documentation to third parties such as insurance providers, who can then double-check the license's issuer and expiration date on their own.

The process of making a DID is simple and open to everybody. There is no history connected to this identity when it is first formed. Over time, the user could link other identifying information to that DID, such as a driver's licence number. Verifying the ownership of a DID by a third party is conceptually similar to confirming the ownership of an email address. An email address, for instance, might be linked to a variety of online profiles. Only the owner of the email account's password would be able to verify the user's ownership by sending them a secret message to their inbox, such as a security code, and having them reply with the code.

The DID would remain in the user's possession, as opposed to the email providers. The owner would additionally protect the password or secret key. An individuality hub, an encrypted database of private information resides somewhere other than the blockchain (expected a mixture of mobile phone, personal computer, cloud data, and offline storage devices), might be used to keep track of a person's identity (Microsoft, n.d.). An identity hub would allow the user to decide what information about themselves should be shared with third parties.

DIDs lower the likelihood of unintended association. When the same identifier is used across many platforms, such as an email address, a correlation issue arises. Without the knowledge or permission of the user, entities can correlate data about a single identity across many platforms. Addresses of email are a key piece of information on virtually all websites. When users reveal the similar email address across many sites, along with possibly other pieces of individual information such as a phone number or physical location, they open the door to the possibility of correlation without even realizing it. In this situation, entities can make cross-site correlations using the data. Users' identities, locations, genders, ages, and interests can all be pieced together by third parties through a combination of tracking cookies and online clicks (Microsoft, n.d.).

Fig. 16.4 shows how a user can engage independently with several online service providers while storing their data in a single, user-managed place. This gives the user the power to decide which providers have access to certain types of data Fig. 16.4.

DIDs allow users to better manage who has access to their data and keep it safe. Using decentralized federated identities blockchains can also improve users' safety when interacting with a variety of online services and resources.

Privacy protection for individuals is essential for identity sovereignty and blockchains make this possible. System Users typically have what is recognized as identity of federated, which can define as unique identifier that a person uses to sign into services or information platforms hosted by different organizations using only a single login and the authentication provided by Single Sign On (SSO). An SSO credential could be used across health care network consisting of various providers such as insurance companies, urgent care clinics

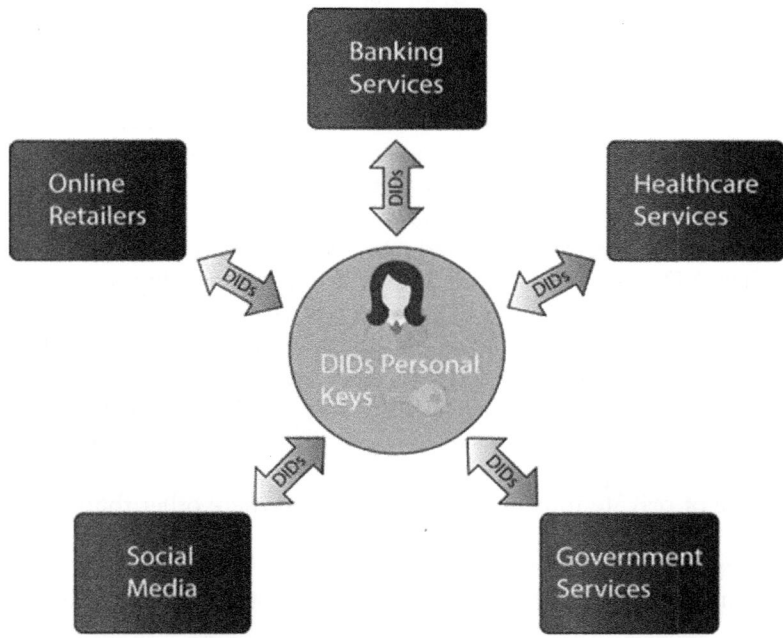

FIGURE 16.4   Online services.

and hospitals. This form of identification is vulnerable to security breaches (ElGayyar et al., 2020) since it is often held and controlled by a service provider.

1. Blockchain Federated Identity

Both private and public blockchain employments are suitable for blockchain based federated identity frameworks, and these frameworks use smart agreements to respond to rule changes that might arise while controlling self-management in the system. Furthermore, these outlines allow users to review and regulate how their identities are used, while also giving the network's business individuals with the capacity to monitor their services are being utilized. This allows for process enhancement and more satisfying user understanding.

2. Zero Knowledge Proof

While still allowing people to keep their privacy and control over their property, zero-knowledge proofs make access to identity and additional crucial data simple. Cryptographic zero-knowledge proofs allow user or "prover" to convince another user or "verifier" that something about the "prover" is true without the "prover" or "verifier" having to provide, expose or share any identifying information. When a client wants to order an alcoholic drink at a bar, the barman will often check IDs to make sure they are at least 21 years old. The customer's full birth date, height, eye color, and home address are all revealed on a driver's licence, all of which could be exploited or stolen.

With the help of cryptographic methods, a prover can show a verifier proposition is true without revealing any info. All drivers above age of 21 are required to submit a secret nickname when applying for a driver's licence. This moniker and ID number hash would then be added to a public catalog of legal drivers over age of 21. You could prove your age at the bar by entering your handle and ID number into hash generator and seeing the output coordinated an existing entry (Lesavre et al., 2019).

Both interactive and noninteractive zero-knowledge proofs exist. Most zero-knowledge protocols involve a dialog between the prover (a person or a computer) and the verifier, who pose series of queries or tasks that, if responded correctly by the prover a sufficient number of intervals will lead verifier to conclude with great probability that the prover's statement is accurate.

An instance of a zero-knowledge proof that can be interacted with is the case of two colored balls that are otherwise identical. There is a red one and a green one. Let's pretend the auditor is totally color blind and can't identify which ball is whose. You need to show the validator that the balls are distinct colors. The certifier hides the balls and pulls out one at random to inspect. The proofer gives a color reference. Again, the verifier ensures this and then enquires if the ball was exchanged. The distinct coloration makes it easy to determine whether or not the ball was exchanged. As the likelihood of repeatedly making an accurate guess decrease toward zero, the existence of two distinct balls becomes increasingly supported by the statistics (Wikipedia, n.d.).

An example of noninteractive proof would be the customer providing the barman with an evidence report that verifies their age but not any other identifying details that would be exposed by showing a photo. An example of this form of evidence is showing which socket value a card in a deck of 52 cards has without also showing its suit. The prover claims to be carrying a "king," but refuses to specify whether it is a "heart," "diamond," "spade," or "club." We may be sure that the prover is in possession of a king of some kind if the string cryptography provides data about the other 48 cards none of which are kings.

4. Cryptography related technologies

4.1 Hash function

Any amount of data can be converted to a hash value of a predetermined length using a hash function, also known as a hash algorithm. Typically, hash values are expressed by meaningless strings of characters (Manikandan and Sakthi, 2020). The fundamental features shared by all hash functions are as follows: if two hash principles are distinct (using the identical hash function), then their respective seed values are distinct. The hash function is described by this property. Therefore, a hash function with this stuff is known as one-way function of hash as depicted in Fig. 16.5. Any appeal change in the input info will end in a substantially dissimilar output hash value which is input sensitivity of hash function. The strongly confused hash function, which is a reflection of the anticollision of hash functions, makes it impossible to calculate whether or not two strings would produce identical output hash values.

4.2 Encryption on asymmetric

Asymmetric encryption differs from encryption based on symmetric in that it encrypts and decrypts using two different keys: a public key and a private key. It is possible to share the encryption technique and public key, while keeping the private key secret (Zheng et al., 2020). As can be seen in Fig. 16.6, the plaintext private key is utilized to encrypt plaintext data once it has been decrypted by the receiving party. The sender's private key is used to sign message digest, while the recipient's public key is used to verify the message digest's authenticity. Figs. 16.6 and 16.7 depicts this procedure.

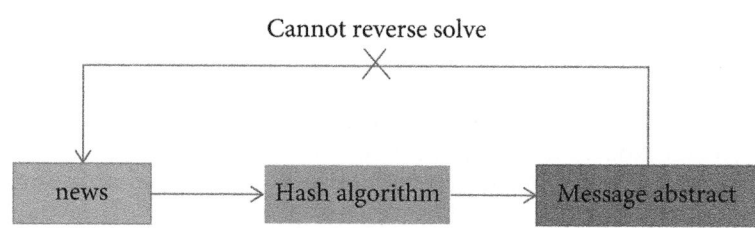

FIGURE 16.5   Process of hash encryption.

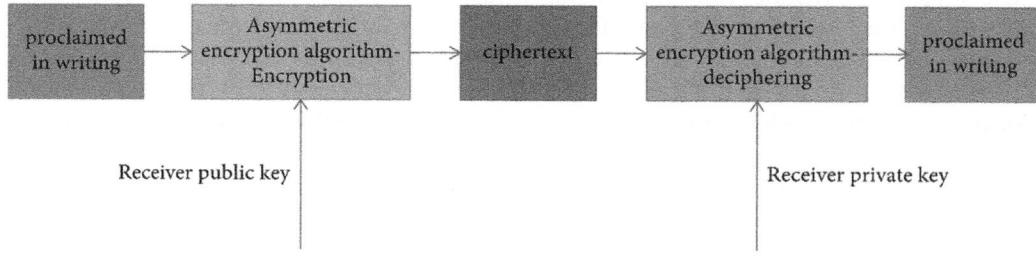

FIGURE 16.6   Procedure of data encryption.

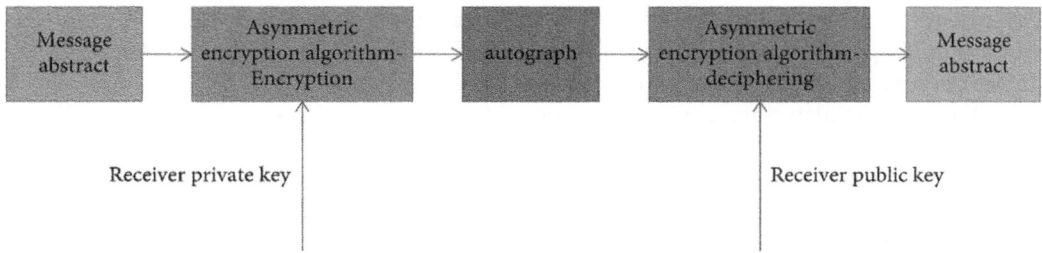

FIGURE 16.7   Procedure of data signature.

5. Benefits of cryptography in blockchain

There are many advantages of using cryptography in a blockchain, but some are listed below.

- **Encryption:** Cryptography relies on asymmetric encryption to safeguard data and communications during transactions on their network from prying eyes. In addition to ensuring that no attacker can develop a valid signature for unposed enquiries from previous queries and their corresponding signatures, the immutability property of cryptography is crucial for blockchain because it allows blocks to link with other blocks strongly and ensures the consistency of data stowed in the blockchain. With the usage of data encryption and public/private key access, cryptography improves the security of transaction records. Due to the cryptographic nature of hashing data in a blockchain cannot be altered. All users can have faith in the veracity of the digital ledger since cryptography renders the transaction irreversible. It enables an infinite number of secure transactions to be documented in the network.
- The **Digital Signature** also offers nonrepudiation function to prevent the sender from denying the delivery of a message. This perk is related to collision resistance since each input value generates a distinct hash function, preventing any potential for messages to be confused with one another.
- **Prevent hackers** from tampering with data by making the digital signature invalid if the data is modified. Using cryptography, it prevents data theft and renders blockchain's encryption unbreakable.

6. Limitations of cryptography in blockchain

The blockchain's use of encryption does have some restrictions, however, as listed below:

- **Difficulty gaining** access to information at crucial junctures: Data that has been heavily encrypted and digitally signed may be inaccessible even to an authorized user. An unauthorized user can launch an assault against the network and effectively disable it.
- One of the most important features of **information security is its reliability**, and this feature cannot be guaranteed by cryptography alone: high availability. Denial of service attacks and total information system failures require additional preventative measures.

- As a result of flaws in the design of protocols, methods, and systems, cryptography is not a shield against the **resulting vulnerabilities and dangers.** The right defense infrastructure design is required to address these problems.
- Cryptography is also **very costly in terms of both time and money.** The public key infrastructure required for public-key cryptography is complex and expensive to set up and keep running. The time it takes to send messages and process data increases when cryptographic techniques are used.
- **Vulnerability:** Cryptographic procedures are only as safe as the mathematical problem they solve. If these mathematical problems are ever solved, it could compromise encryption methods.

7. Artificial Intelligence

Machine learning and cognitive computing are subfields of Artificial Intelligence (AI) in which computers are programmed to follow human cognitive functions such as problem-solving and learning but sometimes with far greater speed and accuracy (Gopichand et al., 2019). Speech and face recognition, medical diagnosis, economic forecasting, disease outbreak monitoring, and other applications are just the tip of the iceberg for the growing use of Artificial Intelligence. Artificial Intelligence (AI) algorithms allow computers to reason and perform actions with the end objective in mind.

AI tools can use blockchain to increase user and stakeholder security by providing new ways to access and learn from data without claiming ownership or control over it. Both the business and its data providers stand to benefit from this measure's potential risk mitigation. Privacy-related AI capability should be built into the structural design of blockchain networks and processed quickly to benefit both blockchain participants and organization or group separately for setting governance procedures and developments.

By combining data from various consumer interactions, businesses using AI are able to form more complete portraits of their clients. Incentives for blockchain players to consolidate client transactions across all blockchain partners will result in more comprehensive databases. While this may be advantageous for blockchain partners, it may have unintended consequences for the customer's privacy and other stakeholders.

Personal info from blockchain contributors and their investors can be combined using AI can protect data security and privacy of personal data, in conjunction with possibilities for identity protection through decentralization. These methods can be used to strengthen the safety of users and stakeholders, as well as data sets and AI models.

Data transparency and privacy processes can have an impact on four groups of people (Sharma et al., 2022): Gathered participant's data (direct and indirect) (Majhi et al., 2022); victims impacted by decisions made by participant data (Aparna, 2021); participant who use user's data in their work; and (Kaur and Ali, 2021) custodians who manage and secure data. All parties win when AI is castoff to oversee data access and create analytical models from that info (Bertino et al., 2019). By strengthening the Blockchain's resistance to attacks, the system's security, and the data's privacy are all enhanced. CI is a branch of AI. Computational Intelligence relies on flexible, adaptable soft computing approaches, while Artificial

Intelligence is grounded in hard computing techniques ("Wikipedia. Computational intelligence. Wikipedia. Available," 1994).

The mixture of blockchain and AI technology allows for development of more secure cryptographic functions and cyphers, making it more difficult for cybercriminals to breach systems over time, despite the fact that their processing power and efforts to do so will only rise. The combination of blockchain technology and AI has been aptly dubbed "blockchain intelligence" (Zheng et al., 2020). Blockchains also allow for the improvement of AI algorithms can monitor the chain's blocks and activity in real-time to determine if an attack is underway. When compared with the native design, this approach significantly improves confidence in the system (Marwala and Xing, 2018; Gopichand et al., 2019).

As blockchain users gain more power over their information, they will be in a better position to determine who sees what information and for what purposes. Participants' consent will be required for data collection to be used in an AI dataset. This allows users to exercise 'opt-in' relatively than 'opt-out' controls and increases the likelihood that their individual information will be used in ways that are in line with their expectations. If organizations implement decentralized identification systems, users may more frequently be offered financial incentives to share their data for use in conventional and AI-powered decision making.

Privacy can also be safeguarded by using smart contracts. Smart bonds can enforce regulations governing the usage of data can manage the yielding and canceling of participant info, which is subject to the permissions supplied by users. Contracts can be scanned with AI to find parties who have, or are likely to have, relevant data for certain purposes.

Effective AI may also be affected by the quantity and quality of data stored in blockchain networks. Compared with databases maintained by individual businesses, the volume and variety of data available for analysis in shared ledgers can be substantially greater. Metadata may be more robust and useful, and more advanced identity-masking processes may be possible with larger datasets.

When compared with data gathered and upheld by numerous organizations in records that are not unchallengeable, the information received from blockchain ledgers is likely to be considerably cleaner and more accurate. This is due to the security, validation, timestamping, and append-only characteristics of blockchain records.

Developers and users of the models will have more faith that they are complying with rules, and the resulting data will be of a higher ethical quality. Organizations can use this info with reduced privacy risk breaches since multi-dimensional user rights can be decided, documented, and in some cases imposed concluded smart contracts. In addition, zero-knowledge proofs allow for the collection of user data, which allows for the performance of complicated analyses that require specific user data, and the acquisition and use of the required information without having access to or possession of PII.

Better quality studies and results are another benefit of using blockchain data and artifacts. The reliability of data rises when it is free of errors and accompanied by descriptive information. Smaller data sets can be more reliable and yield greater insights because of the increased confidence in each data point. When good information is utilized to train models in AI, the resulting algorithms produce better predictions and better decisions. In addition to helping to verify nonblockchain info use in AI models, clean-training information can also valuable for this purpose.

Last but not least, there are a number of ways in which the underlying assumptions of AI models might be enhanced. As a first step, AI designers will need to clearly specify the investigates to be achieved and establish the type and extent of data required for these studies in order to acquire participant authorization, which may cost money. Because of this, designers will have to be more cognizant of the distributed ledgers and personal-data records available to them, as well as the universe of info that could inform these studies. During or before data collection, this could assist spot issues such as the under representation of black faces in photo classification techniques.

Personally Identifiable Information (PII) might be gathered, and it may be used more deliberately with procedure agreements enforced using smart agreements. This opens the door to more moral methods of data collection and administration. Within the bounds of established ethics and regulations, AI models developed with ethically generated and managed data can produce useful outcomes.

8. Computational Intelligence

The term "Computational Intelligence" (also known as "Soft Computing") refers to a novel approach to high-level data processing (Majhi et al., 2022). Computational Intelligence attempts to replicate human intelligence by simulating human abilities such as perceiving, understanding, learning, recognizing, and intelligent. It encompasses study, creation and improvement of computer paradigms inspired by natural language and biological systems. The three mainstays of CI have always been:

- **Artificial Neural Networks (ANN):** ANN are enormously parallel distributed networks with the capability to learn and generalize from instances, taking their cues from the human brain.
- **Fuzzy Systems (FS):** FS model linguistic roughness and handle uncertain situations constructed on a generalization of classical logic, allowing us to do estimated reasoning by drawing inspiration from the human language.
- **Evolutionary Computation (EC):** EC uses the principles of biological evolution to generate, evaluate, and alter a population of potential solutions to optimization problems.

Many computing models, however, have emerged over time that take cues from nature. Computing paradigms such as artificial life, ambient intelligence, cultural learning, social reasoning, artificial endocrine networks and artificial hormone networks are all now included in CI with the aforementioned three basic components are shown in Fig. 16.8, demonstrating that CI is an expanding topic.

8.1 Artificial Intelligence vs. Computational Intelligence

Intelligence displayed by computers, as opposed to humans, is the focus of Artificial Intelligence (AI) research. AI is a subfield of computer science that emphases on generating tools to give machines human-like intelligence. Computational Intelligence (CI) on the other hand is more like a subbranch of AI that focuses on the creation, implementation and improvement

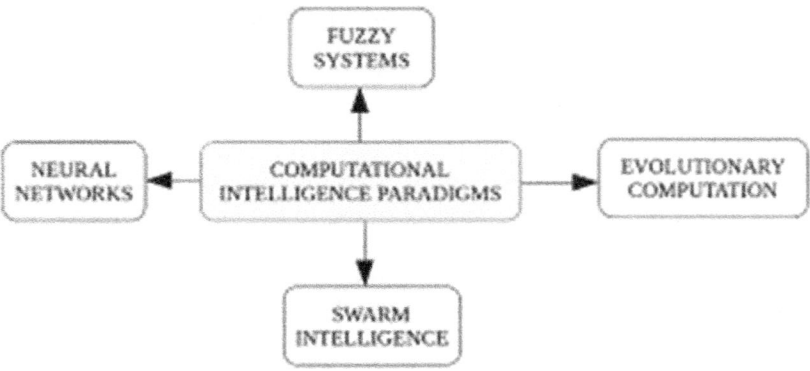

FIGURE 16.8 Components of CI.

of computational models inspired by natural language. Intelligent behavior in dynamic contexts is the focus of the research of adaptive mechanisms.

9. Interlinking of the blockchain technology and CI

Separate but related is the study of Computational Intelligence (CI) and blockchain technology. Both are examples of state-of-the-art technologies with game-changing potential for our daily lives and workplaces. However, they can be fused together to produce innovative, potent answers.

Integrating blockchain with CI using smart contracts is one approach. The conditions of a smart contract are encoded in computer code, making it a contract that can execute itself. When certain circumstances are met, such as the receipt of payment or the completion of delivery, smart contracts created with CI can be automatically executed. This has the potential to drastically cut down on the need of middlemen, which in turn would speed up business dealings.

9.1 Advantages of combining CI with blockchain

There may be several benefits to integrating CI with blockchain technology. The combination of blockchain and CI can provide protection for private information and transactions because each component is intrinsically safe.

- **Transparency** in procedures and choices can be improved by integrating blockchain technology, which is created on a decentralized system with CI.
- With the help of blockchain technology, CI may be used to automate processes and decision-making in a **trustworthy and transparent** manner, leading to increased automation in several areas.
- **Increased efficiency** in data processing and analysis when consuming Blockchain technology for storage of data and management in tandem with CI.
- With the assistance of CI, blockchain technology, which is **inherently scalable**, can facilitate the decentralized management of massive amounts of data and transactions.

9.2 Challenges to the adoption of CI based blockchain solutions

While there are many benefits to implementing blockchain solutions based on CI, there are also potential drawbacks to be aware of. For example:

- **Technical Difficulties:** integrating CI with blockchain technologies might be difficult and call for specialized knowledge. For small and medium-sized businesses, it might be challenging to invest the time and money needed to develop and deploy AI-based blockchain solutions.
- **Scalability:** Unfinished blockchain infrastructure may struggle to accommodate the massive amounts of data and many types of information needed by AI. Because of this blockchain solutions built on CI may be restricted in their applications.
- Organizations may find it challenging to deploy CI and blockchain technologies due to a lack of clear rules governing their use. This **ambiguity** can also hinder progress toward the development of really transformative blockchain-based CI applications.
- **Data privacy:** Because CI models need a lot of data to get better, integrating CI and blockchain can make some people nervous about their data's security.
- **Interoperability:** CI-based blockchain solutions often have trouble communicating and collaborating with one another because to the wide variety of platforms and protocols they use. The scalability and efficacy of such solutions may be stunted by their inability to communicate with one another.
- **Adoption:** Despite the advantages, CI-based blockchain solutions are quiet in the experimental phase of adoption and it may be take some time before businesses fully grasp and accept them.

9.3 Trends for blockchain and CI combo

The following are some emerging tendencies in the intersection between blockchain and CI:

- **Decentralized CI:** blockchain technology can be used to build decentralized CI systems, where information and processing are dispersed throughout a network of computers rather than being centralized in one place. This has the potential to make CI systems more open, safe, and reliable.
- If data is stored on premise and models are trained locally before being shared with a central server, then blockchain can provide Federated learning, which improves data privacy and security.
- Blockchain networks powered by Computational Intelligence: CI may be used to optimize and improve the performance of blockchain networks by doing things such as forecasting network congestion and altering the mining difficulty in response.
- Blockchain technology can be recycled to build a transparent and secure record of AI-based smart contracts, which are created using CI to automatically execute when specific criteria are satisfied.
- **Digital twins** powered by CI can be used to generate digital replicas of physical assets and blockchain technology can be used to keep track of this data in a transparent and safe way.

10. A futuristic approach

The development of decentralized, secure, and unchangeable systems is one of the most promising future consequences of integrating two flexible technologies, Artificial Intelligence and blockchain. Every sector depends heavily on data, and CI and blockchain can improve data handling security. Together blockchain and CI can give rise to innovative, powerful solutions that can revolutionize how sectors such as finance, healthcare, manufacturing, supply chain, and logistics approach trust, transparency, and efficiency.

Better more trustworthy, and more easily auditable systems are just the beginning of what CI and blockchain technology may bring to the business world. When combined AI and blockchain technology can be used to automate tasks, improve forecast accuracy, and streamline decision-making.

Predictive maintenance, intelligent supply chain management and distributed autonomous organizations are just a few examples of what might be made possible by combining these technologies. Trust in CI-based choices and predictions can be boosted by using blockchain technology, which offers a transparent and secure method of managing and protecting sensitive data used for CI models. This has the potential to boost productivity, cut expenses, and open up new income channels for enterprises.

Businesses can benefit from CI-based blockchain solutions because they provide a secure and transparent way to store and handle data, which satisfies regulatory compliance and data protection standards. When combined, CI and blockchain can improve efficiency, security, and transparency in business operations, perhaps leading to a paradigm shift.

11. Conclusion

Rapid advancements in blockchain and CI are opening up previously impossible avenues of data exchange and fusion. Meanwhile, developments in these technologies open up novel avenues for the moral processing of data. Individuals and businesses face a dilemma when sharing personal information; doing so can have significant benefits for both parties, but also expose them to substantial dangers and costs. By using novel mechanisms such as decentralized identities and zero-knowledge proofs, blockchain technology makes it possible to share data in ways that protect user privacy while yet giving them full access and control. These developments have the potential to improve both cybersecurity and the moral treatment of private information. By carefully crafting governance structures and processes blockchain participants can achieve these results.

References

Aparna, G., 2021. A watermark approach for image transmission: implementation of channel coding technique with security. Turkish Journal of Computer and Mathematics Education (TURCOMAT) 12 (3), 3976—3984. https://doi.org/10.17762/turcomat.v12i3.1687.

Bertino, E., Kundu, A., Sura, Z., 2019. Data transparency with blockchain and AI ethics. Journal of Data and Information Quality 11 (4), 1—8. https://doi.org/10.1145/3312750.

Consensys, n.d. Busting the Myth of Private Blockchains.

ElGayyar, M.M., ElYamany, H.F., Grolinger, K., Capretz, M.A.M., Mir, S., 2020. Blockchain-based federated identity and auditing. International Journal of Blockchains and Cryptocurrencies 1 (2), 179—205. https://doi.org/10.1504/ijbc.2020.109004.

Gopichand, G., Sailaja, G., Venkata Vinod Kumar, N., Samatha, T., 2019. Digital signature verification using artificial neural networks. International Journal of Recent Technology and Engineering 7 (6), 467—472.

Grimes, 2019. What Is Personally Identifiable Information (PII)? How to Protect it under GDPR, 2019.

Gupta, A.K., Chakraborty, C., Gupta, B., 2021. Secure transmission of EEG data using watermarking algorithm for the detection of epileptical seizures. Traitement du Signal 38 (2), 473—479. https://doi.org/10.18280/ts.380227.

Heister, S., Yuthas, K., 2020. The blockchain and how it can influence conceptions of the self. Technology in Society 60. https://doi.org/10.1016/j.techsoc.2019.101218.

Kaur, R., Ali, A., 2021. A novel blockchain model for securing IoT based data transmission. International Journal of Grid and Distributed Computing 14, 1045—1055.

Lesavre, L., Varin, P., Mell, P., Davidson, M., Shook, J., 2019. A taxonomic approach to understanding emerging blockchain identity management systems. arXiv. 23318422. https://doi.org/10.48550/arxiv.1908.00929.

Majhi, M., Pal, A.K., Pradhan, J., Islam, S.H., Khan, M.K., 2022. Computational intelligence based secure three-party CBIR scheme for medical data for cloud-assisted healthcare applications. Multimedia Tools and Applications 81 (29), 41545—41577. https://doi.org/10.1007/s11042-020-10483-7.

Manikandan, G., Sakthi, U., 2020. Chinese remainder theorem based key management for secured data transmission in wireless sensor networks. Journal of Computational and Theoretical Nanoscience 17 (5), 2163—2171. https://doi.org/10.1166/jctn.2020.8864.

Marwala, T., Xing, B., 2018. Blockchain and artificial intelligence. arXiv. 23318422 2, 12. https://doi.org/10.48550/arxiv.1802.04451.

Microsoft, n.d. Decentralized Identity.

Sharma, P., Jindal, R., Borah, M.D., 2022. Blockchain-based cloud storage system with CP-ABE-based access control and revocation process. The Journal of Supercomputing 78 (6), 7700—7728. https://doi.org/10.1007/s11227-021-04179-4.

Uribe, D., 2020. Privacy laws, non-fungible tokens, and genomics. The Journal of The British Blockchain Association. 3 (2), 1—10. https://doi.org/10.31585/jbba-3-2-(5)2020.

Wikipedia, n.d. Zero-Knowledge Proof.

Wikipedia, 1994. Computational Intelligence. Wikipedia.

Xiao, Y., Chen, W., 2021. High-fidelity optical transmission around the corner. IEEE Photonics Technology Letters 33 (1), 3—6. https://doi.org/10.1109/LPT.2020.3041482.

Yoo, D.S., Kim, Y., Lee, E.S., Lim, J.S., Hong, S.K., Lee, I.S., Jung, C.S., Yoon, H.C., Wee, S.H., Pfeiffer, D.U., Fournié, G., 2021. Transmission dynamics of African swine fever virus, South Korea, 2019. Emerging Infectious Diseases 27 (7), 1909—1918. https://doi.org/10.3201/eid2707.204230.

Yoon, J., 2019. Democratic Senators Introduce the Consumer Online Privacy Rights Act, 2019.

Zheng, Z., Dai, H., Wu, J., 2020. Blockchain Intelligence: When Blockchain Meets Artificial Intelligence, 2020.

Further reading

Salah, K., Rehman, M., Nizamuddin, N., Al-Fuqaha, A., 2019. Blockchain for AI: review and open research. IEEE (7), 10127—10149. https://doi.org/10.1109/ACCESS.2018.2890507.

Index

Printed and bound by CPI Group (UK) Ltd, Croydon, CR0 4YY

21/10/2024

01776788-0004